Engineering Materials and Processes

Springer

Berlin
Heidelberg
New York
Hong Kong
London
Milan
Paris
Tokyo

Series Editor
Professor Brain Derby, Professor of Materials Science
Manchester Science Centre, Grosvenor Street, Manchester, M1 7HS, UK

Gerd Lütjering · James C. Williams

Titanium

With 335 Figures

Springer

Prof. Dr. Gerd Lütjering
Technical University Hamburg-Harburg
Department of Physical Metallurgy and Materials Technology
21071 Hamburg
Germany

Prof. Dr. James C. Williams
The Ohio State University
Department of Materials Science and Engineering
Columbus, OH 43210
USA

Cataloging-in-Publication Data applied for
Bibliographic information published by Die Deutsche Bibliothek
Die Deutsche Bibliothek lists this publication in the Deutsche Nationalbibliografie;
detailed bibliographic data is available in the Internet at <http://dnb.dd

ISBN 3-540-42990-5 Springer-Verlag Berlin Heidelberg New York

Springer-Verlag Berlin Heidelberg New York
a member of BertelsmannSpringer Science + Business Media GmbH

http://www.springer.de

Printed in Germany

Typesetting: Digital data supplied by authors
Cover-Design: Erich Kirchner, Heidelberg
Printed on acid-free paper 62/3020 Rw 5 4 3 2 1 0

Preface

The authors were motivated to prepare this book by the absence of any recent comprehensive book on titanium. The intent of this book is to provide a modern compendium that addresses both the phyical metallurgy as well as the applications of titanium. Until now the only book on this subject is that by Zwicker which was written in German and published almost 30 years ago. Chapter 1 is an introduction to the subject including some historical aspects of titanium. Chapter 2 is a summary of the Fundamental Aspects of Titanium, Chapter 3 is a summary of the Technological Aspects of Titanium and Chapters 4 through 9 address the specifics of the various classes of titanium ranging from CP Titanium to Titanium Matrix Composites. Finally, Chapter 10 covers "special" properties and applications of titanium.

Our intent has been to address the subject conceptually rather than provide quantities of data of the sort that would be found in a Handbook. It is our intent that this book is useful for materials scientists and engineers interested in using titanium and for students either as a sourcebook or as a textbook. We have attempted to include a representative set of references which provide additional detail for readers interested in specific aspects of titanium. Because of the relatively recent growth of the technological importance of titanium, there is a voluminous literature on titanium. While our references span this literature it has proven impossible to mention every contribution.

We thank the following people for reading one or more chapters and providing suggestions that led to improvements in both the clarity and the content: J. Albrecht, R. G. Baggerly, M. J. Blackburn, J. P. Blank, R. R. Boyer, J. A. Hall, D. Helm, M. C. Juhas, J. M. Larsen, H. A. Lipsitt, S. Lütjering, J. O. Peters, S. R. Seagle.

Mrs. Zimmermann typed and formatted the text of this book and Ms. Wang performed all painstaking work on the figures. Without their dedicated help this book could not have been finished. We would like to thank both of them for their outstanding support.

Finally, we wish to thank our wives, Heide und Joanne, for both their patience and encouragement while we worked on this project.

Contents

1 Introduction

1.1 Purpose of the Book

High strength, low density, and excellent corrosion resistance are the main properties that make titanium attractive for a variety of applications. Examples include aircraft (high strength in combination with low density), aero-engines (high strength, low density, and good creep resistance up to about 550°C), biomedical devices (corrosion resistance and high strength), and components in chemical processing equipment (corrosion resistance).

The relatively high cost of titanium has hindered wider use, for example in automotive applications. To minimize the inherent cost problem, successful applications must take advantage of the special features and characteristics of titanium that differentiate it from competing engineering materials. This requires a more complete understanding of titanium alloys as compared to other, less expensive materials, including the interplay between cost, processing methods, and performance.

The purpose of this book is to provide up-to-date information regarding all aspects of titanium with the intention that further, cost-effective use of titanium will be facilitated. The book covers the basic characteristics of titanium, the complete production route from the ore to finished products, and the numerous opportunities that exist for tailoring the properties to achieve excellent performance in a variety of products. Selection criteria for titanium alloys also are discussed for some applications.

Titanium alloy metallurgy and applications have been the subject of a series of International Conferences on Titanium. These conferences have published proceedings and were held with one exception every four years: 1968 in London [1], 1972 in Boston [2], 1976 in Moscow [3], 1980 in Kyoto [4], 1984 in Munich [5], 1988 in Cannes [6], 1992 in San Diego [7], 1995 in Birmingham [8], 1999 in St. Petersburg [9], the next conference being held 2003 in Hamburg. This book attempts to integrate the information contained in these proceedings and present it as a concise summary. The reader interested in greater detail should refer to the proceedings.

It is intended that the book will be of great interest to all current and potential users of titanium products, to titanium producers, and also to the research and development community in high performance product industries, as well as to students of materials science and engineering.

1.2 History of Titanium

Titanium is present in the earth's crust at a level of about 0.6% and is therefore the fourth most abundant structural metal after aluminum, iron, and magnesium. The most important mineral sources are ilmenite ($FeTiO_3$) and rutile (TiO_2).

The first suspicion of a new, unknown element present in a dark, magnetic iron-sand (ilmenite) in Cornwall (UK) was expressed in 1791 by Gregor, a clergyman and amateur mineralogist. In 1795, Klaproth, a German chemist, analyzed rutile from Hungary and identified an oxide of an unknown element, the same as the one reported by Gregor. Klaproth named the element titanium after the Titans, the powerful sons of the earth in Greek mythology.

Many attempts were made to isolate the metal from the titanium ore using titanium tetrachloride ($TiCl_4$) as an intermittent step. The production of ductile, high purity titanium still proved to be difficult, because of the strong tendency of this metal to react with oxygen and nitrogen. Early demonstrations of the reduction of $TiCl_4$ using either Na or Mg produced small quantities of brittle titanium metal. It was not until well into the 20th century (1937-1940) that a commercially attractive process was developed by Kroll in Luxembourg. This process involved the reduction of titanium tetrachloride with magnesium in an inert gas atmosphere. The resulting titanium is called "titanium sponge" because of its porous and spongy appearance. This famous Kroll process remained essentially unchanged and is the dominant process for titanium production today.

It is noteworthy that the industrial capacity of $TiCl_4$ production existed before the interest in metallic titanium developed. This is because $TiCl_4$ is the feed stock for synthetic, high purity TiO_2 used in paint. Even today, only 5% of the $TiCl_4$ production is used to produce titanium metal.

A more detailed description of the history of titanium can be found in the overview article "Titanium – A Historical Perspective" by Bomberger, Froes, and Morton [10].

1.3 Emergence as Commercial Material

Interest in the properties of titanium started after the Second World War in the late 1940's and early 1950's. Especially in the USA, major U.S. Government sponsored programs led to the installation of large capacity titanium sponge production plants, for example at TIMET (1951) and RMI (1958). In Europe, large scale sponge production started in 1951 in the UK at the Metals Division of Imperial Chemical Industries (later IMI and Deeside Titanium), which became the principal European titanium producer. In France, titanium sponge was produced for several years but discontinued in 1963. In Japan, sponge production started in 1952 and two companies, Osaka Titanium and Toho Titanium, had relatively large capacities by 1954. The Soviet Union started sponge production in 1954 and dramatically increased their capacity. By 1979, the Soviet Union became the world's largest titanium sponge producer. This can be seen by comparing the production capacity in the major titanium sponge producing countries of the world (Table 1.1).

Alloy development in the USA progressed rapidly from about 1950 stimulated by the recognition of the general usefulness of aluminum additions for strengthening. This led in combination with Sn additions to the early development of the α alloy Ti-5Al-2.5Sn[1] for high temperature applications. The addition of Mo as β stabilizing element resulted in the α+β alloy Ti-7Al-4Mo for high strength applications. A major breakthrough was the appearance of the Ti-6Al-4V alloy in the USA in 1954, becoming soon the most important α+β alloy combining excellent properties and good producibility. Today, Ti-6Al-4V is still the most widely used alloy.

In the UK, a somewhat different alloy development route was taken, focusing more on high temperature applications for aero-engines and leading in 1956 to the introduction of the alloy Ti-4Al-4Mo-2Sn-0.5Si (later IMI 550). This marked the first time silicon was added as an alloying element for improving creep resistance.

The first β titanium alloy B120VCA (Ti-13V-11Cr-3Al) was developed in the USA in the mid-1950's as a sheet alloy. Starting in the 1960's, this high strength, age-hardenable sheet alloy was used for the skin of the legendary surveillance airplane SR-71.

Table 1.1. Production capacity of titanium sponge in metric tons [10, 12]

	USA	Japan	UK	USSR	China	Total
1979	20 800	16 200	2 200	39 000	1 800	80 000
1980	25 400	23 200	1 800	42 600	1 800	94 800
1982	27 600	27 300	1 400	44 400	2 300	103 000
1984	30 400	34 000	5 000	47 200	2 700	119 300
1987	25 400	23 100	5 000	49 900	2 700	106 100
1990	30 400	28 800	5 000	52 200	2 700	119 100

Besides these successful alloy developments and the increasing usage of titanium alloys in aerospace application, there was a steady increase in the production of commercially pure titanium (CP titanium), mainly used as corrosion resistant material in non-aerospace applications. Besides in the USA, the production of CP titanium became a major focus in Japan where, in the absence of any domestic major aerospace industry, manufacturing and exporting of CP titanium fabricated products soon became the main area of activity.

[1] Alloy compositions in this book are stated in wt% unless otherwise noted.

1.4
Titanium Industry Status

The development of the world titanium industry can be characterized by two phases. The first phase was dominated by technical progress starting in the 1950's and lasting until the mid-1980's. This phase is described in some detail in the overview article [10] published in 1985 and mentioned earlier in Sect. 1.2. The second and continuing phase can be characterized by the transformation to a commercial industry when the technology was important but economics became a dominant consideration. This second phase is less well documented, but is addressed in a recent overview [11].

The total worldwide capacity for titanium sponge production increased almost steadily from 1980 to 1990 (Table 1.1), showing only a 10% fluctuation around 1987. The traditional "up and down" in the titanium market due to the large dependence on the aerospace industry, especially on the military market, can be seen better on the actual production of sponge leading to values between 41.6% and 86.9% of total capacity, for example in the USA, Table 1.2 [12].

Table 1.2. Sponge production in metric tons and % of capacity in the USA [12]

	1983	1984	1985	1986	1987	1988	1989	1990
Sponge Prod.	12 600	22 000	21 000	15 800	17 900	22 300	25 200	25 000
% of Capacity	41.6	72.7	71.5	57.0	70.4	84.6	86.9	81.2

The total worldwide capacity for the production of titanium sponge dropped sharply by 25% from 1990 to 1995 (Table 1.3) [13], mainly due to lower demand resulting from decreased military budgets in the USA and in the former Soviet Union (CIS) leading to the closing of the RMI sponge plant and to the closing of Deeside Titanium (UK). In the USA, the capacity drop due to the closing of the RMI plant was partially restored shortly afterwards by the opening of a new sponge production plant by TIMET, Henderson, Nevada (capacity about 5000 metric tons/year). It should be pointed out, that the enormous difference in the sponge production capacity between Tables 1.1 and 1.3 for the year 1990 is not real, but is due to the fact, that only after the formation of the CIS, the true capacity numbers became openly available. During the Soviet Union time, these capacity numbers for the USSR were based on estimates by Western experts. As it turned out, these estimates, shown in Table 1.1, were much too low. This interpretation is manifested by the actual production numbers for titanium sponge over the last ten years for the three CIS countries, Japan, and North America (Fig. 1.1). It can be seen that the value in Table 1.3 for the capacity in the USSR in the year 1990 has the correct level and that the actual production of titanium sponge in the USSR during that time period was close to 100% capacity. The worldwide increase in titanium sponge production after the minimum in 1994 (Fig. 1.1) is mainly a consequence of the pickup in commercial airplane sales.

Table 1.3. Production capacity of titanium sponge in metric tons [13]

	USA	Japan	UK	USSR (CIS)	China	Total
1990	30 000	29 000	5 000	91 000	2 700	157 700
1995	15 000	26 000	-	73 000	2 700	116 700

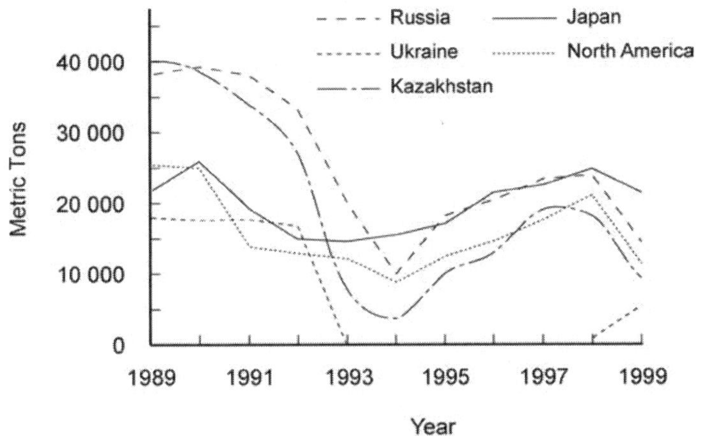

Fig. 1.1. Titanium sponge production in metric tons (source: Materials Bulletin Monthly, March 2000 issue)

The sponge price over the years (Fig. 1.2) also shows this "up and down" due to fluctuations in demand largely because of the variation in aerospace market requirements. The increase in sponge price during the period 1977-1981 is a reaction to the rapid increase in commercial aircraft orders, and the subsequent collapse in aircraft sales in the years 1982-1984 was accompanied by a reduction in sponge price. The smaller fluctuations during the period 1985-1995 are also reflecting the development in commercial airplane sales.

A rough estimate of the added cost attributable to melting and subsequent processing of Ti-6Al-4V can be made as follows. One can start with sponge valued at about 10 US $ per kg and add 4 US $ per kg to create a double-melted ingot. One can also assume as a general rule of thumb that each subsequent processing step doubles the cost. Thus, the price at the mill product stage would be roughly 28 US $ per kg and about 56 US $ per kg for a component.

Large efforts have been made to develop other non-aerospace application areas to lessen the cyclical nature of the titanium industry. This is also true for the former Soviet Union, as shown by Gorynin in his presentation at the World Titanium Conference in San Diego, Table 1.4 [14]. More detailed information about the change in domestic usage of CP titanium mill products in Japan was given by Yamada in 1995, Table 1.5 [15]. It can be seen that the usage of CP titanium mill products increased drastically between 1988 and 1994 in the areas of civil engineering and consumer goods. It should be pointed out that for mill products in Japan CP titanium accounts for nearly 90%, with the remaining 10% being alloyed grades [15].

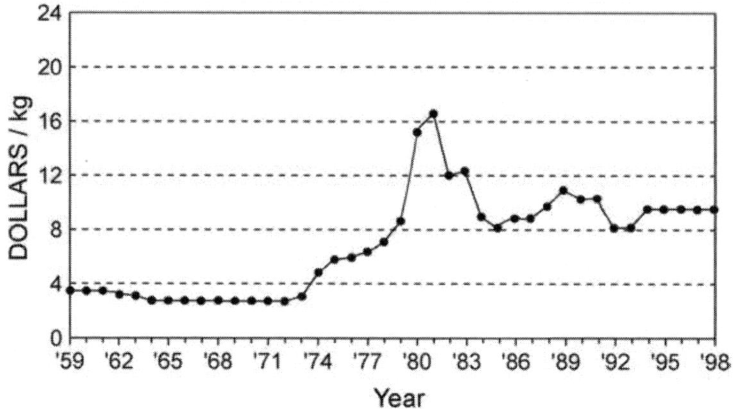

Fig. 1.2. Development of titanium sponge price in US dollars/kg over the years 1959-1998

Table 1.4. Shares of titanium consumption in the USSR [14]

	Before 1967	In the 1970's	In the 1990's (Forecast)
Military Industries	95%	60%	45%
Chemistry and Heavy Industries	-	28%	20%
Domestic Industries	5%	12%	35%

Table 1.5. Domestic usage of CP titanium mill production in Japan in metric tons [15]

	1988	1991	1994
Chemical Industry	1 399	1 338	1 261
Power Industry	611	957	754
Aerospace Industry	19	23	9
Civil Engineering	21	141	320
Automotive Industry	0	14	6
Consumer Goods	0	187	456
Medical Applications	0	9	4
Others	1 011	712	963
Total	3 061	3 381	3 773

In countries or regions with major aerospace industry, for example North America or Europe, the ratio between CP titanium and titanium alloys is different, as can be seen from Fig. 1.3, compiled in 1998 [11]. In sharp contrast to Japan, CP titanium accounts for only 26% of the total USA market. The further breakdown of the remaining 74% for titanium alloys shows that 56% of the market in the USA is covered by the α+β alloy Ti-6Al-4V whereas all β alloys together sum up only to 4%.

Market share numbers for mill products in the USA between 1990 and 1994 are shown in Table 1.6 [13]. It can be seen that during this period the decline was only 12% for non-aerospace products whereas for aerospace applications this decline was much higher. This points out the relative increase in importance of non-aerospace applications over the recent years. This is consistent with the second phase of the world titanium industry mentioned earlier. Economics have a greater influence on titanium utilization in non-aerospace applications. The continued growth of these applications is shown in Fig. 1.4 [11]. This basic change in the titanium markets helps stabilize the volume requirements by diversifying the market for titanium and reducing the reliance on the aerospace industry with its well known cyclical nature.

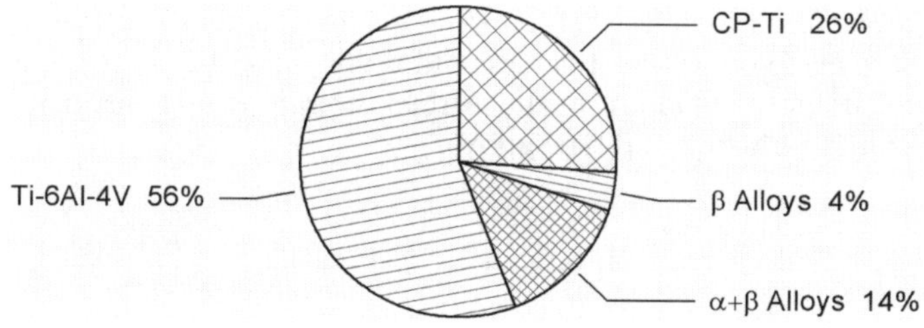

Fig. 1.3. Approximate breakdown of the USA market by alloy type, 1998 [11]

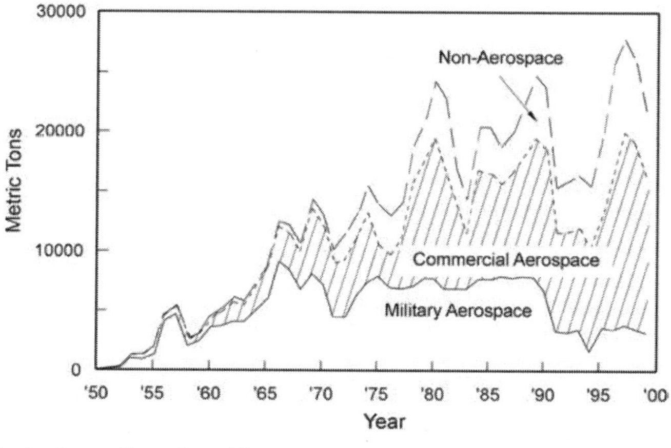

Fig. 1.4. USA titanium mill product shipments by market sector in metric tons [11]

Table 1.6. USA titanium mill products market in metric tons [13]

Market Sector	1990	1994	Change (%)
Defense Aerospace	6 500	3 200	–51
Commercial Aerospace	12 000	7 700	–36
Non-Aerospace	5 400	4 800	–12
Total	23 900	15 700	–35

1.5 Traditional and Emerging Applications

There are two classical application areas for titanium alloys: airframes and aero-engines. These are driven by the superior structural efficiency of titanium alloys as will be described in detail later. The historical growth of titanium usage in the Boeing airframes is shown in Fig. 1.5 [16]. Roughly 10% of the Boeing 777 airframe weight is titanium and it is the first commercial airplane, for which β titanium alloys outweigh the classical Ti-6Al-4V alloy, mainly because most of the landing gear is made out of the high strength β alloy Ti-10-2-3 [16]. For comparison, the total usage of titanium in the Airbus aircraft family lies between 4-5% [17] and includes virtually no β type alloys.

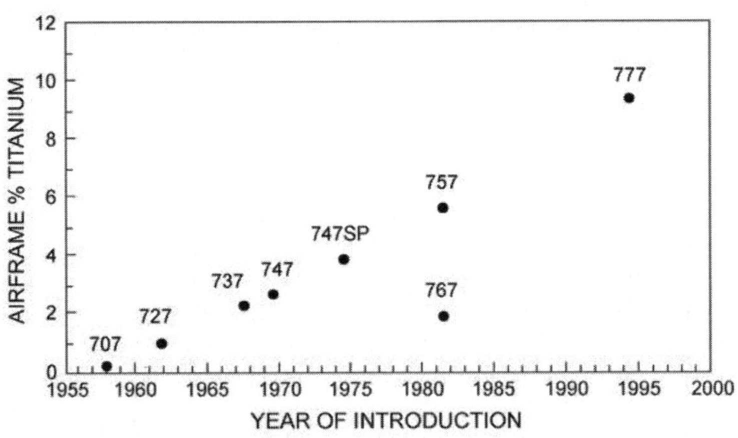

Fig. 1.5. Titanium usage in Boeing airframes [16]

In modern aero-engines, the weight share of titanium is about 25% being applied mainly in the fan and compressor sections for disks and blades which operate at temperatures up to about 500°C. In a recent engine, the GE-90 aero-engine (Fig. 1.6) used for the Boeing 777 aircraft, the large fan blades usually made from Ti-6Al-4V were replaced by polymer composite blades. But, as can be seen from Fig. 1.6, still a large number of titanium parts are used in that engine. The most recent advance in the area of blisk (bladed disk) technology has been made by MTU (DaimlerChrysler) by using linear friction welding to attach the blades to the disk [18]. This allows more efficient material utilization making these blisks attractive economically. Linear friction welding also can be used to replace individual airfoils on the blisks as required to repair in-service damage.

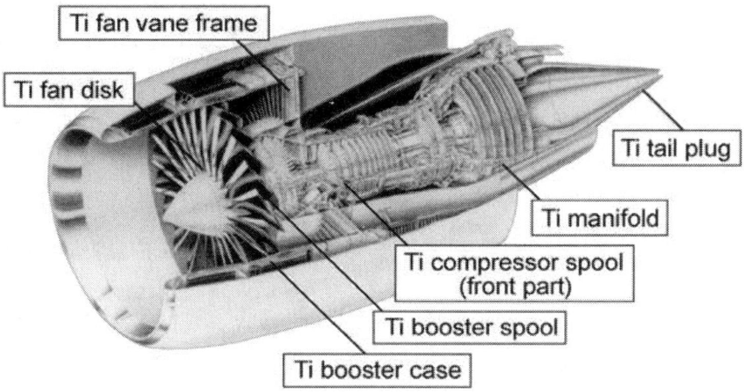

Fig. 1.6. Titanium usage in the GE-90 aero-engine

In the traditional areas of chemical and power industries using titanium mainly as corrosion resistant material, the usage of titanium in offshore structures has become more common in recent years. One example is shown in Fig. 1.7 using titanium as drilling riser string [19]. The principal attractions of titanium for this application are high strength, low density, low modulus (high flexibility), full corrosion resistance to seawater, to drilling mud, and to transported fluids, and seawater fatigue resistance. Because of the requirements for high strength and high fatigue resistance, the Ti-6Al-4V alloy has to be used instead of CP titanium.

Besides the areas mentioned above, building applications such as exterior walls and roofing material have emerged as a new market for titanium. Using CP titanium as building material has become especially popular in Japan [20]. One example is the Fukuoka Dome, built in 1993, which is covered with titanium roofing, retractable for multi-role and all-weather purposes (Fig. 1.8) [20]. Each of these building projects uses large quantities of CP titanium leading to the increased usage in the civil engineering area in Japan (see Table 1.5).

Another "new area" in which titanium use is growing is the area of consumer products, such as spectacle frames, cameras, watches, jewelry, and various kinds of sporting goods. The largest application in the area of sporting goods are golf club heads. Other examples are tennis rackets, bicycle frames, spikes in sprinters' running shoes, etc. [21]. There has been a major effort in the area of bicycle components in the UK by Raleigh Industries Ltd. together with IMI Titanium [22]. The β alloy Ti-15-3 is now being used for handlebars and seat posts on mountain bikes [23]. The real advantage of the handlebars is related to the low modulus of elasticity of the β titanium alloy.

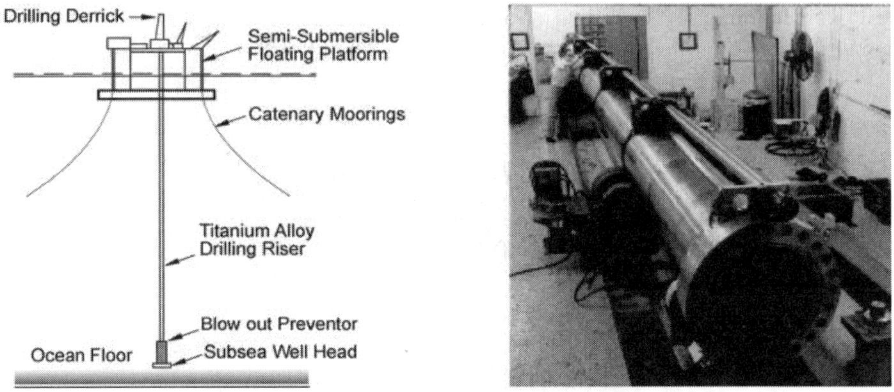

a **b**

Fig. 1.7. Titanium alloy drilling riser [19]: (a) Schematically (b) Riser string joint (courtesy RMI)

Fig. 1.8. Fukuoka Dome with retractable titanium roof (courtesy Y. Kubota, Kobe Steel Ltd.)

The benefits of titanium golf club heads have been widely publicized and in 1997 about 4500 metric tons of Ti-6Al-4V were used in cast driver heads, such as the ones shown in Fig. 1.9. The high strength and the low density of titanium allow a larger club head to be used, which creates a larger area for striking the ball properly while retaining the club head speed [24].

A more established application area of titanium is the biomedical field. Mainly CP titanium and Ti-6Al-4V have been used in the past as implant materials. Vanadium free titanium alloys like Ti-6Al-7Nb and Ti-5Al-2.5Fe were developed because toxicity of vanadium to the human body was suspected. In recent years, efforts have been made to develop β titanium alloys using nontoxic elements such as Nb, Ta, Zr, and Mo as alloying additions [25]. The main advantages of these β alloys over the conventional Ti-6Al-4V alloy are the higher fatigue strength, the lower modulus of elasticity, and the improved biocompatibility.

The last area to be mentioned in this introduction chapter is the application of titanium in mass-produced automobiles. This area also is extremely challenging because of cost sensitivity. Potential parts have been identified, for example valves, valve springs, and connecting rods in the engine area, and suspension springs, bolts, fasteners, and the exhaust system in the car body area. For many years, titanium has been used in high performance vehicles such as Formula-One racing cars or off-road racing trucks [23], but for an application in a family automobile the problem of producing low-cost titanium parts has to be solved. In this context, the cost includes both the raw material and the part fabrication. Successful introduction of titanium in family automobiles will depend on a significant reduction in one or both of these cost elements. For example, today's price for titanium sponge alone is already close to the maximum price that the automobile industry is willing to pay for finished titanium parts [21]. Only completely new ideas, maybe starting already with the extraction process of TiO_2 from the ilmenite ore and with the production process of $TiCl_4$ or avoiding this $TiCl_4$ route altogether, might be able to solve this challenging problem. Alternatively, policy changes that mandate higher fuel economy can create a market for automobiles containing more expensive but lighter materials such as titanium.

Fig. 1.9. Investment cast Ti-6Al-4V golf club heads (courtesy N. E. Paton, Howmet)

2 Fundamental Aspects

2.1 Basic Properties

Some of the basic characteristics of titanium and its alloys are listed in Table 2.1 and compared to those of other structural metallic materials based on Fe, Ni, and Al. Although titanium has the highest strength to density ratio it is the material of choice only for certain niche application areas because of its high price. This high price is mainly a result of the high reactivity of titanium with oxygen. The use of inert atmosphere or vacuum is required during the production process of titanium sponge from titanium tetrachloride as well as during the melting process. Additional major cost elements are energy and the initial high cost of titanium tetrachloride. On the other hand, the high reactivity with oxygen leads to the immediate formation of a stable and adherent oxide surface layer when exposed to air, resulting in the superior corrosion resistance of titanium in various kinds of aggressive environments, especially in aqueous acid environments. The much higher melting temperature of titanium as compared to aluminum, the main competitor in light weight structural applications, gives titanium a definite advantage above application temperatures of about 150°C. The high reactivity of titanium with oxygen limits the maximum use temperature of titanium alloys to about 600°C. Above this temperature the diffusion of oxygen through the oxide surface layer becomes too fast resulting in excessive growth of the oxide layer and embrittlement of the adjacent oxygen rich layer of the titanium alloy (see Sect. 2.9.3).

Table 2.1. Some important characteristics of titanium and titanium based alloys as compared to other structural metallic materials based on Fe, Ni, and Al

	Ti	Fe	Ni	Al
Melting Temperature (°C)	1670	1538	1455	660
Allotropic Transformation (°C)	$\beta \xrightarrow{882} \alpha$	$\gamma \xrightarrow{912} \alpha$	-	-
Crystal Structure	bcc → hex	fcc → bcc	fcc	fcc
Room Temperature E (GPa)	115	215	200	72
Yield Stress Level (MPa)	1000	1000	1000	500
Density (g/cm^3)	4.5	7.9	8.9	2.7
Comparative Corrosion Resistance	Very High	Low	Medium	High
Comparative Reactivity with Oxygen	Very High	Low	Low	High
Comparative Price of Metal	Very High	Low	High	Medium

2.2 Crystal Structure

Pure titanium exhibits an allotropic phase transformation at 882°C, changing from a body-centered cubic crystal structure (β phase) at higher temperatures to a hexagonal close-packed crystal structure (α phase) at lower temperatures. The exact transformation temperature is strongly influenced by interstitial and substitutional elements and therefore depends on the purity of the metal. The hexagonal unit cell of the α phase is shown in Fig. 2.1 indicating also the room temperature values of the lattice parameters a (0.295 nm) and c (0.468 nm). The resulting c/a ratio for pure α titanium is 1.587, smaller than the ideal ratio of 1.633 for the hexagonal close-packed crystal structure. Also indicated in Fig. 2.1 are the three most densely packed types of lattice planes, the (0002) plane, also called basal plane, one of the three $\{10\bar{1}0\}$ planes, also called prismatic planes, and one of the six $\{10\bar{1}1\}$ planes, also called pyramidal planes. The three a_1, a_2, and a_3 axes are the close-packed directions with the indices $<11\bar{2}0>$. The unit cell of the body-centered cubic (bcc) β phase is illustrated in Fig. 2.2 indicating also one variant of the six most densely packed $\{110\}$ lattice planes and the lattice parameter value of pure β titanium at 900°C (a = 0.332 nm). The close-packed directions are the four $<111>$ directions.

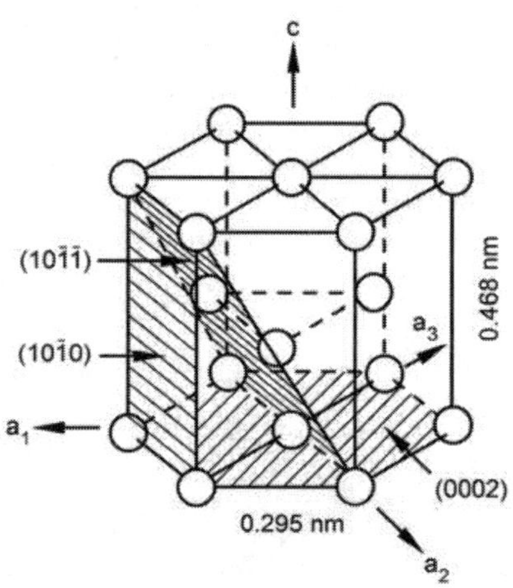

Fig. 2.1. Unit cell of α phase

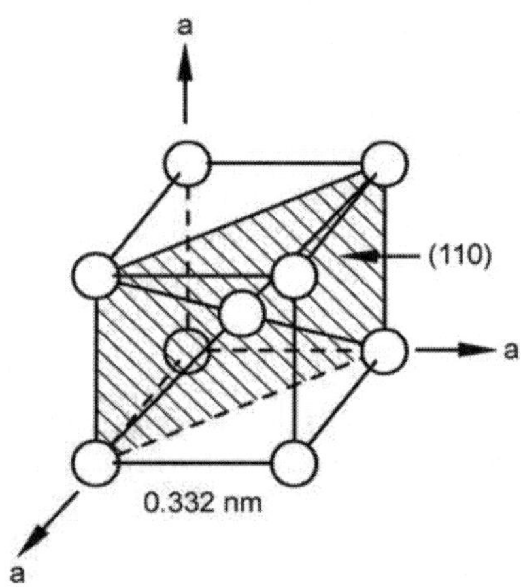

Fig. 2.2. Unit cell of β phase

2.3 Elastic Properties

The intrinsically anisotropic character of the hexagonal crystal structure of the α phase has important consequences for the elastic properties of titanium and its alloys. The variation of the modulus of elasticity E of pure α titanium single crystals at room temperature as a function of the angle γ between the c-axis of the unit cell and the stress axis is shown in Fig. 2.3 [26]. It can be seen that the modulus of elasticity E varies between 145 GPa (stress axis parallel to the c-axis) and 100 GPa (stress axis perpendicular to the c-axis). Similar strong variations are observed for the shear modulus G of single crystals varying between 46 GPa and 34 GPa for shear stresses applied in $<11\bar{2}0>$ direction and in (0002) or $\{10\bar{1}0\}$ planes, respectively. Less pronounced variations in elastic properties are observed in polycrystalline α titanium with crystallographic texture. The actual variations in modulus depend on the nature and intensity of the texture.

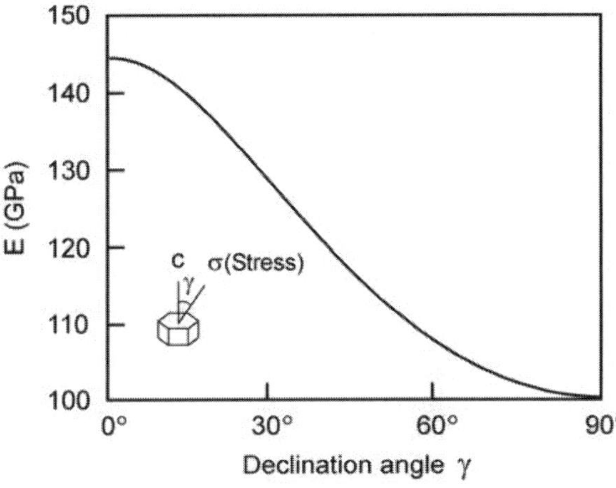

Fig. 2.3. Modulus of elasticity E of α titanium single crystals as a function of declination angle γ [26]

With increasing temperature the modulus of elasticity (E) and the shear modulus (G) decrease almost linearly as shown in Fig. 2.4 for polycrystalline texture-free α titanium [27]. It can be seen that E drops from about 115 GPa at room temperature to about 58 GPa at the β transus temperature, while G is decreasing from about 42 to 20 GPa over the same temperature range.

The modulus of elasticity of the β phase at room temperature cannot be measured for pure titanium because the β phase is not stable. In binary titanium alloys containing sufficiently high concentrations of β stabilizing elements, for example Ti-V alloys with about 20% vanadium, the metastable β phase can be retained to room temperature by fast cooling (see Sect. 2.5). Modulus data for Ti-V alloys in the water quenched condition are shown in Fig. 2.5 [28]. The modulus composition dependence can be discussed in three parts: 0-10% V, 10-20% V, and 20-50% V. From Fig. 2.5, it can be seen that the E value of the β phase increases with increasing vanadium content between 20 and 50% vanadium with a value as low as 85 GPa for 20% V. This shows that the β phase in general has a lower modulus of elasticity than the α phase. The unusual maximum in E at about 15% vanadium is related to the so-called athermal ω phase formation (see Sect. 2.6.1). The steep decline in modulus from 0-10% V is typical for titanium martensite containing β stabilizing elements. Both the minimum and the maximum in the concentration dependence of E disappear upon annealing in the (α+β) phase field (dashed line in Fig. 2.5) and the modulus follows a tie-line between the (α+β) phase field boundaries, as would be expected from the rule of mixtures. Similar concentration dependences of E have been observed for Ti-Mo, Ti-Nb, and other binary alloys involving β stabilizing elements [28]. A commonly used explanation for the steep decline in modulus for martensite containing β stabilizing elements (0-10% range in Fig. 2.5) is the presence of retained metastable β phase which transforms during loading to stress-induced martensite, leading to an apparent low modulus of elasticity of the material [29]. But, as recently shown [30], a Ti-7Mo alloy exhibiting the low E value of 72 GPa was 100% martensitic and did not contain any retained metastable β phase. So, the steep decline in modulus seems to result directly from the fact that the β stabilizing elements severely disturb and reduce the bonding force of the lattice. It is interesting to note, that the martensite in some of these alloys also shows the tendency for spinodal decomposition (see Sect. 2.6.1). In contrast, the most common α stabilizing element (aluminum) increases the modulus of elasticity of the α phase [31]. For solid solutions this concentration dependence of E is atypical. In the case of the Ti-Al system, it seems to be related to the tendency of ordering (see Sect. 2.5) with a concomitant increase in covalent bonding.

In general, commercial β titanium alloys have lower E values than α and α+β alloys. Typical values are 70-90 GPa for the as-quenched condition and 100-105 GPa for the annealed condition of commercial β alloys, 105 GPa for CP titanium, and about 115 GPa for commercial α+β alloys [32].

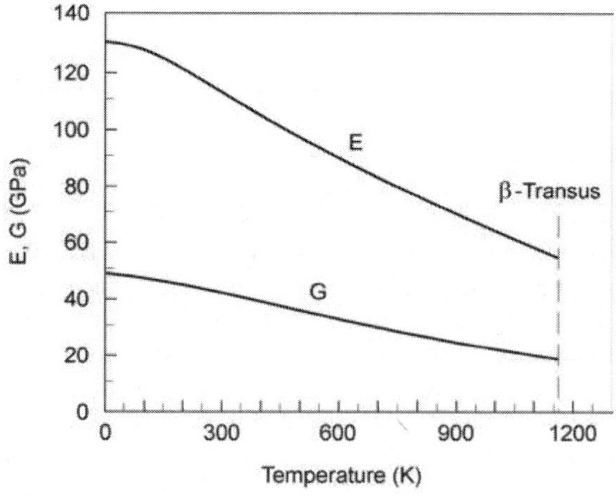

Fig. 2.4. Modulus of elasticity E and shear modulus G as a function of temperature of α titanium polycrystals [27]

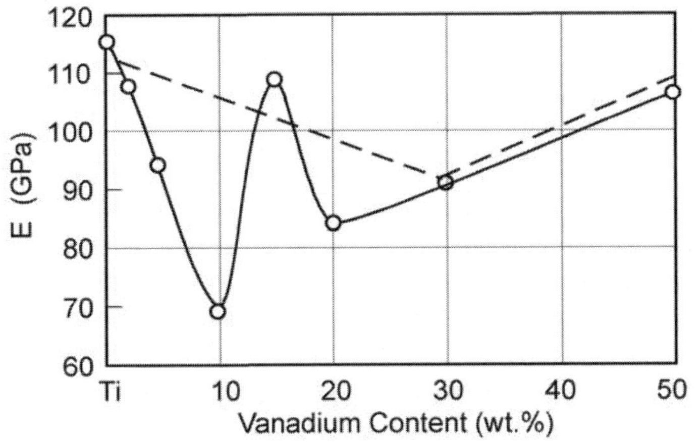

Fig. 2.5. Modulus of elasticity E of Ti-V alloys, solid line: 24h 900°C/WQ, dashed line: annealed at 600°C [28]

2.4 Deformation Modes

The ductile behavior of hexagonal α titanium, especially at low temperatures, results from the activation of twinning deformation modes in addition to conventional slip by dislocations. These twinning modes are important for the deformation behavior of CP titanium and some α titanium alloys. Although twinning is suppressed nearly completely in two phase α+β alloys by the small phase dimensions, high solute content, and presence of Ti_3Al precipitates, these alloys are quite ductile at low temperatures due to their small phase dimensions.

The bcc β phase also shows twinning in addition to slip but the occurrence of twinning in β alloys is limited again to the single phase state and decreases with increasing solute content. In fully heat treated β alloys, which are hardened by the precipitation of α particles, twinning is completely suppressed. In these alloys twinning might occur during the forming operations prior to aging. Some commercial β alloys also can form deformation induced martensite which further enhances their formability (see Sect. 2.6.1). Formation of this deformation induced martensite is very sensitive to alloy composition.

2.4.1 Slip Modes

The various slip planes and slip directions for α titanium are indicated in the hexagonal unit cell in Fig. 2.6. The main slip directions are the three close-packed directions of the type $<11\bar{2}0>$. The slip planes containing this $\vec{a}$ type of Burgers vector are the (0002) plane, the three $\{10\bar{1}0\}$ planes, and the six $\{10\bar{1}1\}$ planes. Among these three different types of slip planes together with the possible slip directions there are a total of 12 slip systems (Table 2.2) [33, 34]. These can be reduced to nominally 8 independent slip systems. However, this number is further reduced to only 4 independent slip systems because the shape changes that are produced by the combined action of slip system types 1 and 2 (see Table 2.2) are exactly the same as those of slip system type 3. Therefore, assuming the von Mises criterion is correct, which requires at least five independent slip systems for a homogenous plastic deformation of polycrystals, the operation of one of the slip systems with a so-called non-basal Burgers vector needs to be activated, either the $\vec{c}$ type with the slip direction $[0001]$ or the $\vec{c}+\vec{a}$ type with the slip direction $<11\bar{2}3>$ (see Fig. 2.6 and Table 2.2). The presence of $\vec{c}+\vec{a}$ type dislocations has been observed by TEM observations in a number of titanium alloys [35, 36]. To justify the presence of these $\vec{c}+\vec{a}$ dislocations it is not so important whether the von Mises criterion is valid or not, but rather to answer the question which slip systems are activated if a grain in a polycrystalline material is orientated such that the applied stress is parallel to the c-axis. In that case, neither a slip system with an $\vec{a}$ Burgers vector nor dislocations with a $\vec{c}$ Burgers vector can be activated because the Schmidt factor for both is zero. From the possible slip planes of the dislocations with a $\vec{c}+\vec{a}$ Burgers vector, the $\{10\bar{1}0\}$ slip planes cannot be activated because they are parallel to the stress axis and from the other possible slip planes (see Figure 2.6) the $\{11\bar{2}2\}$ planes are closer to 45° (higher Schmidt factor) than the $\{10\bar{1}1\}$ planes. Assuming the critical resolved shear stresses (CRSS) are about the same for both types of slip planes, then the slip system with a non-basal Burgers vector which is activated most likely in α titanium is of the type $<11\bar{2}3>$ $\{11\bar{2}2\}$, shown in Table 2.2 as slip system type 4.

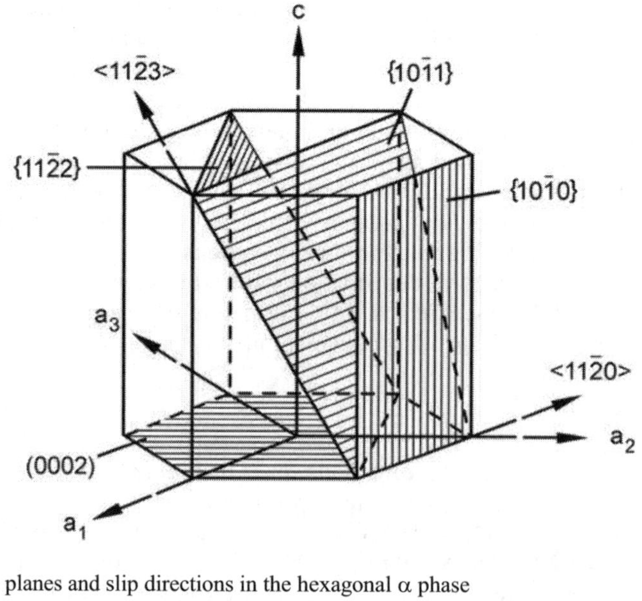

Fig. 2.6. Slip planes and slip directions in the hexagonal α phase

Table 2.2. Slip systems in the hexagonal α phase [33, 34]

Slip system type	Burgers vector type	Slip direction	Slip plane	No. of slip systems Total	Independent
1	$\vec{a}$	$<11\bar{2}0>$	(0002)	3	2
2	$\vec{a}$	$<11\bar{2}0>$	$\{10\bar{1}0\}$	3	2
3	$\vec{a}$	$<11\bar{2}0>$	$\{10\bar{1}1\}$	6	4
4	$\vec{c}+\vec{a}$	$<11\bar{2}3>$	$\{11\bar{2}2\}$	6	5

Taking into account the large difference in CRSS between $\vec{c}+\vec{a}$ slip and $\vec{a}$ slip, which was measured for Ti-6.6Al single crystals (Fig. 2.7) [37], the percentage of grains deforming by $\vec{c}+\vec{a}$ slip will be quite low in α titanium polycrystals without crystallographic texture, because the activation of $\vec{a}$ slip is easier even for an angle of about 10° between the stress axis and the c-axis.

The absolute values of the CRSS are strongly dependent on alloy content and on test temperature (Fig. 2.7). The small differences in CRSS at room temperature between the three types of slip systems with a basal ($\vec{a}$ type) Burgers vector, $\{10\bar{1}0\} < \{10\bar{1}1\} < (0002)$, become even smaller with increasing temperature (Fig. 2.7).

As shown for binary Ti-V alloys [38], the slip systems in bcc β titanium alloys are $\{110\}$, $\{112\}$, and $\{123\}$, all with the same Burgers vector of the type $<111>$, in agreement with the generally observed slip modes in bcc metals.

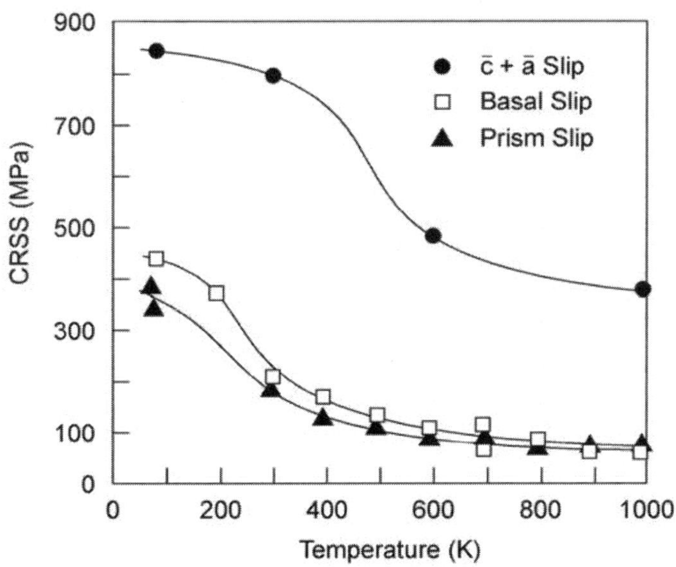

Fig. 2.7. Temperature dependence of CRSS for slip with $\vec{a}$ and $\vec{c}+\vec{a}$ Burgers vectors in single crystals of Ti-6.6Al [37]

2.4.2 Deformation Twinning

The main twinning modes observed in pure α titanium are $\{10\bar{1}2\}$, $\{11\bar{2}1\}$, and $\{11\bar{2}2\}$. The crystallographic elements of these three twin systems in α titanium are listed in Table 2.3 [33]. Twinning modes are especially important for plastic deformation and ductility at low temperatures if the stress axis is parallel to the c-axis and the dislocations with a basal Burgers vector cannot move. In this case, $\{10\bar{1}2\}$ and $\{11\bar{2}1\}$ twins are activated during deformation in tension leading to an extension along the c-axis. The most frequently observed twins are of the $\{10\bar{1}2\}$ type but they have the smallest twinning shear (Table 2.3). The shape change associated with a $\{11\bar{2}1\}$ twin having a much larger magnitude of twinning shear is shown in Fig. 2.8 [33]. Under compression loading parallel to the c-axis $\{11\bar{2}2\}$ twins are activated allowing a contraction along the c-axis (Fig. 2.9) [33]. After compression loading, $\{10\bar{1}1\}$ twins were also observed but only at relatively high deformation temperatures above 400°C [39].

Increasing concentrations of solute atoms in α titanium, such as oxygen or aluminum, suppress the occurrence of twinning. Therefore, twinning as a deformation mode to allow shape changes parallel to the c-axis only plays a major role in pure titanium or in CP titanium with low oxygen concentrations [35].

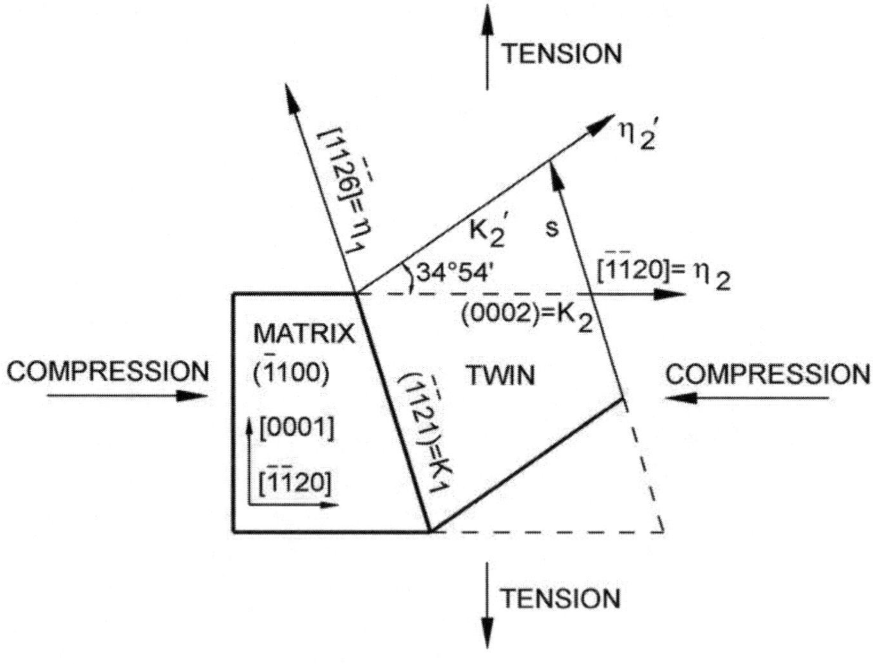

Fig. 2.8. Shape change by $\{11\bar{2}1\}$ twinning [33]

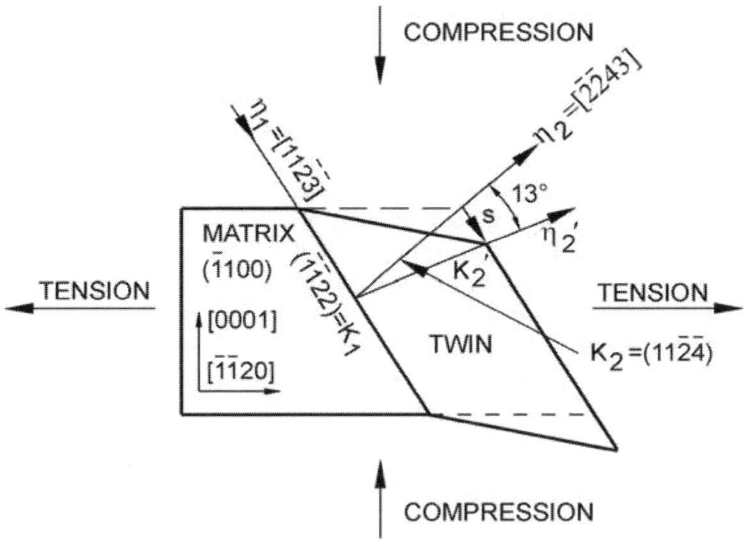

Fig. 2.9. Shape change by $\{11\bar{2}2\}$ twinning [33]

Table 2.3. Twinning elements in α titanium [33]

Twinning plane (1st undeformed plane) (K_1)	Twinning shear direction (η_1)	Second undeformed plane (K_2)	Direction of intersection of plane of shear with K_2 (η_2)	Plane of shear perpendicular to K_1 and K_2	Magnitude of twinning shear
$\{10\bar{1}2\}$	$<10\bar{1}\bar{1}>$	$\{\bar{1}012\}$	$<10\bar{1}1>$	$\{1\bar{2}10\}$	0.167
$\{11\bar{2}1\}$	$<11\bar{2}\bar{6}>$	(0002)	$<11\bar{2}0>$	$\{\bar{1}100\}$	0.638
$\{11\bar{2}2\}$	$<11\bar{2}\bar{3}>$	$\{11\bar{2}\bar{4}\}$	$<22\bar{4}3>$	$\{\bar{1}100\}$	0.225

2.5 Phase Diagrams

Alloying elements in titanium are usually classified into α or β stabilizing additions depending on whether they increase or decrease the α/β transformation temperature of 882°C of pure titanium.

The substitutional element Al and the interstitial elements O, N, and C are all strong α stabilizers and increase the transus temperature with increasing solute content, as can be seen from the schematic phase diagram in Fig. 2.10. Aluminum is the most widely used alloying element in titanium alloys, because it is the only common metal raising the transition temperature and having large solubilities in both the α and β phases. Among the interstitial elements, oxygen can be considered as an alloying element in titanium in those cases where the oxygen content is used to obtain the desired strength level. This is especially true for the different grades of CP titanium. Other α stabilizers include B, Ga, Ge, and the rare earth elements but their solid solubilities are much lower as compared to aluminum or oxygen and none of these elements is used commonly as an alloying element.

The β stabilizing elements are divided into β isomorphous elements and β eutectoid forming elements, depending on the details of the resulting binary phase diagrams. Both types of phase diagrams are shown schematically in Fig. 2.10. The most frequently used β isomorphous elements in titanium alloys are V, Mo, and Nb. Sufficient concentrations of these elements make it possible to stabilize the β phase to room temperature. Other elements belonging to this group that are rarely used or not used at all because of density considerations, are Ta and Re. From the β eutectoid forming elements, Cr, Fe, and Si are used in many titanium alloys, whereas Ni, Cu, Mn, W, Pd, and Bi have only very limited usage. These elements are used only in one or two special purpose alloys. Other β eutectoid forming elements, such as Co, Ag, Au, Pt, Be, Pb, and U, are not used at all in titanium alloys. It should be mentioned, that hydrogen belongs to these β eutectoid forming elements and that the low eutectoid temperature of 300°C in combination with the high diffusivity of hydrogen led to a special process of microstructure refinement, the so-called hydrogenation/dehydrogenation (HDH) process. HDH uses hydrogen as a temporary alloying element, see Sect. 3.5.2. Generally, the maximum hydrogen content in CP titanium and titanium alloys is strictly limited to about 125-150 ppm because of hydrogen embrittlement.

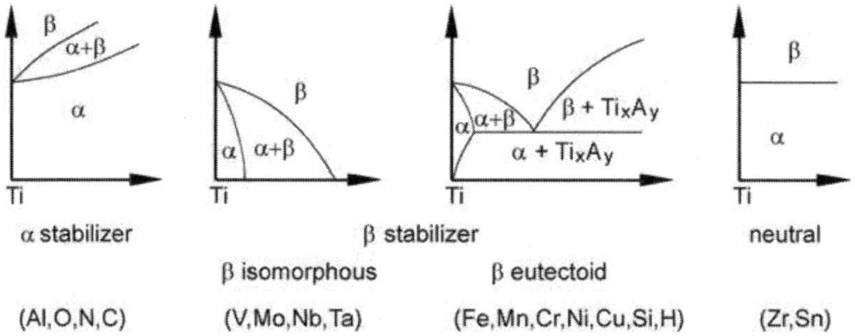

Fig. 2.10. Effect of alloying elements on phase diagrams of titanium alloys (schematically)

In addition, there exist some elements (Zr, Hf and Sn) which behave more or less neutrally (Fig. 2.10), because they lower the α/β transformation temperature only slightly and then increase the transformation temperature again at higher concentrations. Zr and Hf are isomorphous with titanium and, therefore, both exhibit the same β to α allotropic phase transformation. These elements have complete solubilities in the α and β phases of titanium. In contrast, Sn belongs to the β eutectoid forming elements, but has essentially no effect on the α/β transformation temperature. Many commercial multicomponent alloys contain Zr and

Sn but in these alloys both elements are considered and counted as α stabilizing elements. This is because of the chemical similarity of Zr to titanium and because Sn can replace aluminum in the hexagonal ordered Ti_3Al phase (α_2). When present with aluminum, Sn behaves as α stabilizer. This example shows that the interactions between alloying elements make it difficult to understand titanium alloying behavior on the basis of binary Ti-X systems. Rosenberg [40] attempted to express the effect of α stabilizing elements in multicomponent titanium alloys as an equivalent aluminum content by establishing the following equation: $[Al]_{eq.} = [Al] + 0.17\,[Zr] + 0.33\,[Sn] + 10\,[O]$.

A more detailed description of the influence of the various alloying elements on the stability of the α and β phases taking into account electronic and thermodynamic considerations can be found in a summary article by Collings [31].

Although all binary equilibrium phase diagrams of titanium are included in the ASM Handbook on Alloy Phase Diagrams [41], a few of the most important ones will be shown here as illustrations and discussed briefly in the following section.

As already pointed out above, aluminum is the most important α stabilizer and is therefore present in many titanium alloys. The binary Ti-Al phase diagram (Fig. 2.11) shows that with increasing aluminum content the Ti_3Al (α_2) phase will be formed and that the two phase region (α+Ti_3Al) starts at about 5% Al for a temperature of about 500°C. To avoid any appreciable amount of coherent Ti_3Al precipitates in the α phase, the aluminum content in most titanium alloys is limited to about 6%. From Fig. 2.11 it can be seen that for this aluminum level of about 6% the α/β transformation temperature of 882°C for pure titanium is increased to about 1000°C for the two phase region (α+β). In addition to conventional titanium alloys, the Ti-Al phase diagram is also the basis for the so-called titanium-aluminides, which are recently developed alloys based on the two intermetallic compounds Ti_3Al (Alpha-2 alloys and the orthorhombic variant, Ti_2AlNb alloys) and TiAl (Gamma alloys), see Chap. 8.

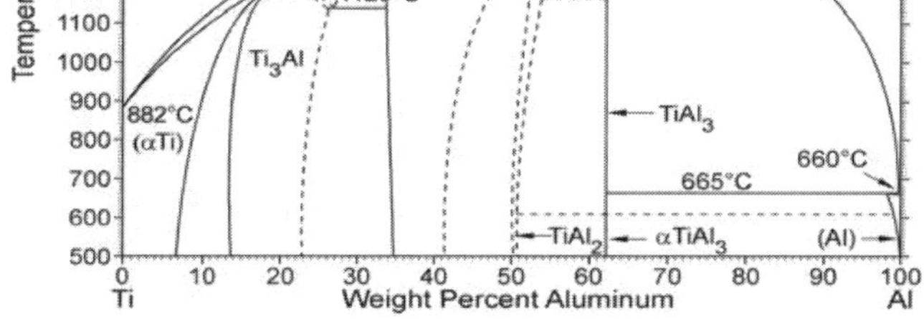

Fig. 2.11. Ti-Al phase diagram [41]

From the three most important β isomorphous elements (V, Mo, and Nb) the Ti-Mo binary phase diagram is selected (Fig. 2.12). This is because it is convenient to calculate an equivalent molybdenum content (as in case of an equivalent aluminum content) for all β stabilizers in multicomponent titanium alloys (see Sect. 2.6.1). For the phase diagram in Fig. 2.12 the older version, published 1958 in the 2nd edition of the Hansen book on Constitution of Binary Alloys [42], is taken rather than the newer version [41] which exhibits a miscibility gap above 20% Mo, in which the β phase separates into two bcc phases (β'+β) outside the (α+β) phase field. The maximum molybdenum content present in conventional titanium alloys is about 15%, so the miscibility gap only adds complexity to the present discussion without helping to understand the effects of alloying additions in the concentration ranges of interest. Indeed, it can be seen from Fig. 2.12 that 15% Mo lowers the β → α+β transformation temperature from 882°C for pure titanium to about 750°C. Furthermore, it can be seen from Fig. 2.12 that the solid solubility of Mo in the α phase is very low (below 1%). The Ti-V and Ti-Nb phase diagrams are qualitatively similar to Fig. 2.12. By adding 15% V, which is also about the maximum vanadium content in conventional titanium alloys, the β → α+β transformation temperature is lowered to about 700°C. The maximum solid solubility of V in the α phase is at 680°C about 3% and therefore much higher as compared to molybdenum. The addition of Nb in conventional titanium alloys is kept within the range of 1-3%, much lower than the maximum amounts of Mo and V. The influence of Nb on the β → α+β transformation temperature is similar to Mo, the transus is lowered to about 750°C by an addition of 15% Nb.

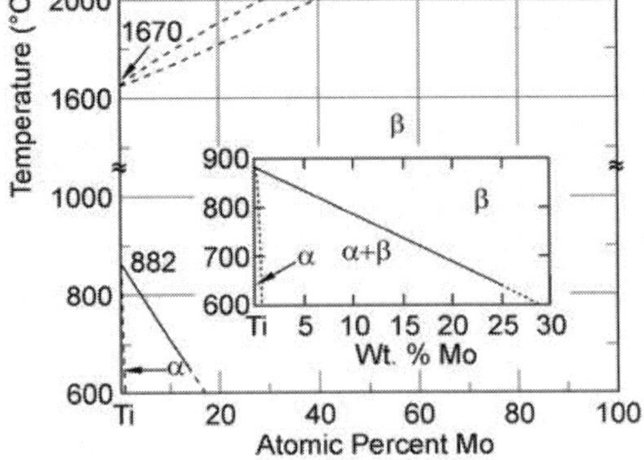

Fig. 2.12. Ti-Mo phase diagram [42]

Among the β eutectoid forming elements, the Ti-Cr binary phase diagram has been chosen and is shown in Fig. 2.13. It can be seen that Cr is an effective β stabilizing element, the eutectoid temperature being 667°C and the eutectoid point lying at about 15% Cr. It is important that the eutectoid decomposition is very sluggish, so that in conventional titanium alloys with Cr contents below 5% the precipitation of the intermetallic compound $TiCr_2$ is avoided. The only exception is the old B120VCA alloy with 11% Cr used in the SR-71 airplane (see Sect. 1.3). This alloy is unstable because it forms $TiCr_2$ precipitates after prolonged exposure to elevated temperatures. This instability is accompanied by a severe loss in ductility. In general, this experience has led to the desire to avoid the formation of the eutectoid compound in this class of alloys. A characteristic of all β eutectoid forming elements is a low solid solubility in the α phase. For example, in the Ti-Cr system (Fig. 2.13) the maximum solubility is only about 0.5%. Consequently, nearly all of the β eutectoid forming element additions partition to the β phase. A second commonly used β eutectoid forming element is Fe which is even a stronger β stabilizing element than chromium, the eutectoid temperature in the Ti-Fe system being about 600°C. The formation of the equilibrium intermetallic compound is avoided in commercial titanium alloys with Fe additions up to a maximum of 5.5% as verified in the recently developed TIMET alloy "low cost beta" (LCB), Ti-1.5Al-5.5Fe-6.8Mo. An exception, where the presence of the compound is desirable, is the β eutectoid forming element Si, which is a common addition to titanium alloys for high temperature application. In this case, the precipitation of the intermetallic compound Ti_5Si_3 is deliberately used for the improvement of creep properties (see Sect. 6.2).

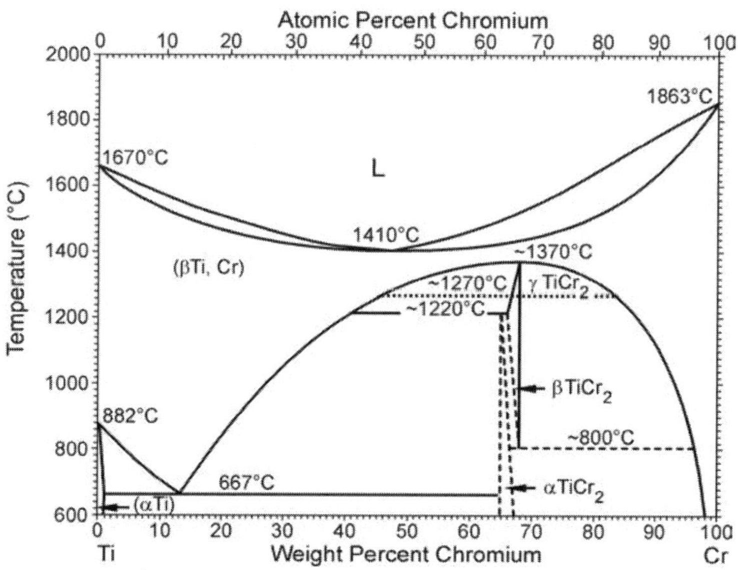

Fig. 2.13. Ti-Cr phase diagram [41]

It should be emphasized, however, that most commercial titanium alloys are multicomponent alloys. As mentioned earlier, the binary phase diagrams can serve only as a qualitative guideline. In principle, ternary or quaternary phase diagrams should be used. This is illustrated in Fig. 2.14 showing isothermal sections through the Ti-rich corner of the Ti-Al-V system at 1000°C, 900°C, and 800°C [31]. In practice, not all the necessary higher order phase diagrams have been determined and judicious use of binary diagrams is the only choice the metallurgists have.

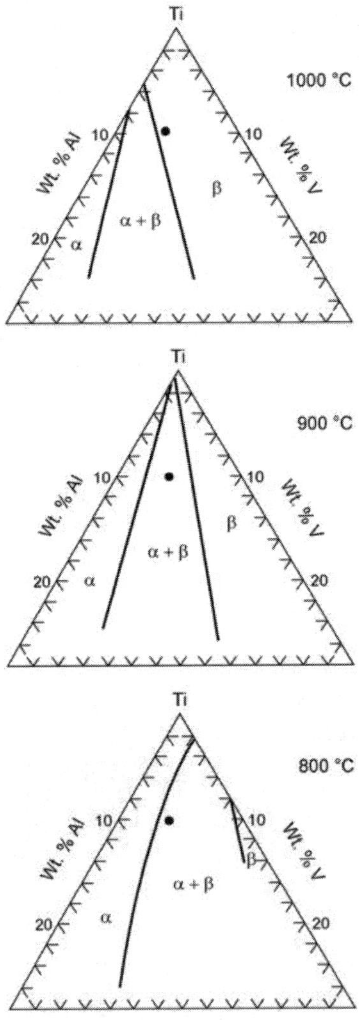

Fig. 2.14. Isothermal sections at 1000°C, 900°C, and 800°C through the Ti-rich corner of the ternary Ti-Al-V phase diagram (solid point: Ti-6Al-4V) [31]

2.6 Phase Transformations

The transformation of the bcc β phase to the hexagonal α phase in CP titanium and titanium alloys can either occur martensitically or by a diffusion controlled nucleation and growth process depending on cooling rate and alloy composition. The crystallographic orientation relationship between α and β has first been studied for zirconium by Burgers [43] and is therefore named the Burgers relationship:

$$(110)_\beta \parallel (0002)_\alpha$$
$$[1\bar{1}1]_\beta \parallel [11\bar{2}0]_\alpha$$

This relationship was confirmed later also for titanium [44]. According to this relationship, a bcc crystal can transform to 12 hexagonal variants, having different orientations with regard to the parent β crystal. This Burgers relationship is closely obeyed for both the martensite transformation and the conventional nucleation and growth process.

2.6.1 Martensite Transformation

The martensite transformation involves the cooperative movement of atoms by a shear type process resulting in a microscopically homogeneous transformation of the bcc into the hexagonal crystal lattice over a given volume. The transformed volume is usually plate shaped or better described geometrically as disk shaped for most titanium alloys. The entire shear transformation process can be reduced to the activation of the following shear systems [45]: $[111]_\beta (11\bar{2})_\beta$ and $[111]_\beta (\bar{1}01)_\beta$ or in hexagonal notation: $[2\bar{1}\bar{1}3]_\alpha (\bar{2}112)_\alpha$ and $[2\bar{1}\bar{1}3]_\alpha (\bar{1}011)_\alpha$. This hexagonal martensite is designated as α' and is observed in two morphologies: massive martensite (other names: lath or packet martensite) and "acicular" martensite [46]. Massive martensite occurs only in pure titanium, very dilute alloys, and in alloys with a high martensitic transformation temperature. "Acicular" martensite occurs in alloys with higher solute content (lower martensitic transformation temperature). The massive martensite consists of large irregular regions (size about 50-100 µm) without any clear internal features visible by light microscopy, but these regions contain packets of small, almost parallel α plates or laths (thickness about 0.5-1 µm) belonging to the same variant of the Burgers relationship. The "acicular" martensite consists of an intimate mixture of individual α plates, each having a different variant of the Burgers relationship (Fig. 2.15). Generally, the martensitic plates contain a high dislocation density and sometimes twins. The hexagonal α' martensite is supersaturated in β stabilizers and, upon annealing in the (α+β) phase field, decomposes to α+β by precipitating incoherent β particles at dislocations or β phase layers at plate boundaries.

With increasing solute content the hexagonal structure of the martensite becomes distorted and, from a crystallographic viewpoint, the crystal structure loses its hexagonal symmetry and must be described as orthorhombic [46]. This orthorhombic martensite is designated α". The α'/α" boundary in terms of solute content is tabulated for some binary titanium systems with transition metals in table 2.4 [31]. For this orthorhombic martensite, the initial stage of decomposition upon annealing in the (α+β) phase field seems to be a spinodal decomposition in solute lean α'' and solute rich α'' regions forming a characteristically modulated microstructure [47], before finally the β phase is precipitated ($\alpha''_{lean} + \alpha''_{rich} \rightarrow \alpha+\beta$). The martensitic start temperature (M_S) of pure titanium depends on the impurity level (oxygen, iron), but lies around 850°C, and increases with increasing amounts of α stabilizers, such as aluminum and oxygen, and decreases with increasing β stabilizer content. Table 2.5 shows the solute contents of some transition metals necessary to suppress the M_S-temperature below room temperature [31]. Using these values of the binary systems, a quantitative rule describing the individual effects of β stabilizing elements in terms of an equivalent Mo content was generated for multicomponent alloys: $[Mo]_{eq.}$ = [Mo] + 0.2 [Ta] + 0.28 [Nb] + 0.4 [W] + 0.67 [V] + 1.25 [Cr] + 1.25 [Ni] + 1.7 [Mn] + 1.7 [Co] + 2.5 [Fe]. Caution should be exercised if attempting to use this equation quantitatively. However, it is a useful qualitative tool which, together with the equivalent Al content derived by Rosenberg [40] (see Sect. 2.5), permits the evaluation of the expected constitution of an alloy with a given chemistry.

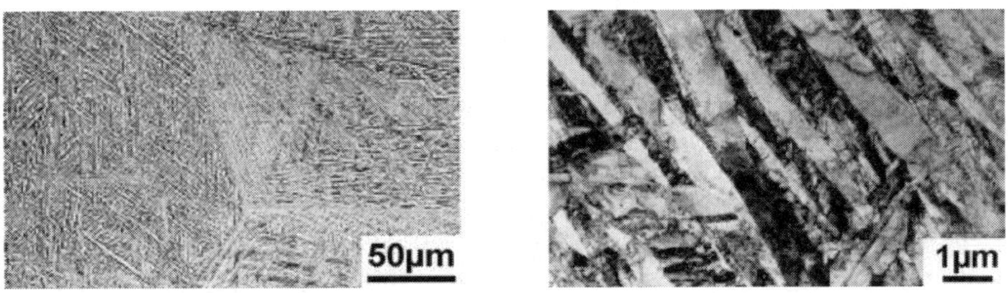

a **b**

Fig. 2.15. "Acicular" martensite in Ti-6Al-4V quenched from the β phase field: (a) LM (b) TEM

Although not being relevant to any practical application of titanium alloys, it should be mentioned that in many alloys in which the martensitic reaction is suppressed, the β phase decomposes upon quenching athermally to the so-called athermal ω phase. The athermal ω phase forms as uniform dispersion of extremely fine particles (size 2-4 nm). These have been suggested to be a precursor to the martensitic reaction because the athermal transformation involves a shear displacement in the $\langle 111 \rangle$ direction of the bcc lattice, see Fig. 2.16 [31] showing the (222) planes of the bcc lattice. From a crystallographic viewpoint, the athermal ω phase has a trigonal symmetry in heavily β stabilized alloys and a hexagonal symmetry (not hexagonal close-packed) in leaner alloys. The transition from hexagonal to trigonal is continuous as a function of alloy content. From a standpoint of dislocation movement in the bcc β structure, the ω particles have a diffuse, coherent interface and the structure is an elastically distorted bcc lattice, i.e. moving dislocations in the β phase can cut all four variants of ω particles. Upon annealing in the metastable (ω+β) phase field, the athermal ω phase grows to form the so-called isothermal ω phase which has the same crystallographic symmetry as the athermal ω but is solute lean with respect to the β phase, see Sect. 2.8.2.

Table 2.4. Composition of the α'/α" (hexagonal/orthorhombic) martensite boundary in some binary titanium systems with transition metals [31]

α'/α" Boundary	V	Nb	Ta	Mo	W
Wt%	9.4	10.5	26.5	4	8
At%	8.9	5.7	8.7	2.0	2.2

Table 2.5. Concentration of some transition metals needed to retain the β phase at room temperature in binary titanium alloys [31]

	V	Nb	Ta	Cr	Mo	W	Mn	Fe	Co	Ni
Wt%	15	36	50	8	10	25	6	4	6	8
At%	14.2	22.5	20.9	7.4	5.2	8	5.3	3.4	4.9	6.6

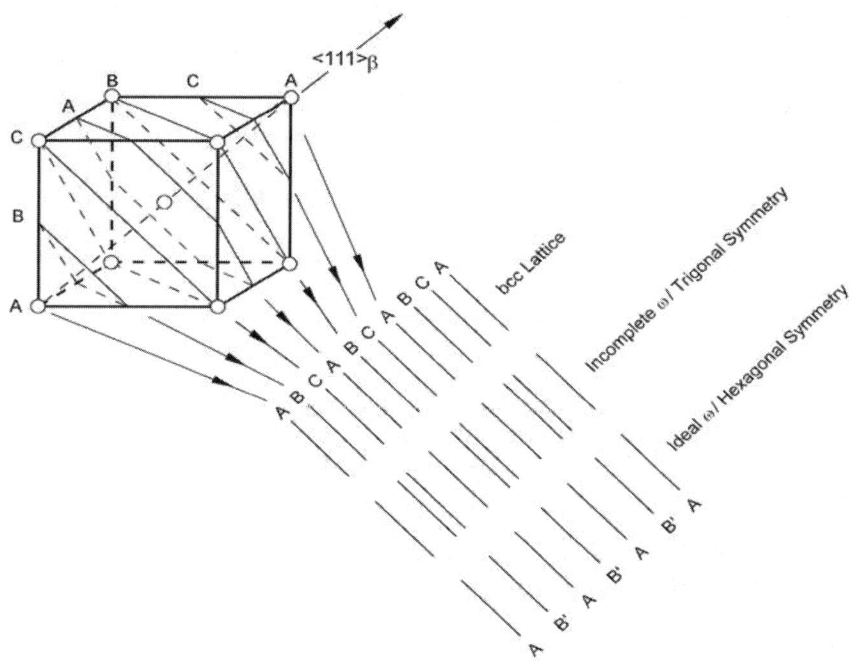

Fig. 2.16. Representation of the β → ω transformation showing the (222) planes of the bcc lattice [31]

2.6.2 Nucleation and Diffusional Growth

When titanium alloys are cooled at sufficiently low rates from the β phase field into the (α+β) phase field, the α phase, which is incoherent with respect to the β phase, first nucleates preferentially at β grain boundaries leading to a more or less continuous α layer along β grain boundaries. During continued cooling α plates nucleate either at the interface of the continuous α layer or at the β grain boundary itself and grow into the β grain as parallel plates [48] belonging to the same variant of the Burgers relationship (so-called α colony). They continue to grow into the β grain interior until they meet other α colonies nucleated at other grain boundary areas of the β grain and belonging to other variants of the Burgers relationship. This process is often called sympathetic nucleation and growth. The individual α plates are separated within the α colonies by the retained β matrix, which are commonly, but incorrectly, called β plates. The α and β plates are also often called α and β lamellae and the resulting microstructure is then designated as lamellar. An example of such a microstructure which was obtained by slow cooling from the β phase field is shown in Fig. 2.17 for the Ti-6Al-4V alloy. In such slowly cooled material, the size of the α colonies can be as large as half of the β grain size.

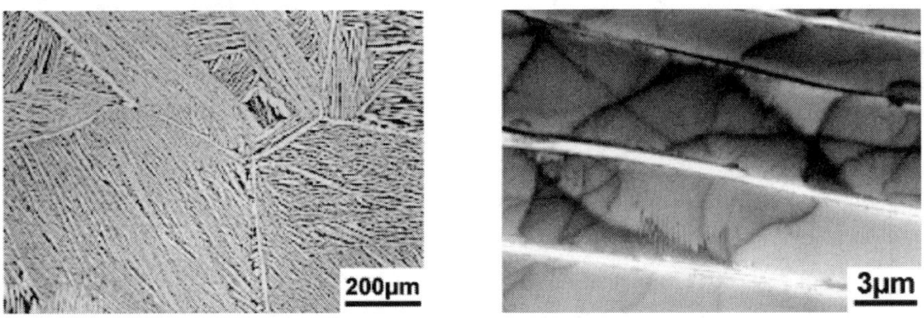

a b

Fig. 2.17. Lamellar α+β microstructure in Ti-6Al-4V slowly cooled from the β phase field: (a) LM (b) TEM

The crystallographic relationship between α and β plates within a colony is schematically shown in Fig. 2.18. It can be seen that the Burgers relationship $(110)_\beta \parallel (0002)_\alpha$ and $[1\bar{1}1]_\beta \parallel [11\bar{2}0]_\alpha$ is strictly obeyed and that the flat surface of the α plates is parallel to the $(\bar{1}100)$ plane of the α phase and parallel to the $(\bar{1}12)$ plane of the β phase. It should be pointed out again, that these flat surfaces are nearly equiaxed shaped (disk shaped) and the diameter is usually called α plate length.

With increasing cooling rate the size of the α colonies as well as the thickness of the individual α plates become smaller. Colonies nucleated at β grain boundaries cannot fill the whole grain interior anymore and colonies start to nucleate also on boundaries of other colonies. To minimize the overall elastic strains, the new α plates nucleating by "point" contact on the broad face of an existing α plate tend to grow nearly perpendicular to that plate. This selective nucleation and growth mechanism in combination with the smaller number of α plates within the colonies leads to a characteristic microstructure called "basket weave" structure or Widmanstätten structure. For a given cooling rate this type of "basket weave" structure is observed more frequently in alloys with higher contents of β stabilizing elements, especially with slow diffusing elements. It should be mentioned that during continuous cooling from the β phase field the incoherent α plates are not nucleating homogeneously throughout the β matrix.

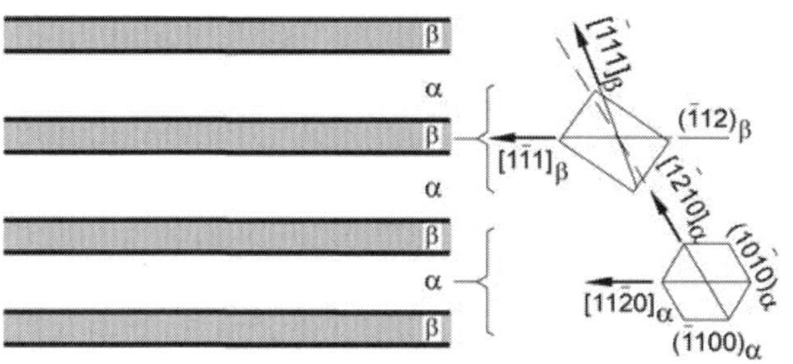

Fig. 2.18. Schematical representation of the crystallographic relationship between α plates and β matrix in α colonies

2.7 Alloy Classification

Commercial titanium alloys are classified conventionally into three different categories (α, α+β, and β alloys) according to their position in a pseudo-binary section through a β isomorphous phase diagram, schematically shown in Fig. 2.19. A list of the most important commercial alloys belonging to each of these three different groups is shown in Table 2.6. In this table the common name, the nominal composition, and the nominal β transus temperature are stated for each alloy.

The group of alloys called α alloys in Table 2.6 consists of the various grades of CP titanium and α alloys which upon annealing well below the β transus contain only small amounts of β phase (2-5 vol%) stabilized by iron. The β phase is helpful in controlling the recrystallized α grain size and improves the hydrogen tolerance of these alloys. The four different grades of CP titanium differ with respect to their oxygen content from 0.18% (grade 1) to 0.40% (grade 4) in order to increase the yield stress level. The two alloys Ti-0.2Pd and Ti-0.3Mo-0.8Ni offer better corrosion resistance than CP titanium. Their common names are grade 7 and grade 12, respectively, and the iron and oxygen limits are identical to grade 2 of CP titanium. Ti-0.2Pd offers better corrosion resistance but is more expensive than Ti-0.3Mo-0.8Ni. The α alloy Ti-5Al-2.5Sn (0.20% oxygen) has a much higher yield stress level (780-820 MPa) than the CP titanium grades (grade 4: 480 MPa). It can be used at service temperatures up to 480°C and in its ELI (extra low interstitials) version with 0.12% oxygen also at low temperatures (–250°C). It is an old alloy, first manufactured in 1950, but still on the market, although it is being replaced by Ti-6Al-4V in many applications.

Table 2.6. Important commercial titanium alloys

Common Name	Alloy Composition (wt%)	T_β (°C)
α Alloys and CP Titanium		
Grade 1	CP-Ti (0.2Fe, 0.18O)	890
Grade 2	CP-Ti (0.3Fe, 0.25O)	915
Grade 3	CP-Ti (0.3Fe, 0.35O)	920
Grade 4	CP-Ti (0.5Fe, 0.40O)	950
Grade 7	Ti-0.2Pd	915
Grade 12	Ti-0.3Mo-0.8Ni	880
Ti-5-2.5	Ti-5Al-2.5Sn	1040
Ti-3-2.5	Ti-3Al-2.5V	935
α+β Alloys		
Ti-811	Ti-8Al-1V-1Mo	1040
IMI 685	Ti-6Al-5Zr-0.5Mo-0.25Si	1020
IMI 834	Ti-5.8Al-4Sn-3.5Zr-0.5Mo-0.7Nb-0.35Si-0.06C	1045
Ti-6242	Ti-6Al-2Sn-4Zr-2Mo-0.1Si	995
Ti-6-4	Ti-6Al-4V (0.20O)	995
Ti-6-4 ELI	Ti-6Al-4V (0.13O)	975
Ti-662	Ti-6Al-6V-2Sn	945
IMI 550	Ti-4Al-2Sn-4Mo-0.5Si	975
β Alloys		
Ti-6246	Ti-6Al-2Sn-4Zr-6Mo	940
Ti-17	Ti-5Al-2Sn-2Zr-4Mo-4Cr	890
SP-700	Ti-4.5Al-3V-2Mo-2Fe	900
Beta-CEZ	Ti-5Al-2Sn-2Cr-4Mo-4Zr-1Fe	890
Ti-10-2-3	Ti-10V-2Fe-3Al	800
Beta 21S	Ti-15Mo-2.7Nb-3Al-0.2Si	810
Ti-LCB	Ti-4.5Fe-6.8Mo-1.5Al	810
Ti-15-3	Ti-15V-3Cr-3Al-3Sn	760
Beta C	Ti-3Al-8V-6Cr-4Mo-4Zr	730
B120VCA	Ti-13V-11Cr-3Al	700

Classifying titanium alloys by their constitution (α, α+β, and β) is convenient but can be misleading. For example, essentially all α alloys contain a small amount of β phase as mentioned above and as will be discussed in more detail in

Chap. 4, Sect. 4.1.1. Perhaps a better criterion for α alloys is the lack of heat treatment response. Following this criterion, the alloy Ti-3Al-2.5V is best classified as α titanium alloy, as shown in Table 2.6. This alloy which is often called "Half Ti-6-4" has excellent cold formability and is used mainly as seamless tubing in aerospace applications and for sporting goods.

The group of α+β alloys (Table 2.6) has a range in the phase diagram (Fig. 2.19) from the α/α+β phase boundary up to the intersection of the M_S-line with room temperature, thus α+β alloys transform martensitically upon fast cooling from the β phase field to room temperature. Alloys which contain in equilibrium only a small volume fraction of β phase (less than about 10 vol%) are also often called "near α" alloys and their main usage is at high service temperatures. Although they belong per definition to the α+β alloys and are listed in that way in Table 2.6, some special features of these high temperature titanium alloys will be discussed in a separate chapter (Chap. 6) distinguishing them from the common α+β alloys (Chap. 5) dominated by the Ti-6Al-4V alloy. The Ti-6Al-4V alloy contains in equilibrium at 800°C about 15 vol% β phase (see Fig. 2.14). This alloy has an exceptional good balance of strength, ductility, fatigue, and fracture properties but can be used only up to temperatures of about 300°C. The ELI version of this popular alloy has especially high fracture toughness values and excellent damage tolerance properties.

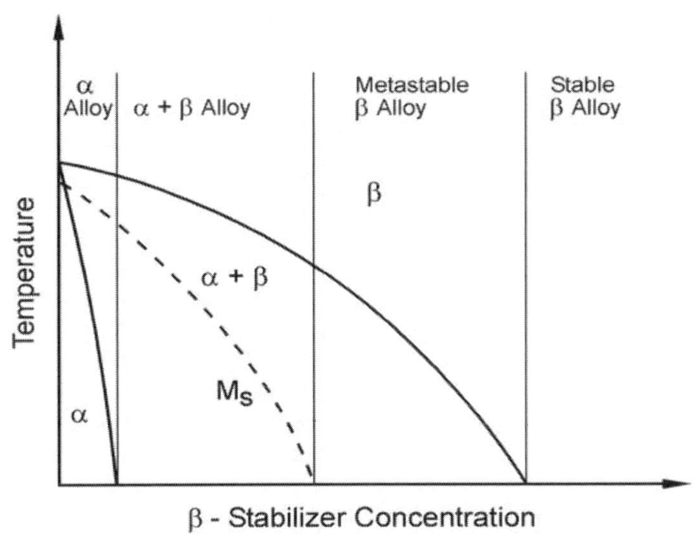

Fig. 2.19. Pseudo-binary section through a β isomorphous phase diagram (schematically)

All alloys in the group of β alloys in Table 2.6 are actually metastable β alloys because they all are located in the equilibrium (α+β) phase region of the phase diagram (Fig. 2.19). Since stable β alloys located in the β single phase field

(Fig. 2.19) do not exist as commercial materials, the expression β alloys is commonly used and also used throughout this book for the metastable β alloys.

The characteristic feature of the β alloys is that they do not transform martensitically upon fast cooling from the β phase field. The two alloys on top of the list in Table 2.6, Ti-6246 and Ti-17, are usually found in the group of α+β alloys [32, 49]. For the Ti-6246 alloy it was recently shown at the Technical University Hamburg-Harburg that all apparent martensite was attributable to artifacts introduced during normal specimen preparation, e.g. stress induced martensite by mechanical polishing for light microscopic and X-ray investigations or thin foil effects in case of TEM observations. There are numerous possibilities to form artifacts in titanium alloys during specimen preparation as will be discussed in more detail in Sects. 3.9.1 and 3.9.2. When the surface layer influenced by mechanical polishing was removed completely by electrolytic polishing, no martensite was found in the Ti-6246 alloy investigated (see Sect. 3.9.1, Fig. 3.79). This material was obtained from an alloy heat containing about 0.10% oxygen. In contrast, heats containing 0.15% oxygen have been reported to form α" martensite during quenching. Based on X-ray diffraction results, this α" martensite was concluded to be a bulk transformation product. For Ti-17, which contains more β stabilizers than Ti-6246, there exists strong evidence in the literature [50] that this alloy does not transform martensitically. Common to all β alloys is that relatively fast cooling from the β phase field and aging in the temperature region of about 500-600°C results in yield stress levels in excess of 1200 MPa. This high yield stress is due to the homogeneous precipitation of fine α platelets from the metastable precursor phases, ω or β', which are formed either during the cooling process to room temperature or during subsequent heating to the aging temperature (see Sect. 2.8.2). By comparison, the maximum yield stress level which can be obtained for α+β alloys, using the same cooling rates and optimum aging treatments, is only about 1000 MPa. This is because relatively coarse α plates form during the cooling process. These coarse α plates are either arranged in colonies or form as individual plates depending on alloy type, as pointed out earlier in Sect. 2.6.2.

Although the number of commonly used β titanium alloys in Table 2.6 is as large as the number of α+β alloys, it should be kept in mind, that the percentage of β alloy usage on the total titanium market is very low, (see Sect. 1.4). However, this percentage of β alloy usage is steadily increasing due to the attractive properties, especially the high yield stress level and for some applications (e.g. springs) the low modulus of elasticity.

2.8 Basic Hardening Mechanisms

From the four different hardening mechanisms in metallic materials (solid solution hardening, strengthening by a high dislocation density, boundary hardening, and precipitation hardening) solid solution and precipitation hardening are present in all commercial titanium alloys and should be discussed separately for the α phase (Sect. 2.8.1) and for the β phase (Sect. 2.8.2). Boundary hardening plays a significant role in α+β alloys cooled at high rates from the β phase field reducing the α-colony size to a few α plates or causing martensitic transformation. In both cases, a high dislocation density is created which also contributes to hardening. It should be pointed out, however, that the martensite in titanium is much softer than the martensite in Fe-C alloys because the interstitial oxygen atoms only cause small elastic distortion of the hexagonal lattice of the titanium martensite. This is in sharp contrast to carbon and nitrogen that cause severe tetragonal distortion of the bcc lattice in ferrous martensite.

2.8.1 Hardening of the Alpha Phase

The α phase is significantly hardened by the interstitial element oxygen. This can best be seen by comparing the yield stress values of the CP titanium grades 1-4 with oxygen levels between 0.18 and 0.40% (Table 2.6). By this increase in oxygen content the yield stress is increased from 170 MPa (grade 1) to 480 MPa (grade 4). In commercial titanium alloys the oxygen content varies between about 0.08% and 0.20% depending on alloy type. Substitutional solid solution hardening of the α phase is caused mainly by the elements Al, Sn, and Zr which have fairly large atomic size differences to titanium and also large solid solubilities in the α phase.

Precipitation hardening of the α phase occurs by coherent Ti_3Al particles above about 5% Al, see Ti-Al phase diagram in Fig. 2.11. These Ti_3Al or α_2 particles have an ordered hexagonal structure, crystallographically a DO_{19} structure, and since they are coherent, they can be sheared by moving dislocations resulting in planar slip and extensive dislocation pile-ups against boundaries. With increasing size, these α_2 particles become ellipsoidal in shape, the long axis being parallel to the c-axis of the hexagonal lattice [51]. This α_2 phase is further stabilized by the elements oxygen and tin, i.e. the ($\alpha+\alpha_2$) phase region is pushed to higher temperatures by these elements. In such cases, Sn substitutes for Al whereas oxygen remains as an interstitial.

Upon annealing of α+β alloys in the two phase region (α+β), significant alloy element partitioning takes place and the α phase is enriched in α stabilizing elements (Al, O, Sn). Substantial volume fractions of coherent α_2 particles can then be precipitated in the α phase by aging for example at 500°C (Ti-6Al-4V, IMI 550), at 550°C (IMI 685), at 590°C (Ti-6242), or even at 700°C (IMI 834). The very high density of these homogeneously distributed α_2 particles in the α phasc is shown in the dark field transmission electron micrograph in Fig. 2.20 for the IMI 834 alloy.

A change from wavy to planar slip is found in pure α titanium with increasing oxygen content [52], similar to that observed for microstructures containing coherent α_2 precipitates,. This observation might indicate that the oxygen atoms are not distributed homogeneously but tend to form zones with short range order. Also, a synergy between oxygen and Al atoms with regard to promoting planar slip has been demonstrated.

Although it has limited relevance to commercial titanium alloys, it should be mentioned that aging of the modulated microstructure, which results from the spinodal decomposition of the orthorhombic α" martensite (see Sect. 2.6.1), leads to a drastic increase in yield stress [30]. This modulated structure can be considered as an array of very small, closely spaced precipitates. In this case, the disordered, solute rich zones become stronger obstacles to dislocation motion as their size and misfit increase with aging. Due to the large volume fraction of these zones in those modulated microstructures, the material becomes brittle on a macroscopic scale. The reason is the intense strain localization that occurs in this microstructure because the modulated zones are destroyed within the slip bands. This leads to the effect that the first slip band in the largest α" martensite plate develops sufficient strain to cause crack nucleation at the plate boundary. The fracture mechanism is microvoid coalescence and growth, not cleavage [30].

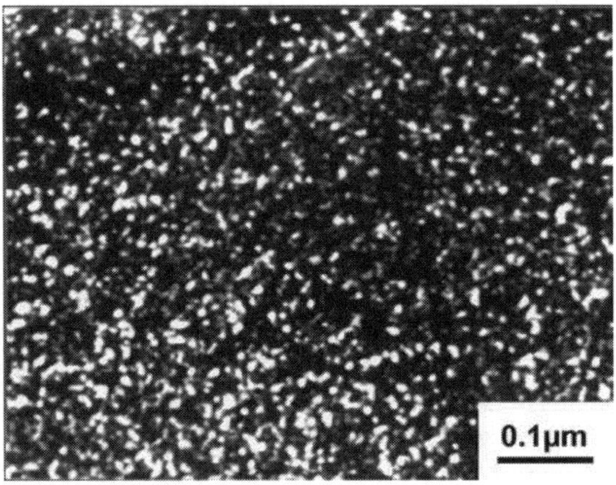

Fig. 2.20. Dark field micrograph of α_2 particles in IMI 834 aged 24 h at 700°C, TEM

2.8.2 Hardening of the Beta Phase

It is difficult to analyze solid solution strengthening of the β phase in a traditional sense because in fast cooled metastable β alloys the precursors of the metastable ω and β' phases cannot be separated from the solute effects. Also, in fully aged microstructures it is difficult to separate the hardening due to α precipitates from the solute effects. In this case, an important solid solution strengthening effect of the β phase is due to alloy element partitioning that accompanies α precipitation. One way to estimate the solid solution strengthening of the β stabilizing elements Mo, V, Nb, Cr, and Fe is to examine the slope of the lattice parameters versus solute content curves for binary alloys pointing out in a qualitative way the size misfit parameter. Data of this type can be found in Pearson's Handbook [53]. From this data it can be seen that the steepest slope is present for Ti-Fe followed by Cr and V, with Nb and Mo having a smaller effect on the lattice parameter.

Precipitation hardening of the β phase is the most effective way to increase the yield stress of commercial β titanium alloys. There are two metastable phases, ω and β', which are observed in β titanium alloys as shown by the schematic phase diagram in Fig. 2.21 [54]. In both cases, it is a miscibility gap into two bcc phases, β_{lean} and β_{rich}, the major difference being the magnitude of distortion of the bcc lattice in the coherent disordered precipitates (β_{lean}) with respect to the bcc lattice of the matrix (β_{rich}). In highly concentrated alloys the magnitude of distortion is very small and the metastable particles are called β' having a bcc crystal structure. In case of less concentrated alloys the distortion of the bcc lattice in the precipitates is much higher and the metastable particles are called isothermal ω having a hexagonal crystal structure from a crystallographic viewpoint (see discussion on athermal ω in Sect. 2.6.1 and Fig. 2.16).

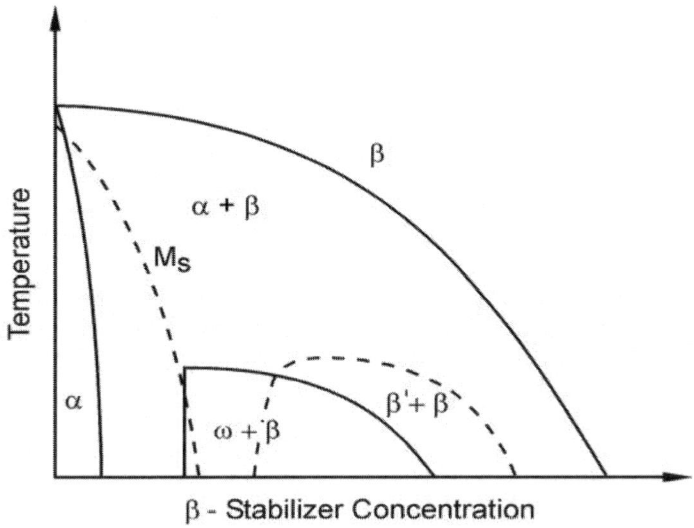

Fig. 2.21. Metastable (ω+β) and (β'+β) phase fields in a β isomorphous phase diagram (schematically)

The isothermal ω particles have either an ellipsoidal or a cuboidal shape depending on the precipitate/matrix misfit [55]. At low misfits, the ω particles are ellipsoidal with the long axis being parallel to one of the four <111> directions of the bcc lattice. An example is shown in the dark field micrograph in Fig. 2.22 imaging one of the four variants of the ellipsoidal ω particles in a Ti-16Mo alloy aged 48 h at 450°C. At larger misfits, the ω particles become cuboidal with the flat surfaces being parallel to the {100} planes of the bcc lattice. An example is shown in Fig. 2.23 for a Ti-8Fe alloy aged 4 h at 400°C.

The β' precipitates have a morphology which varies from spheres or cuboids in Ti-Nb and Ti-V-Zr alloys to plates in Ti-Cr alloys [55], again depending on the magnitude of misfit and coherency strains. One example for the solute lean β' precipitates is shown in Fig. 2.24 for the Ti-15Zr-20V alloy aged 6 h at 450°C.

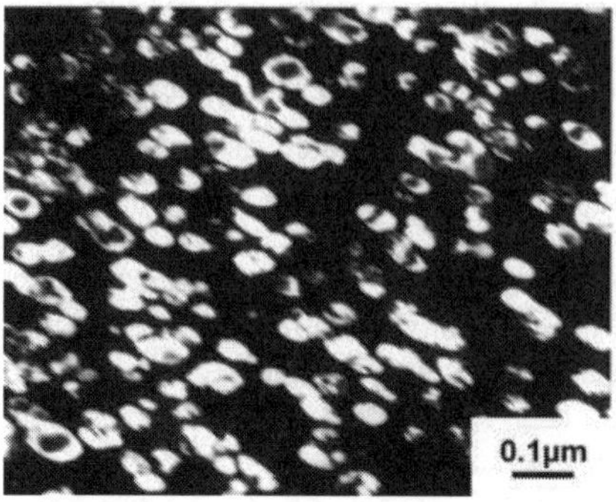

Fig. 2.22. Dark field micrograph of ellipsoidal ω precipitates in Ti-16Mo aged 48 h at 450°C, TEM

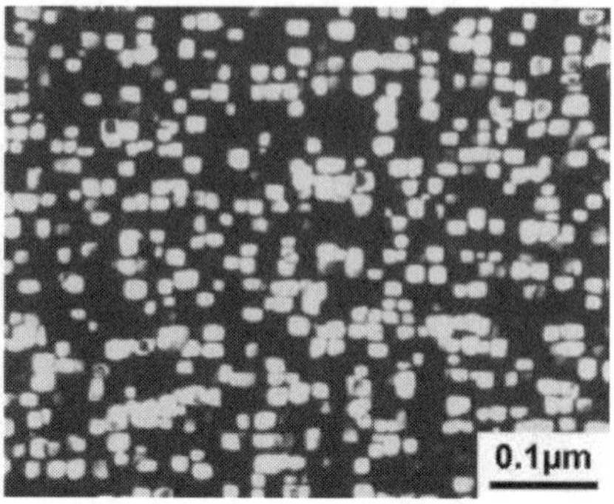

Fig. 2.23. Dark field micrograph of cuboidal ω precipitates in Ti-8Fe aged 4 h at 400°C, TEM

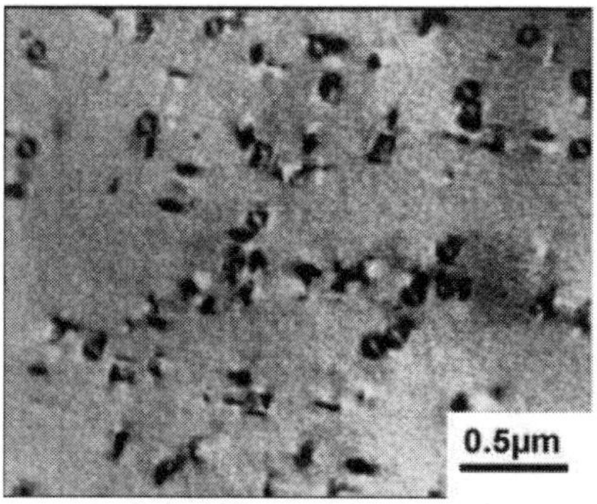

Fig. 2.24. Coherent β' particles in Ti-15Zr-20V aged 6 h at 450°C, TEM

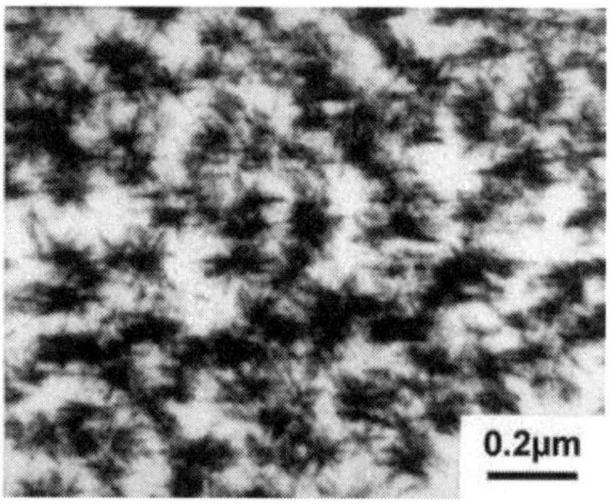

Fig. 2.25. Precipitation of small α platelets at β' particles in Ti-15.6Mo-6.6Al aged 100 h at 350°C, TEM

Both ω and β' are coherent and are sheared by the moving dislocations. This results in the formation of intense, localized slip bands leading to early crack nucleation and low ductility [56]. Therefore, microstructures containing these precipitates are usually avoided in commercial β titanium alloys. Instead, commercial β titanium alloys are aged at slightly higher temperatures in order to precipitate in reasonable aging times incoherent particles of the stable α phase using ω or β' particles as precursors and nucleation sites. Sometimes it is necessary to use a step aging treatment as will be discussed in more detail at the end of this section. Using these precursors, it is possible to obtain a fairly homogeneous distribution of small α platelets. An example of the early stages of α nucleation is shown in Fig. 2.25 for Ti-15.6Mo-6.6Al aged for a long time (100 h) at 350°C [54]. The size and distribution of α platelets in a commercial β titanium alloy, the β-CEZ alloy developed by CEZUS in France, with the recommended aging treatment of 8 h at 580°C is shown in Fig. 2.26. These α platelets also obey the Burgers relationship and the flat surface of the platelets is parallel to one of the {112} planes of the β matrix, as described in Sect. 2.6.2, see also Fig. 2.18. As pointed out already in that section and as can be seen in Fig. 2.26, not all twelve possible variants are nucleated statistically. Instead, in order to minimize the overall elastic strains only two or three variants being nearly perpendicular to each other dominate in a given volume of a β grain.

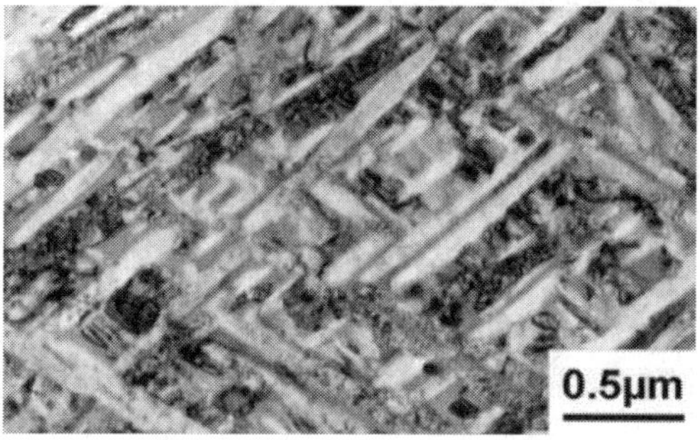

Fig. 2.26. Size and distribution of α platelets in the commercial β titanium alloy β-CEZ aged 8 h at 580°C, TEM

Since these incoherent α platelets are too small to deform plastically they act as hard, undeformable particles. Consequently, high yield stresses can be obtained in β titanium alloys with such microstructures. The yield stress in those alloys can be easily lowered, i.e. adjusted to the desired level, by applying a two-stage heat treatment. In this case, the first annealing step is done at high temperatures in the (α+β) phase field to precipitate a desired volume fraction of large α plates and reducing thereby the volume fraction of small α platelets in the second aging step at lower temperature. The large α plates contribute less to the yield stress than the small α platelets because they are large enough to deform plastically and only boundary hardening is present as a hardening mechanism for the large α plates. However, for all microstructures with α precipitates the dislocation density increases in the β matrix during α precipitation. Therefore, dislocation hardening is also contributing to the yield stress.

It was already pointed out in Sect. 2.6.2 that the α phase always nucleates preferentially at β grain boundaries forming continuous α layers. Especially for β alloys, hardened by small α platelets to high yield stress levels, these continuous α layers become deleterious to the mechanical properties. An example of such a microstructure is shown in Fig. 2.27 for the β alloy Ti-10-2-3. The main intention of the thermo-mechanical processing of β alloys is to eliminate or reduce this negative effect of the continuous α layers on mechanical properties (see Sect. 7.1).

In β alloys with high concentrations of β stabilizing elements it is sometimes difficult to achieve a homogeneous distribution of α platelets by normal aging treatments, especially with aging temperatures above the metastable two phase region, because the formation of the precursors or the α nucleation at the precursors (ω or β') is too sluggish to occur during the heating period to the aging temperature. In such a case, a pre-aging treatment at lower temperature might create a more homogeneous distribution of α platelets, as shown in Fig. 2.28 for the β alloy Beta C [57]. Another possibility is to cold work prior to aging which results in a more homogeneous distribution of α platelets by nucleation at dislocations.

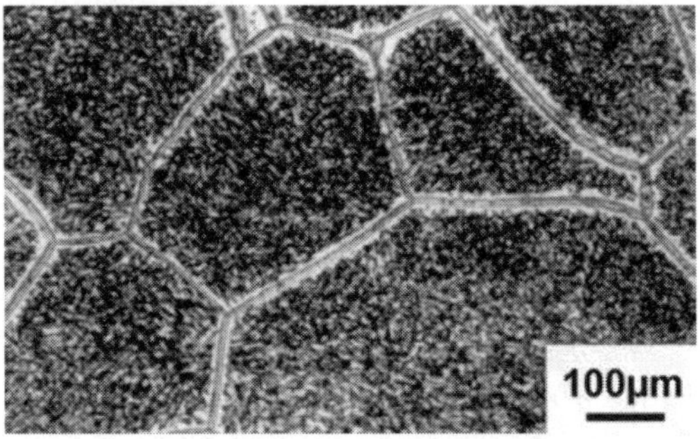

Fig. 2.27. Continuous α layers at β grain boundaries in the β alloy Ti-10-2-3, LM

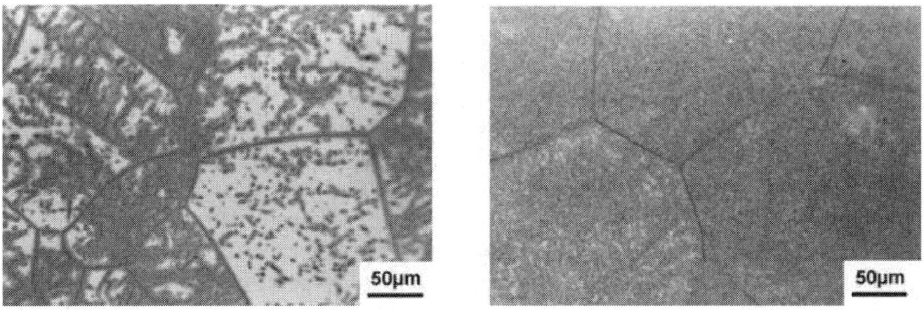

a b

Fig. 2.28. Effect of pre-aging on distribution of α platelets in the β alloy Beta C, LM: (a) 16 h 540°C (b) 4 h 440°C + 16 h 560°C

2.9 Basic Physical and Chemical Properties

For most application purposes the majority of physical and chemical properties of titanium are far less important than the mechanical properties. Noteworthy exceptions are the low density and the formation of the surface oxide layer which imparts very good corrosion resistance. Most of the properties will be discussed in general terms, only a few of them will be treated in some detail. Among these are diffusion (Sect. 2.9.1), corrosion behavior (Sect. 2.9.2), and oxidation (Sect. 2.9.3). The modulus of elasticity was discussed already in Sect. 2.3.

In addition to the basic characteristics, already shown in Table 2.1, other selected physical properties of titanium and its alloys are listed in Table 2.7 [32]. These are compared to other metallic structural materials. The property values for high purity α titanium shown in Table 2.7 are not significantly different than those for the various CP titanium grades. This shows that oxygen concentrations up to 0.40% only slightly affect these properties. On the other hand, if the values for the α+β alloy Ti-6Al-4V and for the β alloy Ti-15-3 are compared with those of pure α titanium, it can be seen that the thermal conductivity and the electrical resistivity vary significantly. Thermal conductivity is lower and electrical resistivity is higher for these commercial alloys, whereas the linear expansion coefficient and the specific heat capacity are only slightly affected. The thermal conductivity and the electrical resistivity both depend on the density and extent of scattering of the conductive electrons. The increase in electrical resistivity with increasing solute content in binary titanium alloys is shown in Fig. 2.29 [31]. It can be seen that there are two branches in the dependency, the upper branch contains the elements which show the tendency for ordering in α titanium and the lower branch consists of the elements showing the tendency for inmiscibility (V, Nb) or behaving completely neutral (Zr). It should be pointed out, that oxygen belongs to the upper branch because the electrical resistivity of CP titanium grade 4 with 0.40% oxygen is 0.60 μΩ m (thermal conductivity: 17 W m^{-1} K^{-1}). In addition to the electrical resistivity of titanium alloys described in Table 2.7, a number of β titanium alloys have been shown to exhibit superconducting behavior. This behavior will be addressed in this book in Chap. 10 (Special Properties and Applications of Titanium).

Table 2.7. Some physical properties of titanium and its alloys as compared to other structural metallic materials

	Linear Thermal Expansion Coefficient (10^{-6} K^{-1})	Thermal Conductivity (W m^{-1} K^{-1})	Specific Heat Capacity (J kg^{-1} K^{-1})	Electrical Resistivity (μΩ m)
α Titanium	8.4	20	523	0.42
Ti-6Al-4V	9.0	7	530	1.67
Ti-15-3	8.5	8	500	1.4
Fe	11.8	80	450	0.09
Ni	13.4	90	440	0.07
Al	23.1	237	900	0.03

Comparing the values for titanium with those of other structural metallic materials it can be seen that the linear thermal expansion coefficient is lower for titanium. Consequently, titanium alloys are an excellent choice for applications requiring a high strength to density ratio and a low thermal expansion. Examples include casings for aero-engines and connecting rods in automobile engines. Unfortunately due to the high price of titanium, the application as connecting rods is limited to high performance, high-priced vehicles. It also should be pointed out, that the linear thermal expansion coefficient for α titanium is about 20% higher parallel to the c-axis than perpendicular to the c-axis. This becomes important if highly textured Ti-6Al-4V material is used as connecting rod material.

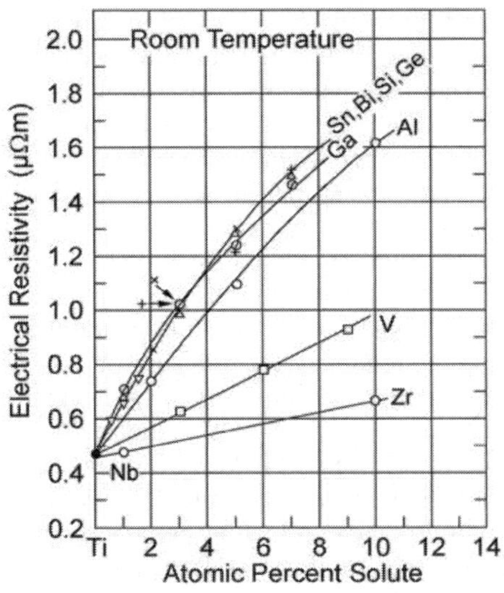

Fig. 2.29. Electrical resistivity of binary titanium alloys [31]

The thermal conductivity of titanium is significantly lower as compared to Fe, Ni, and Al (Table 2.7) which affects for example the obtainable cooling rates after processing and heat treatments, as well as the heating times. The high electrical resistivity of titanium as compared to the other metals in Table 2.7 limits the usefulness of titanium as an electrical conductor. The specific heat capacity of titanium has the same order of magnitude as the other metals listed in Table 2.7.

The strength to density advantage of titanium compared to other structural metallic materials was already illustrated in Table 2.1 for typical α+β titanium alloys (yield stress 1000 MPa, density 4.5 g cm^{-3}). This advantage is not significantly increased for high strength β alloys with typical yield stress values of 1200 MPa, because most β alloys contain heavy elements like Mo and the density of those alloys is increased by about 5% and can be as high as 4.94 g cm^{-3} in the case of the β alloy Beta 21S.

2.9.1 Diffusion

Knowledge of the diffusion rates of interstitial and substitutional alloying elements in the α and β phases of titanium, as well as the self-diffusion rates, is important. Many of the production processes, such as solution and aging heat treatments, hot working, and recrystallization temperatures, are diffusion dependent. So are many application aspects, for example creep, oxidation behavior, and hydrogen embrittlement.

Many diffusivity data were measured especially during the first twenty years of commercialization of titanium and these data are well documented in the German titanium book by Zwicker in 1974 [58]. The intrinsic problem of scatter in the diffusion data, partially due to the measurement method used, was pointed out by Liu and Welsch in their literature survey on diffusivities of oxygen, aluminum, and vanadium in α and β titanium in 1987 [59], done probably with respect to the widely used Ti-6Al-4V alloy. In a recently published overview article [60], new diffusivity data as well as improvements in the understanding of diffusivities, especially for the α phase, are presented.

A selection of diffusivity data is shown in Fig. 2.30 in the form of Arrhenius plots. It can be seen that the self-diffusion of titanium in the β phase is about three orders of magnitude faster than the self-diffusion in the α phase (β-Ti and α-Ti lines in Fig. 2.30). The diffusion rates of substitutional elements in the β phase can be either slower or faster than the self-diffusion of titanium [58]. From the group of slow diffusing elements, Al and Mo are shown as examples in Fig. 2.30. Among the other important alloying elements belonging to this group, V and Sn are close to Al, and Nb lies in between Al and Mo. Among the group of fast diffusing elements, Fe is shown as an example in Fig. 2.30. Ni is even slightly faster, whereas Cr and Mn fall in between Fe and the β-Ti self-diffusion line. From the various diffusion rates measured for oxygen in β titanium, the upper set of curves, shown in [59], is taken for Fig. 2.30 because, as pointed out in [59], the slope (activation energy) of the lower set of curves seems to be too high.

From the diffusivity data for α titanium, the results obtained for oxygen all show reasonable agreement [59], and the line shown in Fig. 2.30 seems to be well established. Only a limited number of diffusivity measurements for aluminum in α titanium exists and, apart from the scatter, most of these data lie close to the oxygen line in Fig. 2.30 [59]. This is unusually high for substitutional elements. It is well established that the group Fe, Ni, and Co exhibits an abnormally high diffusivity in α titanium (see Fe line in Fig. 2.30). This is explained by an interstitial diffusion mechanism of these elements [60]. This fast interstitial diffusion mechanism also increases the vacancy diffusion rate for self-diffusion and Al in α titanium, if the measurements are done on materials with normal Fe impurity content [60]. Measurements on ultrahigh purity α titanium with respect to Fe, Ni, and Co impurities [60] resulted in very low diffusivity rates for self-diffusion and Al in α titanium (Fig. 2.30) being characteristic of vacancy controlled diffusion. Besides Al, the elements Zr, Hf, Au, In, and Ga are also normal diffusing elements (vacancy mechanism) and their diffusivities are close to the Al line in Fig. 2.30. Cr and Mn probably belong as Fe, Ni, and Co to the fast diffusion elements (interstitial mechanism) although their diffusivity is about two orders of magnitude slower than Fe, Co, and Ni. According to [60], this is still much too fast for a vacancy mechanism.

It should be emphasized once more, that the diffusivity data of substitutional elements in α titanium apparently depend strongly on the Fe impurity level of the material. For commercial titanium alloys, this effect is particularly important for the diffusion of aluminum in the α phase.

The interstitial element hydrogen exhibits very high diffusion rates in the β phase as well as in the α phase (Fig. 2.30), which has serious consequences for applications of titanium alloys in aqueous or humid gaseous environments, especially under high static loads (stress corrosion cracking) or fatigue loading (corrosion fatigue), because of hydrogen embrittlement.

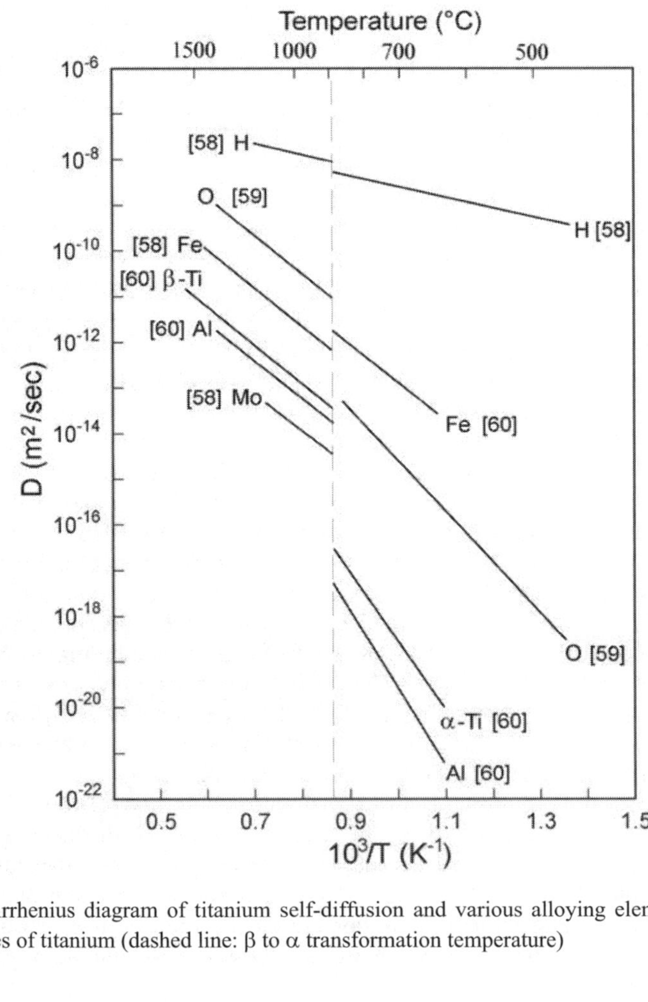

Fig. 2.30. Arrhenius diagram of titanium self-diffusion and various alloying elements in the β and α phases of titanium (dashed line: β to α transformation temperature)

2.9.2 Corrosion Behavior

In the galvanic series of metals, titanium has a standard potential of −1.63 V which is close to aluminum. Therefore, titanium cannot be considered as being intrinsically noble. Yet, the excellent resistance of titanium to general corrosion in most environments is well-known. This is the result of a stable protective surface film, which consists basically of TiO_2. This thin oxide film passivates titanium as long as the integrity of the film is maintained. In general, this is the case in most oxidizing environments, for example in salt solutions, including chlorides, hypochlorites, sulfates, and sulfites, or in nitric and chromic acid solutions. On the other hand, titanium is not corrosion resistant under reducing conditions, where the protective nature of the oxide film breaks down. Consequently, the corrosion resistance of titanium in reducing environments, such as sulfuric, hydrochloric, and phosphoric acids, is not good [61]. For example, titanium dissolves very rapidly in hydrofluoric acid, mainly because this acid causes the oxide film to break down, exposing reactive metallic titanium. This is why a HF-HNO_3 mixture is used to chemically mill and to pickle titanium during its production. In many commercially used reducing environments, the stability and integrity of the protective titanium oxide film can be improved by adding inhibitors (oxidizing agents) to these environments [61].

Passivated titanium is very corrosion resistant in flowing seawater at room temperature and in this environment has a potential close to that of Hastelloy, Inconel, Monel, and passive austenitic stainless steels [62]. In addition, titanium normally does not contain inclusions such as oxides, carbides, and sulfides. Therefore, titanium has a much greater resistance to pitting corrosion than the materials just mentioned.

The good general corrosion resistance of unalloyed titanium is also observed for α+β and β alloys. From an economic standpoint (cost, formability, weldability) CP titanium grades are preferred in applications that do not require higher strength levels. The corrosion resistance of CP titanium grade 2 in reducing acids can be improved significantly by small additions of noble metals, such as 0.2% Pd, grade 7 (Table 2.6), or to a smaller extent by adding 0.3% Mo+0.8% Ni, grade 12 (Table 2.6). This is shown in Table 2.8, where CP titanium grade 2 and the alloy grades 7 and 12 are compared with respect to the acid concentration limits resulting in a corrosion rate of about 125 μm/year [63]. Table 2.8 represents a summary of data acquired by Covington and Schutz, TIMET. A more detailed coverage of these data can be found in the Metals Handbook [61]. The data in Table 2.8 show that the highest acid concentration limit in all three acids is obtained for grade 7 with 0.2% Pd, followed by grade 12, and then by CP titanium grade 2. The addition of 0.2% Pd shifts the corrosion potential in the acids towards the increasingly noble (positive) direction, where the protective surface oxide film is stable, and full passivity can be achieved in dilute reducing acids [64].

The resistance to general corrosion of the higher strength α+β and β titanium alloys has been evaluated in a reducing acid environment by Schutz [65]. These results show that alloy contents of more than 3% Mo and 8% Zr are exceptionally beneficial, vanadium is of minor importance, but Al levels above 3% are increasingly detrimental. Table 2.9 shows the active-to-passive transition acid concentrations in boiling HCl at a corrosion rate of about 125 µm/year for the α+β alloy Ti-6Al-4V and for various β alloys. It can be seen that the β alloy Beta 21S with 15% Mo exhibits the best corrosion resistance in reducing acid environments, but some of the advantage is lost if used in the high strength aged condition.

Table 2.8. Acid concentration limits (wt%) for CP titanium grade 2 and alloy grades 7 and 12 in reducing acids for a corrosion rate of about 125 µm/year [63]

Acid	Temperature	Grade 2	Grade 7	Grade 12
HCl	24°C	6	25	9
	Boiling	0.6	4.6	1.3
H_2SO_4	24°C	5	48	10
	Boiling	0.5	7	1.5
H_3PO_4	24°C	30	80	40
	Boiling	0.7	3.5	2

Table 2.9. Active-to-passive transition acid concentrations at a corrosion rate of about 125 µm/year for Ti-6Al-4V and various β alloys in boiling HCl [65]

Alloy	Annealed Condition (% HCl)	Aged Condition (% HCl)
Ti-10-2-3	-	0.08
B120VCA	0.10	-
Ti-15-3	0.12	0.08
Ti-6-4	0.12	0.13
Beta C	1.1	0.87
Beta 21S	5.0	1.5

The resistance to pitting corrosion of titanium is generally very high because of its protective oxide film and, as already mentioned, the virtual absence of inclusions. The resistance to pitting is measured electrochemically either by the anodic breakdown potential or by the repassivation potential [61]. A comparison of the repassivation potential (also known as critical pitting potential, E_{pit}) between CP titanium and various titanium alloys in boiling 5% NaCl solution is shown in Table 2.10 [65]. It can be seen that CP titanium exhibits the highest value (6.2 V) and has therefore, in this comparison, the highest resistance to pitting corrosion. Although the potential values are lower for all titanium alloys listed in Table 2.10, the titanium alloys are still considered to be resistant to pitting corrosion because the repassivation potential values are higher than 1 V [61]. The values in Table 2.10 show that some β alloys, including Beta C and Beta 21S, exhibit a better resistance to pitting corrosion than the α+β alloy Ti-6Al-4V.

Crevice corrosion is often the critical mode of attack for titanium in chloride, fluoride, or sulfate containing solutions when temperatures exceed 75°C. Within crevices oxygen depleted reducing acid conditions can develop with pH values as low as 1 or below, because of the small and restricted volume of solution involved. Figure 2.31 illustrates the combination of pH level and temperature at which crevice attack is observed for various titanium grades (Table 2.6) in NaCl-rich brines [61]. It can be seen that the crevice corrosion resistance of CP titanium grade 2 can be improved by adding 0.3Mo+0.8Ni (grade 12) or even more by the addition of Pd (grade 7). The resistance to crevice corrosion of Ti-6Al-4V is similar to that of CP titanium grade 2, whereas the β titanium alloys Beta C and Beta 21S show a better resistance to crevice corrosion in most aggressive environments [65]. A further discussion of the relation between environmental effects and alloy performance in specific applications is included in Sect. 4.2 of the chapter "Commercially Pure (CP) Titanium and Alpha Alloys".

Table 2.10. Repassivation potential (also known as critical pitting potential, E_{pit}) in boiling 5% NaCl solution of CP titanium, Ti-6Al-4V, and various β alloys in the annealed condition [65]

	CP-Ti Grade 2	Ti-6-4	Ti-15-3	B120VCA	Beta 21S	Beta C
Repassivation Potential (V) vs. Ag/AgCl	6.2	1.8	2.0	2.7	2.8	3.0

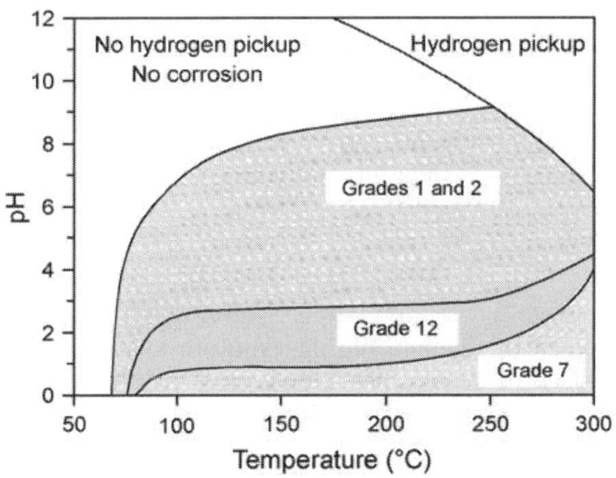

Fig. 2.31. Temperature and pH limits for crevice corrosion of various titanium grades in NaCl-rich brines (attack in shaded areas) [61]

The combination of corrosive environment and applied stress can cause the degradation of some important mechanical properties. The tensile ductility can be reduced if crack nucleation is shifted to the specimen surface, cracks connected to the surface can propagate under constant load conditions (stress corrosion cracking), and in fatigue loading surface cracks can nucleate and propagate at lower stress amplitudes as compared to inert environment (corrosion fatigue). The magnitude of these effects will be shown and discussed in this book in the relevant materials sections, for example for α+β alloys in Chap.5. The present discussion will be limited to a brief description of the basic mechanism which is responsible for the above mentioned effects.

Hydrogen is the fastest diffusing element (Fig. 2.30) and the most detrimental acting environmental species in combination with an applied stress (hydrogen embrittlement). In general, there are two possible sources for hydrogen, the internal hydrogen of the material or external hydrogen from the environment. The negative effects due to internal hydrogen are well under control in titanium by strictly limiting the maximum hydrogen content in CP titanium and titanium alloys to 125-150 ppm. Still, hydrogen related problems can occur in the presence of sharp notches, as will be discussed in Chap. 4.

The external hydrogen from the environment can be swept into the material interior by moving dislocations if the slip steps at the surface are higher than the thickness of the protective oxide layer. In this way, the hydrogen concentration within the slip bands can locally reach such a high level that the fracture stress within the slip bands is reduced leading to easier crack nucleation and crack propagation. For the hexagonal α phase, it is observed that this hydrogen induced fracture preferentially takes place on the basal planes. Therefore, a pronounced effect of crystallographic texture on the magnitude of degradation of the relevant mechanical properties is observed for α+β titanium alloys (Sect. 5.2.6). The reason for the preferred fracture along basal planes is still unknown. In contrast to α and α+β alloys, β titanium alloys are less sensitive to hydrogen embrittlement especially in the annealed condition. This beneficial behavior is somewhat reduced for the aged condition with a higher volume fraction of α phase. This higher tolerance of β alloys to hydrogen is attributed to the bcc crystal structure of the β matrix and to the higher solid solubility of hydrogen in the β phase as compared to the α phase.

2.9.3 Oxidation

The oxidation product of titanium during exposure to air is TiO_2 which has a tetragonal rutile crystal structure. This oxide layer is often called scale and is an n-type anion-defective oxide, through which the oxygen ions can diffuse. The reaction front is at the metal/oxide interface and the scale grows into the titanium base material. The driving force for the rapid oxidation of titanium is the high chemical affinity of titanium to oxygen which is higher than for titanium and nitrogen. During the oxidation process, the high affinity of titanium to oxygen and the high solid solubility of oxygen in titanium (about 14.5%) results in the simultaneous formation of the scale and an adjacent oxygen rich layer in the base metal. This oxygen rich layer is called α-case because it is a continuous layer of oxygen stabilized α phase. As mentioned in Sect. 2.8.1, an increasing oxygen level strengthens the α phase and changes the deformation behavior of α titanium from a wavy to a planar slip mode. Therefore, the hard, less ductile α-case can result in the formation of surface cracks under tension loading. The low local ductility and the large slip offsets at the surface can cause low overall ductility or early crack nucleation under fatigue loading conditions. The high temperature application of conventional titanium alloys is therefore limited to a temperature regime below about 550°C. The diffusion rates through the scale (oxide layer) below 550°C are slow enough to prevent excess oxygen contents being dissolved in the bulk material, resulting in no significant α-case formation.

In order to decrease the diffusion rate of oxygen through the scale, various additions of alloying elements have been investigated [58]. Improvements were found by adding Al, Si, Cr (> 10%), Nb, Ta, W, and Mo. These elements form either thermally stable oxides (Al, Si, Cr) or have a valency greater than four, for example Nb^{5+}. By substituting for the Ti^{4+} ions in the TiO_2 structure niobium reduces the number of anion vacancies and therefore reduces the oxygen diffusion rate. Based on this effect, a β titanium sheet alloy (Beta 21S) with the composition Ti-15Mo-2.7Nb-3Al-0.2Si (see Table 2.6) was developed [66]. This β alloy has a higher oxidation resistance but a lower high temperature strength and creep resistance than the α+β high temperature alloys Ti-6242 and IMI 834. However, increasing amounts of aluminum are much more effective in lowering the diffusion rates, because aluminum forms a very dense and thermally stable α-Al_2O_3 oxide. The resulting scale consists of a heterogeneous mixture of TiO_2 and Al_2O_3 underneath the TiO_2 surface oxide layer, as shown schematically in Fig. 2.32 [67].

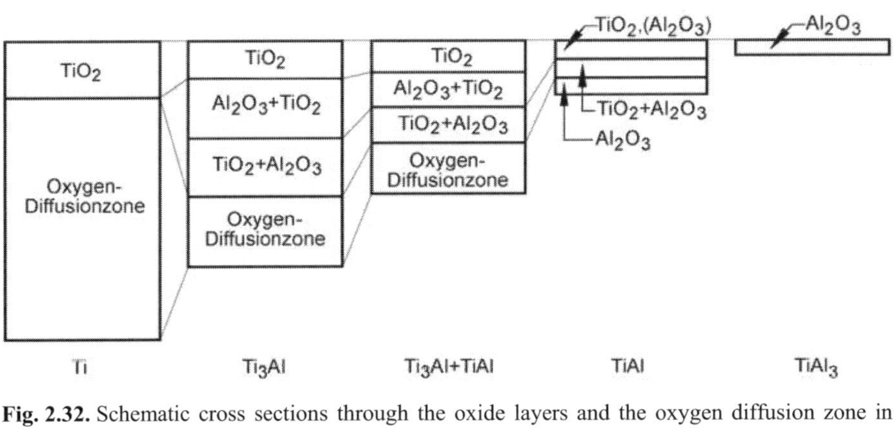

Fig. 2.32. Schematic cross sections through the oxide layers and the oxygen diffusion zone in titanium and titanium-aluminides [67]

The improved oxidation resistance of titanium aluminides, such as Ti_3Al or γ-TiAl based alloys, results from an increased volume fraction of Al_2O_3 in the scale (Fig. 2.32). The amount of Al_2O_3 increases with aluminum concentration and the Al_2O_3 layer becomes continuous around 40 at% Al. Consequently, γ-TiAl exhibits a better oxidation resistance than alloys based on Ti_3Al. This is because TiO_2 is not stable on titanium alloys at high temperatures and the Al_2O_3 layer is not continuous on Ti_3Al (Fig. 2.32), whereas the Al_2O_3 layer is continuous on γ-TiAl and stable up to much higher temperatures. This improved oxidation resistance can be used for the development of surface coatings for conventional titanium alloys, such as IMI 834, to allow application temperatures above 550°C. Many different coatings have been investigated, for example Pt, NiCr, Si, Si_3N_4, Al, MCrAlY, silicates, SiO_2, Nb [68], but the most promising results are obtained by sputtering Ti-Al coatings. This is shown in Fig. 2.33 for the high temperature alloy Ti-1100 [69]. Although this alloy is no longer produced by TIMET, the results are still useful, because Ti-1100 exhibits at 700°C a similar oxidation behavior as IMI 834 [70]. From Fig. 2.33 it can be seen that the Ti-Al coating produced a better oxidation resistance than Si or Pt coatings. Furthermore, the Ti-Al coated material exhibited a better oxidation resistance at 750°C than the uncoated material at 600°C.

A special case of oxidation resistance is the ignition and burn resistance. In normal atmospheric air environment, all titanium alloys are generally resistant to ignition and burning, but under special conditions, such as in gas turbine aero-engine compressors (high pressures, high air flow velocities), ignition and burning is possible for many titanium alloys [64]. This special case will be discussed in more detail in Chap. 10 (Special Properties and Applications of Titanium), together with some alloying approaches that have been used to mitigate this problem.

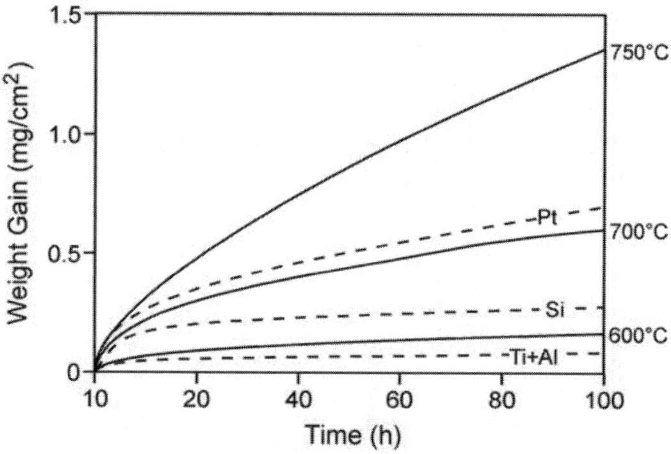

Fig. 2.33. Oxidation behavior of Ti-1100 material at different temperatures in comparison to coated material at 750°C (dashed curves) [69]

3 Technological Aspects

This chapter addresses the various aspects that are associated with producing titanium as a commercial material (hence “Technological Aspects”). It commences with a short discussion of the production of metallic titanium (titanium sponge), then continues with a discussion of all aspects of titanium production ranging from melting, including alloying (Sect. 3.2), to processing into useful product forms (Sect. 3.3). Included are products, such as billet, bar, plate, and sheet. Section 3.4 describes shaping processes for the manufacture of components. The emphasis in Sect. 3.5 is on near net shape processes, since these are one means of reducing the cost of using titanium alloys, which is a principal constraint to their increased use. Section 3.6 addresses the joining methods most commonly used for titanium and its alloys, and in Sect. 3.7 various surface treatment processes are described. Section 3.8 illustrates some of the inspection methods used during production of titanium mill products and components. Especially those used for high performance applications are described. Finally, it concludes in Sect. 3.9 with a discussion of characterization methods, which are particular to titanium alloys.

3.1 Sponge Production

Metallic titanium, as obtained from the ore, is called sponge. This is because it is porous and has a sponge-like appearance. Titanium as a chemical species is very abundant. It is the fourth most prevalent metallic element in the earth’s crust (only exceeded by Al, Fe, and Mg). The starting ore for the production of titanium is either rutile (TiO_2) or ilmenite ($FeTiO_3$). The extraction of metallic titanium from these ores occurs in five distinct stages or operations:

- Chlorination of the ore to produce $TiCl_4$.
- Distillation of the $TiCl_4$ to purify it.
- Reduction of the $TiCl_4$ to produce metallic titanium (the Kroll process).
- Purification of the metallic titanium (the sponge) to remove by-products of the reduction process.
- Crushing and sizing of the metallic titanium to create a suitable product for subsequent melting of CP titanium and titanium alloys.

The chlorination process starts with relatively impure rutile. If the ore is ilmenite instead of rutile, the starting material is TiO_2 enriched slag that is a by-product of the electromelting of ilmenite with carbon to produce iron. Chlorination occurs in a fluidized bed containing TiO_2, carbon (coke), and impurities that accompany the rutile into the chlorinator as shown schematically in Fig. 3.1. As shown, Cl_2

(gaseous) is introduced at the bottom of the chlorinator and contacts the (impure) TiO_2 and carbon reactants. The reaction products are metal chlorides (MCl_x), CO_2, CO, and gaseous $TiCl_4$ (the boiling point of $TiCl_4$ is 136°C). These products are removed at the top of the reactor vessel and go directly into the fractional distillation unit (Fig. 3.2).

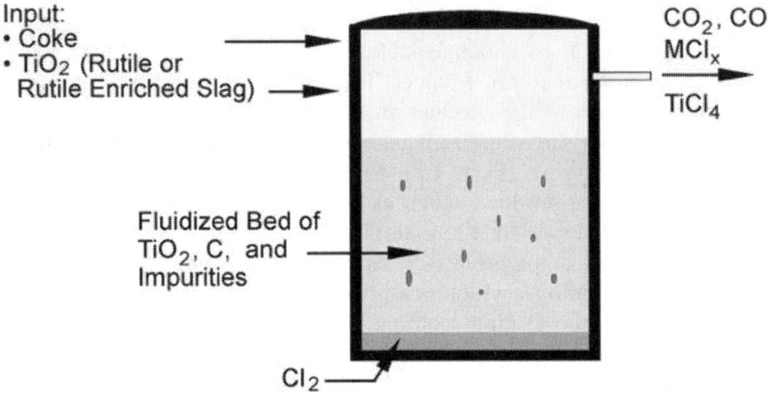

Fig. 3.1. Schematic drawing of a fluidized bed chlorinator used for producing $TiCl_4$ (courtesy J. A. Hall)

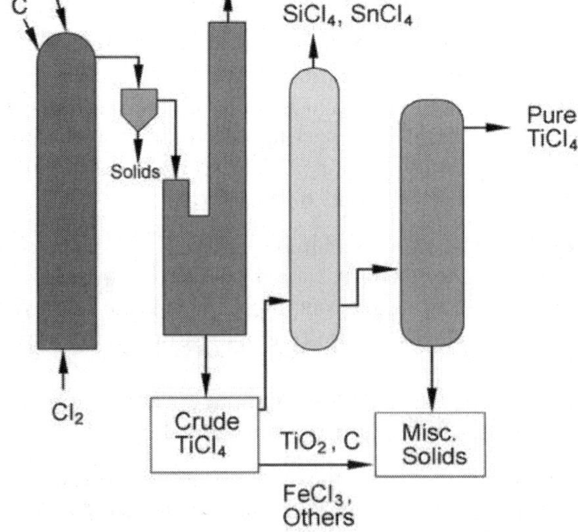

Fig. 3.2. Schematic drawing of a chlorinator on the left feeding the fractional distillation unit (two columns in the center) and a holding vessel on the right (courtesy J. A. Hall)

The basic chlorination reactions are as follows:

$$TiO_2 + 2Cl_2 + C \rightarrow TiCl_4 + CO_2$$

and

$$TiO_2 + 2Cl_2 + 2C \rightarrow TiCl_4 + 2CO.$$

The second step in the production route is the distillation process because the starting grade of $TiCl_4$ that comes from the chlorination process requires further purification. This is accomplished by fractional distillation of the $TiCl_4$ as shown in Fig. 3.2. Here it can be seen that a two step distillation process is used. The first step removes the low boiling point impurities such as CO and CO_2 and the second removes the higher boiling point impurities such as $SiCl_4$ and $SnCl_4$. The purified $TiCl_4$ is stored under inert cover gas until it is used.

The next stage in the production route is the reduction of the $TiCl_4$, the Kroll process. The purified $TiCl_4$ is put into a reactor filled with inert gas and already containing metallic Mg and heated to 800-850°C to drive the following overall reduction reaction:

$$TiCl_4 + 2Mg \rightarrow Ti + 2MgCl_2.$$

This actually occurs in two steps as follows:

$$TiCl_4 + Mg \rightarrow TiCl_2 + MgCl_2$$

followed by

$$TiCl_2 + Mg \rightarrow Ti + MgCl_2.$$

A schematic of a Kroll reaction vessel is shown in Fig. 3.3 with the vessel on the left coupled to a vacuum distillation vessel on the right. The reduction reactions were originally studied by Kroll [71] in the late 1930's and the reduction of $TiCl_4$ by Mg is still known as the Kroll process. The metallic titanium, that is the product of the final reduction shown above, is in itself quite pure, but occurs as a mixture of pure metal and $MgCl_2$. Most of the $MgCl_2$ is removed continually as the Kroll reaction proceeds but there is some residual amount that must be removed during the metallic titanium extraction step described later.

Because the reduction reaction is exothermic, the $TiCl_4$ is added to the vessel containing Mg at a rate that allows the temperature to be managed. This is necessary to prevent the solid reaction product from becoming so dense that the volatile products are trapped inside. This reaction product is a solid mass of intermingled mixture of metallic titanium and $MgCl_2$. This is called a "sponge cake" and is the product of the Kroll process.

Earlier (1910), Hunter [72] had demonstrated that $TiCl_4$ could be reduced using molten Na and this method of making sponge is called the Hunter process. During the 1960-1995 period significant quantities of titanium sponge were produced using this process. Today, there are no large scale titanium production operations left that use this process. This is mainly because the economics of using Mg as the reducing agent are more attractive than using Na.

The next step in the production route is the extraction of the metallic titanium from the sponge cake by removal of the residual $MgCl_2$. Separation of the $MgCl_2$ can be done by one of several methods: acid leaching, inert gas sweep, or vacuum distillation. The former of these processes utilizes the preferential solubility of $MgCl_2$ in acidic solution, allowing removal of the $MgCl_2$ from the crushed sponge cake in a separate leaching operation. This process is no longer used extensively. The other processes have the advantage of removing the $MgCl_2$ in situ in the Kroll reactor vessel. These processes utilize the high vapor pressure of $MgCl_2$ to selectively remove it by evaporation and then recondense it for Mg and Cl_2 recovery away from the sponge. The inert gas method uses argon as a carrier gas to transport the $MgCl_2$ vapor.

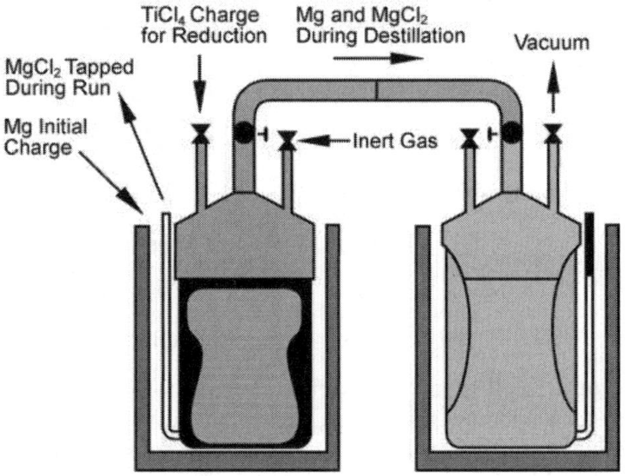

Fig. 3.3. Schematic of a Kroll reaction vessel on the left coupled with a collection vessel on the right for the Mg and $MgCl_2$ that are removed during the vacuum destillation (courtesy J. A. Hall)

The vacuum distillation process (VDP) is shown schematically in Fig. 3.3. In this process, the sponge cake is heated in situ in the Kroll reactor vessel on the left under vacuum. This allows the volatile $MgCl_2$ and excess metallic Mg to be extracted by evaporation and recondensed in another vessel (the one on the right in Fig. 3.3). This vessel becomes the Kroll reaction vessel for the next reduction run after additional Mg is added. The left hand vessel in Fig. 3.3 containing the metallic titanium sponge cake is then replaced by an empty one. This process is a semi-continuous one which is economically advantageous. The resulting vacuum distilled sponge has the lowest volatile content of the three sponge purifying processes. Because of the high temperature (700-850°C) at which VDP is conducted, the sponge does pick up small amounts of Fe and Ni from the stainless steel reaction vessel. The Ni is especially undesirable in high temperature alloys, since Ni reduces the creep strength when exceeding specific limits. There also is some sintering of the sponge cake.

In both processes (inert gas sweep and VDP), the Mg and Cl_2 are recovered and recycled. Today, Mg reduced titanium sponge production is nearly a closed loop batch process with only modest amounts of "make up" Mg and chlorine being required from batch to batch.

The last stage in the production route is the crushing and sizing of the titanium sponge. After removal of the excess Mg and $MgCl_2$, the sponge mass is crushed to produce granules of metallic titanium. After crushing and sizing, the coarser sponge granules are further sheared to reduce their size. These crushing and shearing operations are conducted in air but require care. Titanium is potentially pyrophoric and any fires that occur during this operation can contaminate the sponge with nitrogen rich regions that later result in melt related defects. Higher VDP temperatures reduce the ease of subdividing the sponge cake. Unless otherwise specified, titanium sponge producers typically do not strive for average sponge particle sizes less than 3-5 cm. This eliminates the cost of further crushing or shearing operations and avoids the threat of incurring sponge fires during these operations. The desired or specified sponge particle size depends on the end product that is being produced. Coarser granules (up to 2.5 cm) can be used for commercially pure titanium (CP titanium) and standard grades of most alloys, but for high performance applications, such as aircraft engine rotors, smaller sizes (1 cm maximum) are typically required. This is because of the concern for interstitial stabilized defects in the melted product for rotor grade material. Examples of these sponge particles are shown in Fig. 3.4.

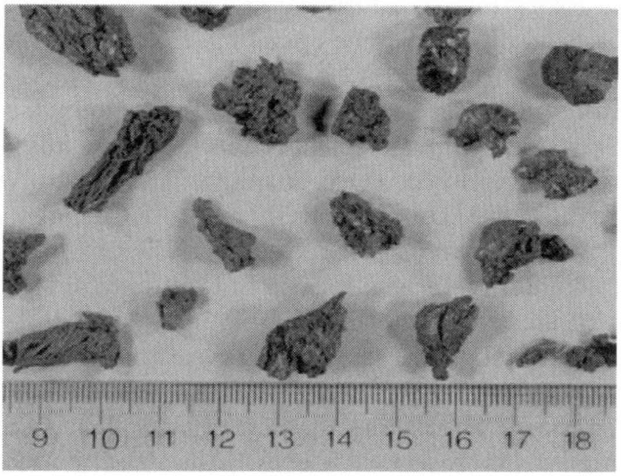

Fig. 3.4. Low magnification photo showing individual sponge particles (courtesy J. A. Hall)

The cost of producing titanium sponge can be conveniently separated into five components or cost elements. These are labor, equipment maintenance, utilities, and the two main ingredients (Mg and $TiCl_4$). A pie chart is shown in Fig. 3.5 that identifies the relative contributions of each of these elements to the overall cost. It can be seen that $TiCl_4$ comprises more than 50% of the cost, so efforts aimed at reducing the cost of titanium sponge must address this matter.

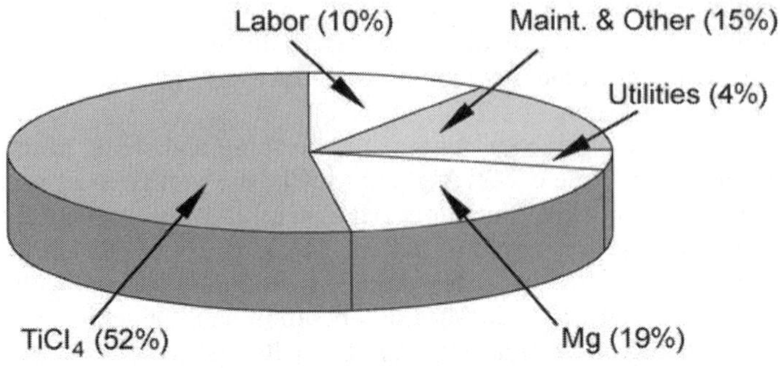

Fig. 3.5. Relative proportions of major cost elements for titanium sponge production (courtesy J. A. Hall)

Other processes for producing metallic titanium have been under investigation for years. Most of these have been directed to reducing the cost of sponge, generally without economic success. Electrolytic production (also called electrowinning) of titanium is one example that appears attractive and in the 1975-1985 time frame Dow-Howmet successfully demonstrated a pilot scale operation in the USA [73]. The down market for titanium at this time resulted in a decision not to proceed to full-scale operation. Consequently, the practicality of sustaining a reliable and affordable operation of a large capacity electrolytic reduction cell has not been demonstrated. The issues that remain to be demonstrated are the ability to seal a large cell in order to maintain a pure operating environment and the long-term stability of the electrodes.

Other recent efforts to produce very high purity titanium by electrolytic refining have been quite successful, both technically and economically [74]. Electrolytic refining starts by dissolving lower purity titanium in an electrolyte and redepositing it as high purity titanium. Through careful control of the deposition conditions and electrolyte purity a very high purity product can be obtained. This high purity metal is made into sputtering targets for use in electronic device fabrication. The economic success of electrolytically refined titanium is because the users of this high purity material use relatively small quantities in high value products, so the economics are completely different from structural applications.

There is a new process for making titanium sponge that is currently under intensive investigation. This process is known as the Electro-Deoxidation Process (EDO)™ [75]. The EDO process converts a pressed and sintered TiO_2 cathode to titanium in situ by electrolytically separating the oxygen from the titanium ions using a molten $CaCl_2$ bath and a graphite anode. This leaves porous metallic titanium in place of the original cathode. In principle, this process also has the capability to make pre-alloyed sponge if oxides of the desired alloying elements are blended into the oxide cathode and are electrolytically reduced along with the TiO_2. While the results obtained using this process are quite limited and the potential for scale up is still a matter of analysis and subsequent demonstration, the potential for such a process is exciting for several reasons. First, the ability to make pre-alloyed sponge would permit elimination of the sponge and master alloy blending and mechanical compaction steps used to form a first melt electrode for ingot metallurgy melting, as described in Sect. 3.2.1. This may result in significant cost savings. Second, the ability to introduce alloying elements into titanium (e.g. W, Cu) that are difficult to introduce using conventional ingot metallurgy practice as discussed later. This new process opens a number of alloy synthesis options that have not been previously explored because of the melting constraint. The EDO process has been proven to be technically feasible but there are many details ranging from repeatability to product cost after scale up that require detailed study and analysis. Because it is somewhat revolutionary in its capability, the EDO process is mentioned here even though it's future as a commercial reality is still unclear.

3.2 Melting

This section describes the procedures used to formulate titanium alloys and the melting technology used to produce ingots which are the starting materials for both mill products and remelt stock for titanium castings. This process is commonly referred to as "melting", but the resolidification of the molten metal is the key to obtaining homogeneous, high quality ingots for conversion to mill products.

A significant portion of this section is devoted to the discussion and characterization of melt related defects. These defects must be minimized for titanium to perform at a level that justifies its cost. It is because of the potential for these defects to be formed and the severe consequences of their presence that the elaborate and expensive methods are used to melt titanium and produce ingots. While the cost of preventing these defects is high, this strong, lightweight material would not be available for the most demanding applications if these defects could not be eliminated. The detailed nature of the defects is discussed later in Sect. 3.2.3, but it is useful here to outline the types of possible defects to emphasize the reasons for the approach that is taken for melting titanium. There are five principal types of defects in titanium. The primary source of these is melting. There are interstitial stabilized defects, known as type I defects or referred to as high interstitial defects (HIDs), tungsten rich inclusions, known as high density inclusions or HDIs, alpha stabilizer rich regions called type II defects, beta stabilizer rich regions called "beta flecks", and voids that occur during the solidification of the ingot. In contrast to other classes of metallic materials, where melting is used to eliminate defects, melting can introduce defects in titanium. Once formed, these defects can be difficult to eliminate through all subsequent processing steps, including remelting. In all cases, the entire spectrum of causes of these defects is not understood and the severity of the performance degradation associated with their presence is different for each type, as will be discussed later in this section. Table 3.1 is a summary of the types of known defects in titanium and some of their possible causes.

Molten titanium is very reactive, therefore, special means are required to produce ingots of both unalloyed titanium (CP titanium) and the various titanium alloys. Titanium and its alloys are melted either in a vacuum arc remelt (VAR) furnace or in a cold hearth melting (CHM) furnace. In either case, the melting is done in a manner that prevents molten titanium from contacting furnace refractories such as those used in vacuum induction melting furnaces or from being exposed to air. Production of titanium and titanium alloys has been done by vacuum arc melting since titanium has been a commercial product. Cold hearth melting has only become commercially feasible for rotor grade titanium since about 1985.

Table 3.1. Melt related defects known in titanium and their possible causes

Defect Type	Possible Causes
Type I ("Hard Alpha"), also called High Interstitial Defect (HID)	Sponge production – Fires during handling or shearing First melt electrode production – Fires during compaction – Improperly conditioned scrap – Contaminated master alloy – Contamination during welding Melting and remelting – Small water leak – Air leak – Aggressive grinding during ingot conditioning
High Density Inclusions (HDIs)	Scrap additions – Tungsten welding electrodes – Tool bits mixed into turnings
Beta Flecks	Melting segregation Conversion too close to transus (including adiabatic heating effects)
Type II (Alpha Stabilized)	Improper final melt phase (excessive pipe formation) Improper ingot top removal Al-rich "drop-ins" during EBM
Voids	Incorporation of shrinkage pipe during conversion Improper conversion practice

3.2.1 Vacuum Arc Remelting (VAR)

Vacuum arc remelting is actually a misnomer in that it is the initial melting process used in the production of titanium. This is in contrast to the production of nickel base alloys and specialty steels where the first melt process is vacuum induction melting, followed by vacuum arc remelting. Vacuum arc remelting is the most commonly used process for making titanium, but the use of cold hearth melting is growing as will be described later. Over time, the vacuum arc remelting process has been used to successfully make larger and larger ingots. The ingot size (diameter and weight) has increased due to the improved capability of melting larger diameter ingots of CP titanium grades and alloys such as Ti-6Al-4V. Larger ingots are more economical because losses during conversion of the ingot to the final product are smaller and the melting time including reloading of the furnace is shorter. Both of these factors plus minimizing the number of VAR units required for production result in lower product cost. Today, it is common to melt ingots of these materials as large as about 100 cm in diameter and weighing as much as 10 000-15 000 kg. Other titanium alloys are more difficult to melt because of a higher propensity for alloy element segregation during solidification leading to both beta flecks and type II defects. The minimization of beta flecks occurs if the segregation prone alloys are produced in smaller ingots, which impacts material costs. Type II defects are avoided by a melt practice that eliminates or minimizes shrinkage pipe at the ingot top.

The vacuum arc melting begins with a first melt electrode that is made up of mechanically compacted blocks of sponge and alloying elements, where each block has the desired nominal alloy composition. Sponge and alloying elements are blended together in a twin cone blender. This mixture is then placed in a die and mechanically compacted at room temperature into blocks using a hydraulic press. The as-compacted blocks have adequate "green strength" to remain intact during handling and melting. These blocks are welded together in an inert gas welding chamber to create the first melt electrode or "stick". Because of the high cost of winning titanium, there is a strong economic incentive to recycle and reuse titanium scrap (often called revert) by reincorporating it into ingots during melting. This reuse is accomplished in unalloyed grades and non-rotor alloys by adding scrap of the same composition to this electrode during the electrode welding operation. This scrap is carefully controlled with regard to its origin and cleanliness. For example, the use of scrap that has been flame cut is generally not allowed. This is because experience has shown that the N and C enriched regions along the flame cut edges are not always refined out during melting. This can leave interstitial stabilized defects in the final product. Turnings generated during the machining of titanium parts are also used in the electrode make-up, but these also are subject to special controls. The turnings must be cleaned to remove any residual cutting fluids and X-rayed to ensure that they contain no broken WC cutting tools or other high density inclusions that can end up in the ingot. The usage of revert material is limited for various applications by different specifications. A picture of a first melt electrode is shown in Fig. 3.6. This figure shows the individual briquettes of compacted sponge and master alloy and the titanium straps that are welded to them to hold the electrode together during the first vacuum arc melt. This electrode is held in the VAR furnace by a stub. A stub configuration can be seen in Fig. 3.6 at the left. Once the first melting operation is complete, the ingot is removed from the copper mold. Figure 3.7 shows a large titanium alloy ingot after the VAR process is complete. Beside the ingot on the right is the vacuum jacket for the VAR furnace which is about 125 cm in diameter. This ingot is inverted and melted again.

Rotor grade VAR materials are typically triple melted so the ingot is once again inverted and the remelting process is repeated a second time in this case.

Fig. 3.6. First melt VAR electrode with welded individual briquettes and the stub on the left (courtesy RMI)

Fig. 3.7. VAR ingot (on left) after first melt (courtesy J. A. Hall)

VAR production of homogeneous, sound ingots of titanium alloys requires care and the detailed melting procedure depends on the particular alloy. Over the past 30 years dozens of improvements have been made to the process, some major and some minor. All of these have been directed toward reducing the possibility of defects and the extent of variation in the ingots. Figure 3.8 shows schematically the VAR process. This figure shows the VAR furnace, the electrode being melted, and the water-cooled copper crucible containing the new ingot with the molten pool at its top. In the figure the molten pool at the top of the new ingot is inside the solid line drawn near the top of the new ingot. There are a number of parameters that must be monitored and controlled during the final melting operation. This is required to ensure homogeneity and soundness of the ingots. Important parameters that require attention and that are often monitored during melting include the following:

- The vacuum in the furnace is continuously monitored as this ensures that no air or small water leaks are occurring to contaminate the melt with nitrogen or oxygen (major water leaks create a serious explosion hazard).
- The melt rate is continuously adjusted to control the size of the molten pool at the top of the ingot (see Fig. 3.8). The propensity for freezing segregation to occur varies with alloy type. The melt rate and molten pool depth are controlled accordingly. This is largely based on experience. In segregation prone alloys such as Ti-17 or Ti-10V-2Fe-3Al, it is common to reduce the ingot diameter to about 75 cm and melt at lower rates (5-6 kg/min versus 8-10 kg/min). This modified melt practice creates a smaller, shallower molten pool at the top of the ingot. The lower melt rates use correspondingly lower power settings (200-275 versus 400-500 kVA).
- Most VAR furnaces are equipped with electrical coils at the top of the ingot mold that create an electromagnetic field used to stir the molten metal. This is done to achieve improved ingot homogeneity. The extent to which stirring is employed varies between titanium producers and between alloys. There is no general agreement on the benefit it produces or, even, the extent to which it is necessary.
- As the final part (25-35%) of the ingot is approached, the melt rate is reduced by reducing the power in several steps. In the VAR process this is the same as the hot topping operation practiced during conventional ingot metallurgy melting of Ni or Fe base alloys. This procedure minimizes the extent of shrinkage pipe formation and other defects such as type II at the ingot top. Minimization of shrinkage pipe reduces the loss of metal during conversion and helps eliminate defects that can be created when this pipe is inadvertently incorporated into the product.

The techniques used to control melting are quite empirical and are equipment dependent. Consequently, there is a significant "art" content involved in the melting operation. This makes experienced melt furnace operators (known as "melters") a valuable resource to all titanium producers. Eventually, the use of better process controls coupled with knowledge based systems may eliminate this dependence on individuals with a great deal of experience.

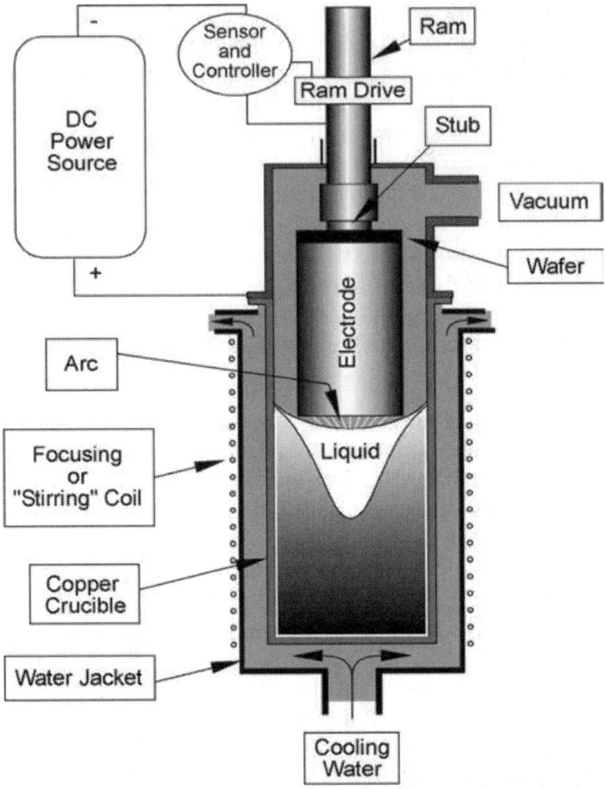

Fig. 3.8. Schematic of VAR furnace and ingot during a second melt, the electrode being remelted is at the top and the new ingot is at the bottom (courtesy J. A. Hall)

3.2.2 Cold Hearth Melting (CHM)

Cold hearth melting (CHM) is a newer melting method that seems to have several advantages over the VAR process for rotor grade material [76, 77]. A schematic of a cold hearth furnace is shown in Fig. 3.9. This method utilizes a water cooled copper vessel (the hearth) which contains the molten titanium. Cold hearth melting is conducted in either a plasma arc or an electron beam melting furnace. In both cases, the heat input from the heat source (the electron beam or the plasma torch) is balanced against the rate of heat extraction from the water-cooled copper hearth. This maintains a thin layer of solid titanium alloy (called the "skull") in contact with the hearth, so the molten titanium alloy only contacts the solid titanium alloy. This prevents any contamination by the hearth. The potential advantages of cold hearth melting include the following:

- It permits the residence time of the titanium alloy in the molten state to be controlled independently of the volume of molten metal solidifying as an ingot. In principle, this creates the opportunity for refining the alloys through dissolution of any nitrogen or oxygen rich defects without incurring a large, deep molten metal pool as in the VAR process that can cause solute segregation.
- It automatically introduces gravity separation of high density inclusions such as WC tool bits or tungsten welding electrode tips that are introduced along with the revert. The high density inclusions become trapped in the mushy zone of the skull and are not transmitted to the ingot. This is in complete contrast to the VAR process, where all of the material in the electrode ends up in the ingot.
- It allows direct casting of non-axisymmetric shapes, such as slabs or bars. These cast products are much better suited for conversion to flat mill products (plate, sheet, and strip) than large round ingots. Consequently, the conversion losses are lower and products made this way can be more cost competitive. This capability has proved to be a particularly attractive method for making sheet and strip from alloys that are readily rolled into coil without reheating. Of special interest are the various grades of CP titanium listed in Table 2.6.
- As noted earlier, the cost of titanium is a deterrent to its expanded use. Cold hearth melting is the most efficient method for reusing all forms of revert.
- In contrast to the physical environment in the VAR furnace chamber, the cold hearth furnace is more conducive to the use of online sensors. Consequently, this process is more amenable to real-time process controls and detection of process variations during the melting process.

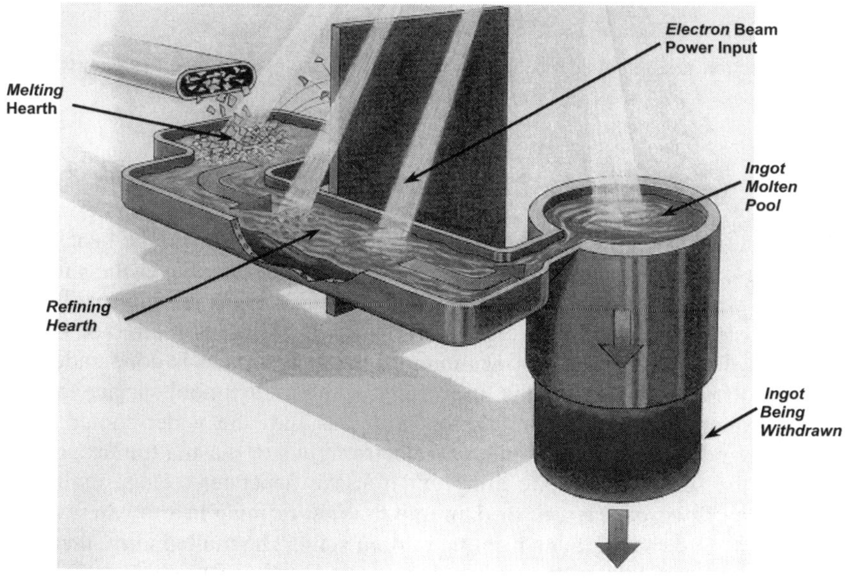

Fig. 3.9. Schematic of the hearth in an electron beam melting furnace (courtesy THT-TIMET)

There are two types of cold hearth melting furnaces in use today. One uses plasma arc melting (PAM) technology and the other uses electron beam melting (EBM) technology. These furnaces obviously differ in heat source (plasma torches vs. electron beam guns). Also, electron beam melting furnaces operate in vacuum, whereas plasma arc melting furnaces operate in a partial pressure of argon. Apart from these differences, the remaining physical arrangements of these furnaces are very similar. Production cold hearth melting furnaces are typically large. Figure 3.10 shows a man standing atop the vacuum chamber that contains the hearth of an electron beam melting furnace. This furnace is capable of melting about 10^6 kg of titanium per year.

Fig. 3.10. Electron beam cold hearth furnace showing the size of such facilities (courtesy Teledyne Allvac)

The overall hearth in a cold hearth melting furnace usually consists of several compartments, Fig. 3.9. From Fig. 3.9, it can be seen that the input material is fed into the first chamber (the melting hearth) where the energy from the heat sources melt it. These heat sources are continuously moved or scanned over the surface of the molten pool. In plasma arc melting furnaces, the torches are physically reoriented whereas in electron beam melting furnaces, the beam is redirected by electromagnetic deflection coils. This scanning of the heat sources is done under computer control, which optimizes the uniformity of the molten pool surface temperature, for example by compensating for heat loss into the water-cooled copper hearth at the pool edges. The number of electron guns or plasma torches, the heat input, and the scan patterns are alloy and furnace dependent. Once molten, the titanium alloy flows over a weir or dam into the first refining hearth where another array of heat sources maintains it in the molten state. The molten alloy then flows over a second weir into the second refining hearth where another array of heat sources keeps it molten and permits further refining. The molten alloy finally flows over the exit lip of the second refining hearth into an ingot mold. In the current example the ingot is round, but rectangular shapes also can be directly cast as previously mentioned. A large rectangular slab from an electron beam cold hearth melting furnace is shown in Fig. 3.11. This slab weighs approximately 12 000 kg. The principal disadvantage of cold hearth melting is the relatively smaller superheat that can be achieved in the molten alloy as it enters the mold, compared to the VAR process. Here, superheat is defined as the ΔT of the molten alloy above the liquidus temperature. This leads to a rougher ingot surface (roughness also is called wall quality) which can lead to increased conditioning requirements and added cost compared to VAR. In the case of slab casting, the geometric advantage of the shape for flat rolled products overwhelms any penalty due to ingot wall quality.

Fig. 3.11. Large titanium slab directly cast from a cold hearth melting furnace (courtesy Teledyne Allvac)

During the time the molten metal is in the two refining hearths, there is a significant opportunity to remove any interstitial stabilized inclusions by dissolution in the molten titanium alloy (ergo alloy refining). A thermodynamic driving force exists for this dissolution, however, the kinetics are relatively slow. Thus, the promise of complete refining (removal) of unwanted interstitial stabilized inclusions by cold hearth melting has not been realized. It is clear that refining does occur, but the required residence time in the hearth to achieve complete refining of relatively large inclusions would make the cold hearth melting process economically unattractive. It is clear, however, that cold hearth melting does allow complete removal of high density inclusions (HDIs). This is a major benefit of cold hearth melting of rotor grade material. This is particularly true during periods when titanium is in short supply, so there is a strong incentive to reuse all available turnings. Increased turning use increases the probability of having HDIs present even in rotor grade materials. Extensive production experience, supported by modeling, has shown that HDIs are not carried along over the weirs and into the ingot in the cold hearth melting case. This is because the density differences between titanium and the HDIs cause them to sink and stick in the mushy zone between the solid skull and the molten alloy.

Current cold hearth melting practice for rotor grade alloys requires application of a final VAR step to the ingot after it is produced in a cold hearth melting furnace [78]. This is because both electron beam and plasma arc ingots made by cold hearth melting can have process related defects that must be removed to qualify the material as rotor grade. In the electron beam melting process, aluminum evaporates from the melt and deposits on the cold roof and walls of the furnace vacuum chamber. These can drop into the melt at the latter stages of melting and result in Al-rich regions that are not adequately homogenized before solidification. The final VAR step re-homogenizes the alloy. In the plasma arc melting process, small inert gas bubbles from the argon ion plasma are formed in the molten metal and are trapped in the ingot. Since argon is essentially insoluble in titanium, these bubbles remain as pores. The final VAR step allows these bubbles to escape. There are two incentives to eliminate the final VAR operation, one is economic and the other is concern that this step creates an opportunity to reintroduce interstitial stabilized inclusions if a vacuum or water leak or other melt furnace malfunctions should occur. It appears that the most promising means of achieving this objective is with a partial pressure plasma arc melting process. Under such melting conditions, evaporation and condensation of aluminum is minimized and the partial pressure allows elimination of the argon filled pores. Realization of this goal is a number of years away because of the necessity of producing substantial quantities of material and inspecting it to statistically establish confidence in the "hearth only" process capability. Until qualification for rotor grade, the material produced by an unapproved melt practice must be used for non-rotor applications. There are many noncritical applications for some alloys such as Ti-6Al-4V and none for others such as Ti-17 or Ti-6246. Consequently, the qualification procedure for Ti-6Al-4V is more affordable because the material made during qualification can be sold for these non-rotor applications. This minimizes the nonrecurring cost of this qualification procedure by allowing all except the incremental cost of the CHM process to be recovered through the sale of the product made during the qualification process. Thus, there is an added economic penalty for qualifying a new CHM source for Ti-17 or Ti-6246. This added economic penalty extends the time required to achieve qualification and complicates the qualification process.

3.2.3 Melt Related Defects

Frequent reference already has been made to melt related defects in titanium alloys (Table 3.1). Experience has shown that, once formed, these defects are very difficult to eliminate and can have an extremely detrimental impact on material performance. Consequently, a separate section is devoted to the subject of melt related defects.

Melt related defects can be categorized as either intrinsic or extrinsic, depending on their origin. Extrinsic defects are caused by inadvertant introduction of impurities during the preparation of the electrode or during the melt process. Intrinsic defects are those that can be present if the ingot solidification occurs without proper control. The nature and origin of melt related defects in VAR material are discussed in [79].

As previously mentioned, the solidification of titanium ingots must be controlled to ensure homogeneity. The degree of difficulty involved in achieving homogeneous solidification depends significantly on the alloy. Those alloys that contain β eutectoid forming elements, such as Fe, Cr, Mn, Ni, and Cu, typically have depressed freezing temperatures resulting in solidification over a significant temperature range. This situation can lead to solute segregation during ingot solidification. The most common source of depressed freezing is a eutectic reaction in the (liquid+solid) portion of the phase diagram, such as in the Ti-Fe, Ti-Mn, and Ti-Cu systems. The Ti-Fe phase diagram is shown as example in Fig. 3.12. The presence of the eutectic extends the freezing range of the alloy and causes the last liquid to solidify to be enriched in solute. This creates the possibility of long range solute segregation during ingot solidification. Alloys that only contain β isomorphous alloying additions, such as Mo, V, and Nb, do not have similar depressed solidification temperatures and these alloys are much less prone to freezing segregation (see Ti-Mo phase diagram in Chap. 2, Fig. 2.12). Segregation of Fe or Cr during freezing results in regions that have a lower β transus temperature. These regions exhibit a different microstructure than the surrounding material in the final product. These solute rich regions sometimes become clearly visible in materials heat treated below, but near the nominal β transus and are known generally as "beta flecks". An example of a beta fleck in the alloy Ti-10V-2Fe-3Al is shown in Fig. 3.13 [80]. Here the large β grains and the lower volume fraction of α precipitates within the beta fleck are clearly visible. Beta flecks are the direct result of remaining alloying element freezing segregation. The segregated areas typically occur on a scale ranging from a few hundred micrometers to a few millimeters. These solidification related defects can occur in any titanium alloy but, as mentioned earlier, alloys containing eutectoid forming elements, such as Cr, Fe, or Ni, are considerably more susceptible to the creation of beta flecks. These beta flecks shown in Fig. 3.13 are detrimental to the fatigue strength because they are weaker and deform preferentially leading to early crack nucleation.

The reactivity of titanium also creates the possibility of formation of interstitial stabilized inclusions. These are known as type I defects and are most frequently the nitrogen rich compound TiN. The nitrogen stabilized type I inclusions are very hard and brittle and, therefore, are often called "hard alpha". Consequently, they fracture at relatively low stresses leading to incipient cracks in the material. Type I inclusions also can have high concentrations of oxygen and/or carbon, but this is less common. A nitrogen stabilized type I inclusion in a forging is shown in Fig. 3.14a. Because of the propensity to crack at low strains, the presence of a type I inclusion can seriously decrease the fatigue capability of the material. Thus, great care is taken to minimize their occurrence. Over the past 25 years numerous restrictions have been placed on both input material and on melt practice with the goal of minimizing the presence of type I inclusions in titanium alloy products. This has reduced the frequency of these defects by 10 to 100 times. Today, the frequency of type I defects detected in rotor grade titanium alloys is less than one defect per every 500 000 kg of material melted. Since the aircraft engine industry uses over 1 000 000 kg of titanium alloys each year this means that defect detection and elimination after melting is still essential. The most effective inspection method is ultrasonic inspection. This will be described in greater detail later in Sect. 3.8. Here, it is important to mention that the basis for ultrasonic detection of type I defects is the void that usually accompanies the defect, as can be seen in Fig.3.14a. The cause of this void is the strain incompatibility between the matrix and the rigid type I inclusion. The reduced ductility of the nitrogen rich region of the matrix adjacent to the inclusion may also be a factor, depending on the temperature at which the strain is introduced. In principle, the TiN should be detectable, because it has a modulus about 30% higher than the average value of the titanium alloy matrix. In practice, this modulus difference is about the same as the elastic anisotropy of α titanium. Thus, any ultrasonic technique sensitive enough to detect such differences also will detect regions of texture or preferred orientation and lead to many false calls during ultrasonic inspection. Causes of "false calls" are discussed in Sect. 3.8.1. Disposition of false calls consumes time and resources and is a significant factor in adding cost to the production of rotor quality material.

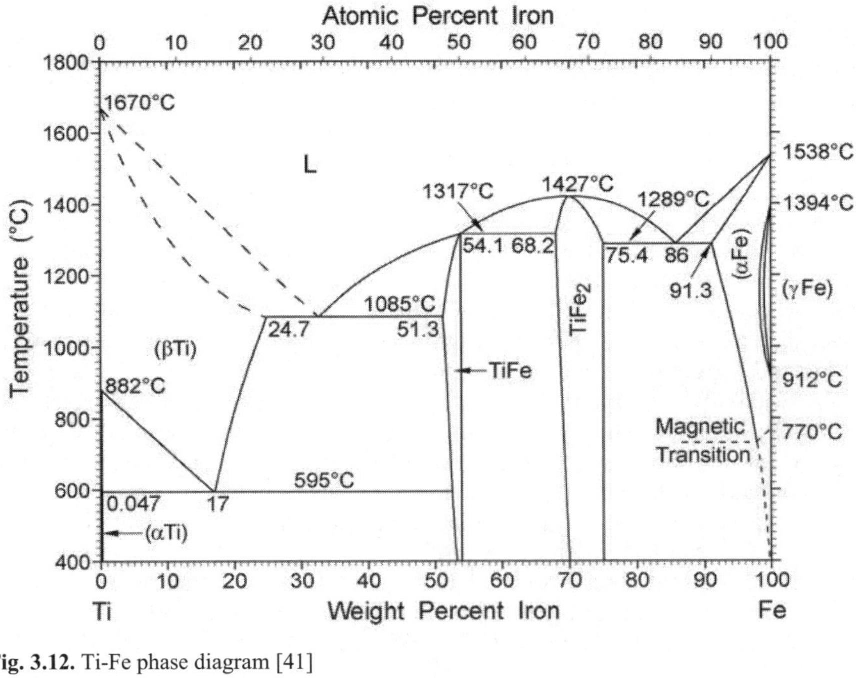

Fig. 3.12. Ti-Fe phase diagram [41]

Fig. 3.13. Fe enriched region, known as beta fleck, in Ti-10V-2Fe-3Al, LM (courtesy R. R. Boyer, Boeing)

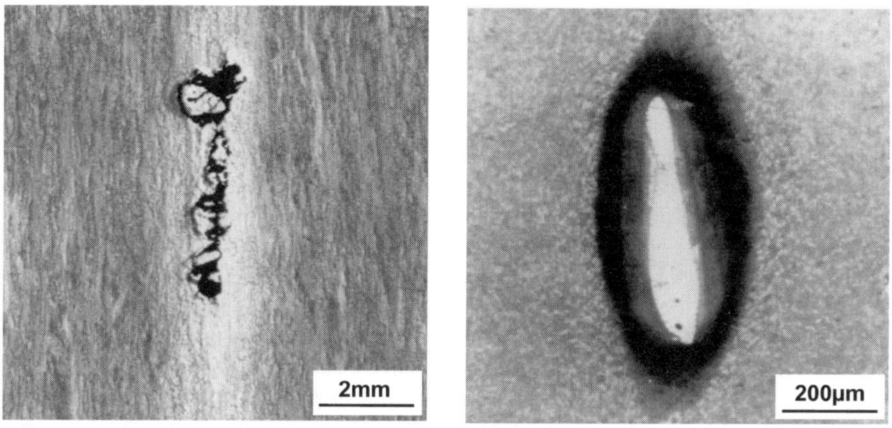

a b

Fig. 3.14. Melt related defects in Ti-6Al-4V forgings, LM: (a) Type I nitrogen stabilized inclusion (b) High density, tungsten rich inclusion (courtesy C. E. Shamblen, GE Aircraft Engines)

There also is another melt related defect that is Al-rich and usually is a result of incorporation of Al-rich regions from the shrinkage pipe near the top of the ingot into the product. These Al-rich regions are known as type II defects. Type II defects are less detrimental to properties than type I defects, but in high strength alloys such as Ti-17, these defects do not respond to heat treatment (aging) as much as the surrounding matrix and remain softer. Consequently, they will deform preferentially in fatigue situations, leading to earlier crack nucleation. Type II defects are eliminated by proper melt practice and by cropping the portion at the ingot top that contains the shrinkage pipe so it does not appear in the product.

Alloy segregation effects (beta flecks and type II defects) also can be minimized by ingot homogenization.

It has been mentioned that the use of titanium turnings is a common means of recycling titanium and is done to reduce the cost of the product and to increase the available quantity of titanium alloys. This is particularly helpful during periods when the titanium demand is high relative to capacity. Other types of scrap than turnings and the mill scrap internally generated by the titanium producer are also used. For example, pieces of sheet and plate generated by fabricators of titanium equipment, e.g. heat exchanger, also are separated by grade and reused. One of the issues that the reuse practice creates is the possibility that tungsten rich, high density inclusions (HDIs) can be introduced into the product. These tungsten rich inclusions can come from two principal sources: tungsten inert gas (TIG) welding electrodes that are embedded in welds and WC from broken tool bits. The high melting temperature of tungsten and WC make these inclusions relatively stable during vacuum arc melting and subsequent remelting. Thus, they are incorporated into the ingot with little or no modification. An example of a WC inclusion in a titanium alloy forging is shown in Fig. 3.14b. The dark etching matrix adjacent to the inclusion contains fine α precipitates in the β phase. This β phase is enriched in W due to diffusion.

The use of scrap, or revert as it is often called, including turnings varies widely by alloy, by material grade, and by user specification. The various engine manufacturers have different permissible levels of turnings that can be used in rotor grade material ranging from none to as much as 50%. Other (non-rotor) grades have more liberal allowances and CP titanium has no limitation. The incorporation of turnings and revert into an ingot is considerably easier and more economical in the case of cold hearth melting, because they can be fed directly into the melting hearth without prior consolidation.

This discussion of titanium alloy melting shows the complexity of producing commercial quantities of mill products from reactive materials for high performance applications. Significant technological advances have been made over the past 40-50 years in which titanium alloys have been in existence as commercial products. While the processes still can be improved, the reliability of titanium components, as affected by melt practice, has increased greatly. This improvement is an ongoing effort. The value of continuous improvement must be recognized by both users and producers.

3.3 Primary Processing

Once the ingot is melted and conditioned, it is generally given a homogenization anneal in the β phase field prior to working. Not all titanium producers use a homogenization treatment, and not all producers use this for all alloys. When a homogenization treatment is used, the times and temperatures are alloy dependent, but are typically 200-450°C above the β transus for times of 20-30 hours. It is important to emphasize that homogenization does not remove HDIs and type I (hard alpha) defects.

Irrespective of melt practice, after the final melt process is complete and before hot working, the ingot undergoes conditioning. Conditioning creates a smoother ingot surface that is free from stress concentrators that can induce cracking during the ingot breakdown and conversion operations. This conditioning can be done by grinding or by lathe turning of round ingots, or by grinding slabs. Grinding often is done manually and care must be exercised to control the local surface temperature increase during grinding. If not done carefully, these temperatures can reach high enough levels to create interstitial stabilized regions in situ. These can subsequently be embedded in the final product.

The initial working is done on a forging press at a temperature about 150°C above the β transus. During this operation, the initial round ingot is converted to a square or a round cornered square piece. The amount of initial strain before the first reheat also is somewhat alloy dependent, and dependent on whether the ingot has been homogenized, but is typically 28-38% (e.g. 90 cm diameter is reduced to 63-68 cm square). The work piece is fan-cooled after this initial working operation. The piece is then reheated to 35-50°C below the β transus and given a further reduction of about 30-40% to recrystallize it and to refine the structure in preparation for continued hot working. After this working operation in the (α+β) phase field, it is air-cooled. The piece is then reheated to about 50°C above the β transus and worked another 30-40% followed by a rapid cool (Ti-6Al-4V is water-quenched, other alloys such as Ti-17 or Ti-10-2-3 are aggressively fan-cooled). The fundamental intent of this initial heating, working, cooling, reheating, working, and cooling is to improve the compositional homogeneity and thus the structural uniformity of the alloy. This improves its subsequent response to thermomechanical processing operations. The remaining hot working operations typically are all done in the (α+β) phase field and a minimum of 65% additional reduction is typically used to obtain uniform structures that will respond to forging or heat treatment and are more amenable to ultrasonic inspection. Working in the (α+β) phase field is necessary to refine the macrostructure for subsequent microstructure control and for inspectability. In segregation prone alloys, where the initial ingot size is smaller, an upsetting operation may be used to provide a large enough diameter work piece that will yield a large diameter billet and still maintain the reductions outlined earlier. At this point, the remaining working steps are mainly determined by the shape of the product form that the ingot is destined to become (billet, plate, sheet, or bar). Before continuing, the work piece also is inspected for surface cracks or tears and any sharp discontinuities that could propagate during continued working. If present, these are blended out during a conditioning operation.

There are basically four types of titanium alloy mill products: billet, flat rolled product (plate and sheet), bar, and casting electrode. Billet is typically round and is used as the input material for forgings and for rolled rings. Sheet is flat rolled product up to a thickness of about 25 mm. Plate is flat rolled product that is thicker than about 25 mm. Bar can be round, square, or it can be shaped during rolling to meet a special need. Casting electrode is remelted by the casting producer, so it is more like billet in shape, but it is shaped to the desired diameter and cut to length without regard for the need to create a particular microstructure since it will be remelted. The relative amounts of each of these product forms are shown in Fig. 3.15. From this figure it can be seen that billet and flat rolled products comprise about 84% of the market. Billet and large diameter bar are made using a rotary general forging machine (called GFM) or a forging press. Since billet is a semifinished product used for making die forgings, the billet surface finish must be adequate to permit high sensitivity ultrasonic inspection. Depending on the alloy and the billet forging temperature, rough grinding or machining may be required. Bar, on the other hand, is a finished product. As such, bar always requires grinding or turning to impart an acceptable surface finish and uniform diameter of the desired dimensions and shapes. The surface finish and concentricity of the GFM product is better than the forged bar from a forging press, making the GFM product a more economical product. The other benefit of these machines is that they introduce uniform work into the product and require fewer, if any reheats. Figure 3.16 shows a rotary forging machine making a round product. Rotary forging of titanium alloys has the benefit of being economical and is a process that is relatively easy to control, and which yields a product that has more uniform hot work. This is beneficial when developing the desired microstructure in forgings made from the billet. Smaller diameter bar is finished in a rod mill that has a series of shaped rolls having successively smaller openings (a Kocks mill is an example of such a mill). In either case, the working temperature is in the (α+β) phase field, typically 50-70°C below the β transus. Titanium alloys have relatively poor thermal conductivity and relatively high flow stresses at all α+β working temperatures. At high rates of deformation, the thermal conductivity is too low to allow dissipation of the heat generated during working. Consequently, titanium alloys exhibit adiabatic heating during working. Adiabatic heating can lead to localized flow softening in active shear bands, resulting in further strain localization (this can be thought of in terms of positive feedback as used in control theory). The resulting intense shear bands cause inhomogeneous strain distributions and non-uniform microstructures after recrystallization of the final product. This effect is amplified at lower working temperatures because the flow stress is higher and more heat is generated in the active shear bands. Similarly, higher deformation rates increase the rate of heat generation. Under extreme circumstances the critical strain for ductile fracture is locally exceeded in the shear bands and small voids are formed. These voids are known as strain induced porosity (SIP). An example of strain induced porosity in Ti-6Al-4V is shown in Fig. 3.17. Once this porosity is created in forging billet, it is not always healed during subsequent forging operations. Consequently, it can be carried over into a finished component as a defect that can act as an early fatigue crack initiation site. Strain induced porosity also can be created in plate, particularly if the working temperature is allowed to drop. Alternatively, adiabatic heating, if it is properly managed, can extend the working time and deformation extent between reheats. Under these circumstances adiabatic heating is beneficial. In alloys that are prone to beta fleck formation, adiabatic heating must be carefully managed to avoid exceeding the local β transus in solute rich regions of the work piece.

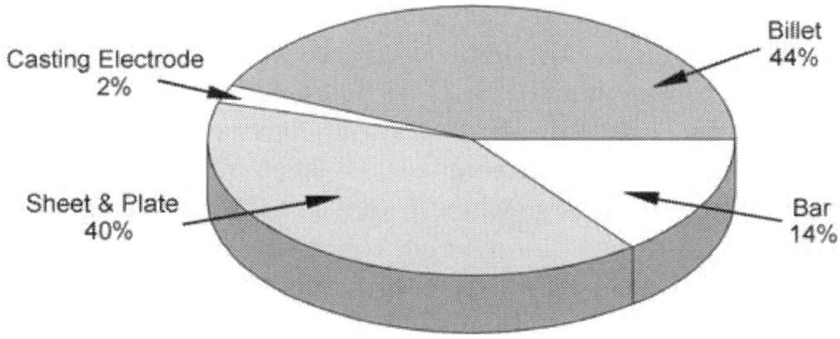

Fig. 3.15. Distribution of titanium production by product form (courtesy RMI)

Fig. 3.16. Rotary general forging machine (GFM) reducing the diameter of a round billet (courtesy Teledyne Allvac)

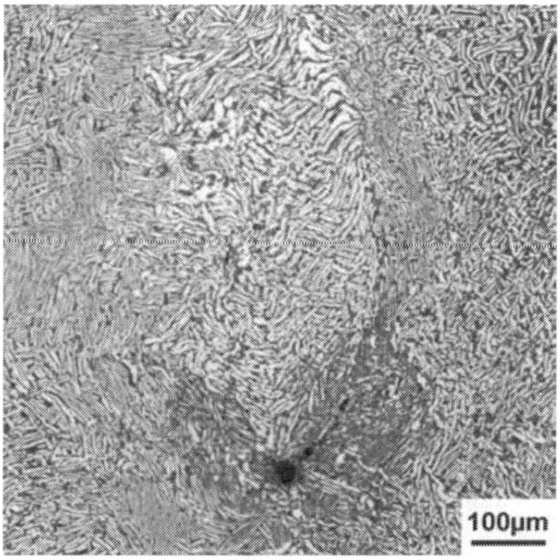

Fig. 3.17. Strain induced void at a β grain boundary in Ti-6Al-4V billet, LM (courtesy P. Wayte, GE Aircraft Engines)

As seen in Fig. 3.15, plate and sheet comprise 40% of titanium products. Plate, sheet, and small diameter bar (including rod) are all made in a rolling mill with either flat or shaped rolls. Figure 3.18 shows a large, "four-high" rolling mill (each work roll is backed by a second roll, so the roll stack consists of four rolls, hence "four-high") used for hot rolling titanium alloy sheet and plate. The starting material for plate and sheet is a forged bloom that is an intermediate product of the ingot conversion process. The bloom is hot rolled into a slab, which then receives further hot reductions. For plate, this final hot reduction stops at the desired thickness. The plate is then given an annealing treatment before it is surface conditioned by grinding and/or pickling. The hot rolling operations are also done in the (α+β) phase field, typically at about 50-100°C below the β transus. Flat rolled products require reheating more frequently because the plane strain imposed during rolling increases the incidence of edge cracking. The final annealing treatment is typically more of a stress relief than a true anneal. For example, Ti-6Al-4V is often shipped to users in the so-called mill-annealed condition. Mill-annealing means that the plate has been held at about 700°C for times as short as 1 hour or as long as 8 hours. At higher annealing temperatures the yield stress can be lower than the initial residual stresses. This can cause relaxation of internal residual stresses by gross plastic deformation of the plate causing it to loose flatness during the annealing treatment. Holding for longer times will result in a recovery of flatness due to creep flattening. Creep flattening occurs when the temperature is high enough that the product undergoes time dependent plastic deformation due to the weight of the product which results in an improvement in flatness. Thus, the flatness of the product after annealing is a function of the amount of work and the annealing time and temperature. As will be discussed in Chap. 5, the properties of mill-annealed plate are quite variable because this is not a well-controlled microstructural condition. Figure 3.19 shows a number of plates that have been processed and are awaiting shipment. Such plates are most frequently used for machining into aircraft parts.

Fig. 3.18. "Four-high" plate mill shown in the process of rolling a large plate, about 4 m long (courtesy RMI)

Fig. 3.19. Ti-6Al-4V plates after hot rolling, annealing, and surface conditioning (courtesy J. A. Hall)

Titanium alloy sheet is typically pack rolled to avoid surface oxidation. In pack rolling, a group of sheet blanks are sealed in a steel retort and rolled as a group. The individual blanks have an inert "parting agent" between them to prevent the sheets from bonding to one another during rolling. After hot rolling is complete, the pack is cut open and the finished sheets are extracted, pickled, and creep flattened or annealed in fixtures, depending the required flatness. Some final cold reduction also can be done to meet gage and flatness requirements. Sheet packs are often cross rolled to widen the sheet. Cross rolling also reduces the intensity of texture (preferred crystallographic orientation) and the texture symmetry in the final product. Texture can be very important to the forming response of sheet, especially in $\alpha+\beta$ alloys such as Ti-6Al-4V and Ti-8Al-1V-1Mo. In any case, the ratio of longitudinal to transverse strain introduced during hot working is large in flat rolled products. Further, the working temperatures happen to be in the range where basal/transverse type of textures are readily developed (see Sect. 5.1.2 and Fig. 5.8). Consequently, these products always have fairly intense textures. This makes some properties (e.g. yield stress and modulus of elasticity) anisotropic. If this is recognized and accounted for in product designs, it does not create a problem.

Strip is sheet-like product, but strip is typically narrower than sheet and very long. Strip is essentially unidirectionally rolled and coiled after the final rolling operation. The majority of strip is produced from either one of the grades of CP titanium or the alloy Ti-3Al-2.5V. The production of strip has enabled the economical production of welded CP titanium tubing for use in the chemical industry. The early stages of strip production are identical to those used for sheet and plate. However, at the slab stage, the strip is hot rolled, annealed, pickled, surface conditioned by grinding, and coiled as hot band, which is the intermediate product used for cold rolling. Figure 3.20 shows coils of annealed hot band that are awaiting cold rolling to final gage. It then is cold rolled to gage in a multi-stand mill such as a Steckel or a Sendzimir mill. These mills use several rolls to back a pair of very small diameter work rolls to ensure flatness of the thin, cold rolled product. A schematic of a Sendzimir mill is shown in Fig. 3.21. After cold rolling, the strip is annealed again and coiled for shipment. Because of the unidirectional working of strip, it always has a significant degree of texture, but because most of the material that is produced as strip is CP titanium, this typically is not a limitation.

Fig. 3.20. Coiled CP titanium hot band ready for cold rolling on a Steckel or Sendzimir mill into finished coils of strip (courtesy J. A. Hall)

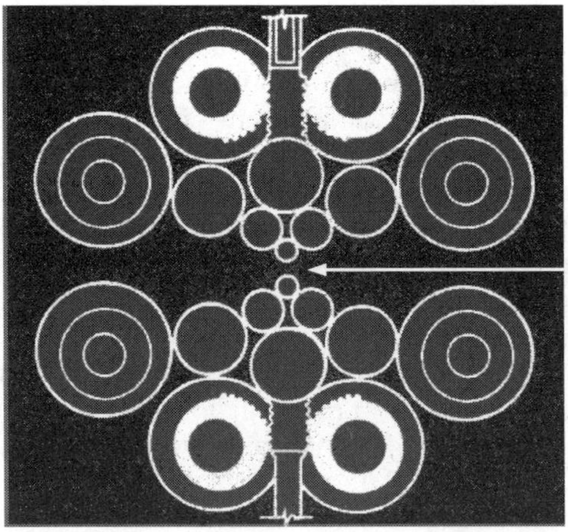

Fig. 3.21. Schematic of a Sendzimir mill used for cold rolling thin gage sheet, arrow: small work rolls (courtesy J. A. Hall)

3.4 Shaping into Components

Shaping of titanium mill products into actual components is accomplished by mechanical deformation operations such as forging or ring rolling to create rough shapes with the desired properties. These shapes are finished by mechanical metal removal (milling, turning, and drilling) to improve feature definition and reduce weight of the final component. This section describes these operations.

3.4.1 Forging

Forging is the principal shaping process used for making titanium alloy components [81, 82]. This is consistent with billet being the largest percentage of titanium alloy products sold each year (Fig. 3.15). Over the past two decades, titanium forging technology has progressed dramatically, particularly with regards to two aspects. First, the ability to achieve shapes that are closer to those of the desired component (near net shape) has greatly improved. Second, the ability to manipulate and control the microstructure to tailor the properties has become accepted as standard. Near net shape capability affects the cost of forged components whereas microstructure control affects the component performance. The cost of forgings always is a limitation in their use and has become a central issue in the competition between forgings and castings. The capability of titanium castings will be discussed in the next section of this chapter.

Titanium alloys are forged using hammers, either drop or steam driven, and large presses that are driven by either mechanical screws or hydraulic actuators. The tendency is to require press forging for high performance applications and for alloys that require tighter controls over the forging parameters (strain, strain rate, and temperature). Titanium alloys have much higher flow stresses during forging than aluminum alloys or alloy steel, so the required forging pressures are greater. This places limitations on the number of available facilities where large titanium forgings can be made. At present, there are two large presses in USA and one in Europe. These have maximum force capacity of 50 000 metric tons. There also is a 75 000 metric ton press (the largest in the world) at the Verkhnyaya Salda Metallurgical Production Operation (VSMPO) in the former Soviet Union. These are very large, expensive machines, so it is not surprising that there are so few of them. One of these large presses is shown in Fig. 3.22, together with a large titanium forging that is being produced.

Titanium alloys are forged using either open die or closed die methods. In open die forgings the work piece is not constrained laterally by the forging die, whereas it is in a closed die forging. Clearly, modeling these two forging processes requires different boundary conditions. The closed die process is more difficult to model, but once modeled, it is easier to control. In principle, the preferred forging method depends on the required degree of microstructural control as is dictated by the criticality of the intended application. In practice, selection of the forging method also is influenced by other economic factors. Perhaps the most important of these is the estimated total number of forgings that are to be made, since this determines the cost per piece that accrues from making the more complex and expensive dies used for closed die forging. A typical airframe or turbine engine forging is made in several steps. These steps are captured in a forging production plan that is approved by the customer before the first forging is produced. The first step is a potting or roughing operation that creates a work piece with the desired shape and aspect ratio for the first rough forging step, often called the blocker. Potting is a simple upsetting operation of small diameter billet to create a somewhat larger diameter work piece. This is done in cases where small diameter billet is required to ensure adequate hot work has been done for reasons of microstructural control. In the case of highly irregular shapes, the blocker is used to create nonuniform cross sections by gathering material where large, local cross sections are required, e.g. the root and mid-span shroud regions of an aircraft engine fan blade. This is followed by several more rough forging steps and a finish forging operation to produce the final shape. The number of forging steps is determined by the size of the forging, by the complexity of the final shape, and by the workability of the alloy being forged. In all cases, the work piece is reheated after each processing step. Most $\alpha+\beta$ alloys such as Ti-6Al-4V are forged in the $(\alpha+\beta)$ phase field, but β processing is used in some cases where damage tolerant properties are the design limiting consideration. The effect of processing on properties is discussed in detail in Chap. 5.

Fig. 3.22. Photo of a forging press (50 000 metric tons) with a large aircraft bulkhead forging (courtesy R. G. Broadwell, Wyman Gordon)

Today, the production plan for all forgings intended for high performance applications is created with the aid of computer modeling. This modeling addresses the strain, strain rate, and working temperature applied during each forging step. This is done to assure that the completed forging will have a microstructure including the necessary residual work stored in it, to allow the desired response to the final heat treatment. The results of the use of modeling is the ability to rapidly and accurately produce a wide range of forged shapes that have the required microstructure after forging and heat treatment. This modeling is still empirical with respect to the constitutive expressions that are used to relate external variables (strain, strain rate, and temperature) and the microstructural evolution. Thus, while the use of modeling has advanced the state-of-the-art in thermo-mechanical processing, the ability to model processes using first principles has yet to be realized. The actual solid mechanics modeling has become quite sophisticated but standard, user friendly software packages such as DEFORM and ABAQUS are widely available now. These programs can be run on a high-end desktop computer, so the cost and complexity of this modeling from a computational standpoint is quite nominal.

The shape making capability of hot die press forging is quite remarkable. These complex forgings also have well-defined microstructural requirements to assure that they exhibit the necessary mechanical and physical properties, for example strength, modulus, and ultrasonic clarity, to meet design intent. The size of these forgings also has increased over the past 30-35 years. Figure 3.23 shows a landing gear beam forging for a Boeing 747 aircraft prior to machining. This forging weighs about 950 kg and is one of the largest titanium alloy forgings produced today. Titanium forgings for aerospace applications are extensively machined to create complex, lightweight components. The result of this machining is removal of a substantial fraction of the original forged weight. In the manufacture of complex shaped aerospace components from forgings, it is quite common to have a finish machined part that weighs less than 10% of the initial forging weight. This practice is so common that the ratio of the forging weight to the machined part weight is called the “buy to fly” ratio. This requirement for machining is the result of the inability to make nearer net shape forgings with the desired microstructure and, in some cases, the need to create a forging that has a rectilinear shape for ease of precise ultrasonic inspection. This rectilinear shape is often referred to as a “sonic shape”. This requirement adds considerable cost to finished parts made from forgings. The sources of the additional cost are, therefore, twofold: first, heavier forgings are required at the outset and, second, it typically costs more per unit weight to remove the excess material by machining than the initial forging costs per unit weight.

The reasons for removing such large quantities of material are twofold. First, the emphasis on low weight requires machining of deep pockets and cutouts in the forging that cannot be created during the forging process. Second, critical forgings, particularly those used in turbine engines, require detailed, accurate ultrasonic inspection. This inspection is best accomplished on a rectilinear shape, as mentioned above. This requirement places practical limits on the level of shape complexity that can be included in the forging envelope. Today, this is often the limiting consideration rather than the intrinsic shape making capability of the forging process. Figure 3.24 shows a machined fan disk for a large turbine engine. This finished disk weighs about 170 kg, whereas the original forging from which it is machined weighs about 1000 kg. Large airframe components machined from forgings often have even lower utilization factors expressed as a fraction of the initial forged weight (sometimes $< 5\%$). An example of a bulkhead forging used in a twin engine military aircraft is shown in Fig. 3.25. Here the deep pockets that have been machined into the forging can be seen and these significantly contribute to the material removal requirements.

Despite the cost of making and machining forgings, they are the most common product forms used in high performance applications. This is because forgings have the best properties and because the thermo-mechanical processing that is an integral part of forging production, creates an opportunity to balance the critical properties for a specific application.

Fig. 3.23. Photo of a landing gear beam forging of Ti-6Al-4V for a Boeing 747 (courtesy R. G. Broadwell, Wyman Gordon)

Fig. 3.24. Large commercial aircraft engine fan disk that has been machined from a forging (courtesy GE Aircraft Engines)

Fig. 3.25. Bulkhead for a twin engine military aircraft machined from a large forging such as the one shown in Fig. 3.22 (courtesy R. G. Broadwell, Wyman Gordon)

3.4.2 Ring Rolling

The creation of round shapes is most easily accomplished by the ring rolling process. Ring rolling is used to make both rings and cylinders. The distinction between these is the ratio of diameter to axial height, with rings having the larger ratio. Rolled rings are made by piercing a hole in the center of a piece of billet to create a thick walled cylinder. This cylinder is heated and placed in a ring rolling machine which consists of two rolls. These rolls apply a through-thickness pressure and turn to rotate the cylinder, which increases in diameter while the wall thickness is decreased. A photo of a ring rolling machine making a cylinder is shown in Fig. 3.26. Due to the lateral constraint of the rolls, changes in the axial dimension of the cylinder can be controlled during ring rolling.

Rolled rings are more accurately round, are seamless, and have good properties in the circumferential direction. They can be joined together by fusion welding or friction (inertia) welding to form cylinders that are used for casings for jet engines and rocket engines, among other things. An example of a seamless rolled ring being machined into a fan case for an aircraft engine is shown in Fig. 3.27. Rolled rings can be large and diameters greater than 300 cm can be made, although 50-100 cm are more common sizes. Large cylinders for pressure vessels with axial heights as large as 150 cm are commonly used. A large ring rolled cylinder of CP titanium is shown in Fig. 3.28.

Fig. 3.26. Ring rolling machine producing a large seamless cylinder (courtesy D. Furrer, Ladish Co.)

Fig. 3.27. Ti-6Al-4V seamless rolled ring during machining to make a fan casing for a large aircraft engine (courtesey D. Furrer, Ladish Co.)

Fig. 3.28. Large CP titanium ring rolled cylinder ready for shipment, the starting billet for such a cylinder is also shown in the right foreground (courtesy D. Furrer, Ladish Co.)

3.4.3 Metal Removal (Machining)

Titanium and titanium alloys are generally considered difficult to machine in comparison to steel or aluminum alloys for all conventional machining methods [83, 84]. Included in these methods are milling, turning, end milling, drilling, and reaming. Titanium alloys have low thermal conductivity which reduces the rate of dissipation of the heat generated at the tool/work piece interface. This causes significant reductions in tool life. This problem becomes more severe at higher rates of metal removal, in terms of tool/work piece surface speed. Slow speeds and heavy depths of cut are commonly used in the machining of titanium alloys. Heavy cuts are used in part to offset the slow speeds and in part to be certain that each successive cut is deeper than any work hardened layer from the previous cut. In addition, the intrinsic reactivity of titanium causes the freshly exposed surface to react with the tool, further accelerating the rate of tool wear. Improved cutting tool materials such as carbides and ceramics ameliorate the tool wear problem, but these tools are more expensive. Consequently, the cost of removing a unit volume of material is not reduced as much as it appears just from extended tool life when the tool cost is included in the calculation. In machining components, that have large fatigue limited surfaces, the state of residual stress induced by machining is important. In such cases, tool lift-off, breakage, or change are often disallowed, making the use of long life tools more economically attractive. Improved coolants and the use of high coolant volumes during machining help with heat removal during milling and turning. In the case of drilling, it is difficult to deliver high coolant volumes to the bottom of a deep hole that is being drilled. Special drill bits, called coolant fed drills, that have coolant passages along their axis, provide improved coolant supply, but the coolant volume and recirculation rate is still low compared to a milling or turning operation. In fatigue limited parts, holes are typically created in three steps: drilling, followed by reaming, and then honing. This is expensive, but is the most reliable method for making high aspect ratio holes (depth more than five times the diameter) that are free from damage and, therefore, have the best possible fatigue characteristics. The allowable feeds and speeds used depend both on the specific machining operations and on the titanium alloy being machined. CP titanium can be machined about twice as fast as the higher strength alloys such as Ti-6Al-4V. By comparison to Ni base alloys or steel, titanium alloys are less stiff (50% lower modulus of elasticity) and this can cause deflection of the work piece if it is not adequately supported by tooling. It is important to recognize this issue and address it during final machining operations, especially if tight dimensional tolerances are required. For repetitive large volume jobs, the cost of additional tooling is not that significant but for small lot sizes it can add significant cost.

Perhaps the most serious aspect of the tool life issue, apart from the intrinsic cost issue, is the potential for introducing surface damage if machining is done with a dull or damaged tool. It is well known that abusive grinding of carbon steels and low alloy steels can cause untempered martensite to form, which is brittle and causes significant shortening of fatigue life. Surface damage in titanium alloys is less obvious but still has serious consequences with regard to fatigue. Detection of surface damage in titanium is more difficult, thus elimination of the possibility of damage introduction in critical locations of fracture critical components is accomplished through process control. These controls include monitoring the torque on the machine tool spindle during drilling or milling and alerting the operator if the torque rises to a level known to be consistent with a damaged or worn tool. Another type of control is strict limits on tool use, e.g. a maximum number of holes for each tool. Establishing safe limits for torque or tool use is a highly empirical process, making machining process parameter control imperative. A fundamental understanding of the detailed processes that occur during metal removal would allow higher confidence limits to be established faster.

Because machining damage has serious consequences on component performance, inspection methods to ensure no damage is being introduced have been developed. These inspection methods will be described in detail later in Sect. 3.8, but it is important to mention that blue etch anodize (BEA) inspection is currently considered the most reliable method for detecting surface damage. There is growing interest in the use of eddy current inspection, but this is not a mature method of today.

3.5 Near Net Shape Processes

The cost of titanium alloys has been identified as a significant factor in the decision to use them or not. Because titanium alloys are expensive, it is essential to make the most efficient use of them. One approach to realizing high material utilization has been the development of net shape processing methods, each driven by the need to minimize the amount of wasted metal created during the realization of a titanium alloy component. The following sections describe four processes that have been examined as possible means of producing near net shape components.

3.5.1 Casting

The usage of titanium components made by casting has grown over the past ten to fifteen years. This growth has occurred as the cost of castings has decreased and as the quality and capability of titanium castings have improved [85, 86]. This allows castings to be used in a wider range of components with a broader range of requirements that are derived from design intent. The growth in the use of castings has resulted in fewer fabrications. A fabrication is a component that is built up from a number of smaller parts, i.e. it is a fabricated component. In some cases, castings have displaced parts that were previously machined from forgings or machined out of heavy plate or billet. The substitution of castings for components made by other methods has been enabled by three factors. First, improvement of net shape casting technology, second, improvement of the fatigue properties of castings due to the use of hot isostatic pressing (HIP) to heal internal porosity that can act as crack initiation sites, and third, decreased metal-mold reaction. Examples of two investment cast frames for an aircraft engine are shown in Fig. 3.29. These one piece castings replaced fabricated frames that were welded and mechanically fastened. Each of these frames, when used as a fabrication, contained more than 100 individual parts including the fasteners.

Fig. 3.29. Investment cast frames for aircraft engines (courtesy GE Aircraft Engines)

Casting of titanium alloys for production applications is done using one of two methods. The first is conventional casting using rammed graphite as the mold material (as opposed to sand for ferrous and bronze castings). Rammed graphite has a minimal tendency to react with the molten titanium and this results in castings that have better surface condition. Rammed graphite titanium castings can have quite intricate shapes and have good surface finish after the typical finishing operations such as tumble cleaning and chemical milling. Figure 3.30 shows a pair of articles that have been cast using rammed graphite molds. Investment casting is the second and more widely used method for making titanium alloy castings. This method is more costly to execute, but its shape making capability makes the resulting castings economically attractive because they displace very expensive forged and machined or fabricated parts. The investment casting method is capable of producing highly intricate shapes with essentially a net shape. An example of such casting was shown in Fig. 3.29. In this method, a pattern of the part is made and this is used to create a mold for making wax shapes. These shapes are made by injection molding the wax into the die made from the pattern, using a standard injection molding machine. After removal from the mold, the wax pattern is coated with a nonreactive face coat intended to minimize the reaction between the ceramic shell and the molten metal during casting. After applying the face coat, the wax pattern is repeatedly dipped into a ceramic slurry to create a shell that becomes the vessel into which the molten titanium alloy is poured to make the casting. An example of the shell used to cast a frame similar to that in Fig. 3.29 is shown in Fig. 3.31. It is essential to maintain a uniform wall thickness in the shell, both for mechanical strength and for heat transfer during solidification. The extensive use of robots to manipulate these shells has led to advances in uniformity of the shells. Once the shell is thick enough to have the necessary mechanical strength, it is put into a low temperature furnace to remove the wax by melting. The shell is then fired at a higher temperature to improve its strength. Both the green (unfired) and fired shells are quite fragile and the use of robots has improved the consistency of shell properties which has reduced shell losses due to handling in the foundry. Investment castings can contain internal passages and these are created by inserting a core in the shell and supporting it to avoid any movement during casting. These cores are removed after casting by chemical dissolution or leaching. Consequently, there must be a passage from the surface of the casting to the core passage to allow entry of the chemicals and to allow the reacted solution to escape. The molten titanium alloy is then introduced into the shell under vacuum and the shell is allowed to cool in a controlled manner to minimize the occurrence of shrinkage pore formation. A schematic diagram of a vacuum casting furnace for titanium alloys is shown in Fig. 3.32. The reactivity of titanium requires the use of a nonreactive melting technique. The methods available include non-consumable electrode and consumable electrode VAR and cold walled, split crucible induction melting. As in the case of cold hearth melting described earlier, these methods are capable of achieving only a limited degree of superheat of the molten titanium as compared, for instance, to vacuum induction melting of superalloys. The limited superheat in turn limits the ability to fill narrow passages in the casting.

Fig. 3.30. Rammed graphite castings made from CP titanium grade 2 (courtesy J. A. Hall)

Fig. 3.31. Shell for a large structural investment casting being prepared for placement into the casting furnace (courtesy N. E. Paton, Howmet)

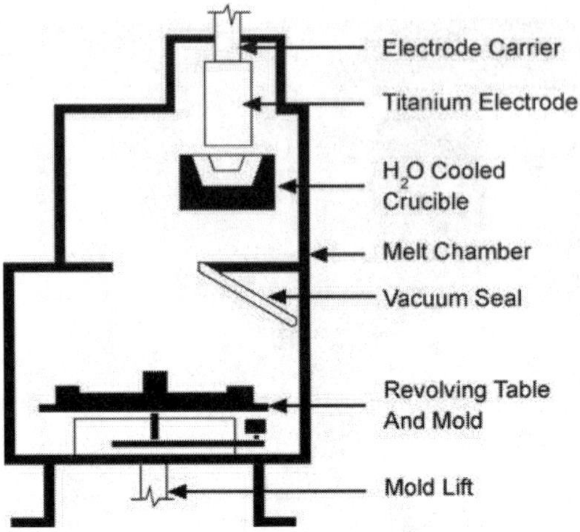

Fig. 3.32. Schematic of a vacuum casting furnace used for making titanium castings (courtesy N. E. Paton, Howmet)

There is a great deal of experience based know-how in the design of the gates through which the molten alloy is introduced into the shell. The gates must be located to ensure that the shell cavity is filled before solidification is very advanced. The use of risers that provide supplemental molten metal helps this to be achieved. This is particularly important for titanium casting because of the lower superheat of the molten metal. These gates and risers are removed and recycled which results in significant cost saving. There is an increasing use being made of a cold wall, split crucible induction furnace for melting titanium. As its name implies, this type of furnace has no continuous path for induced current in the crucible, so only the charge is heated and melted. Among other uses, this is an effective way to remelt and to directly reuse this recycled titanium. Heat and fluid flow models are used to assist in the placement of gates and risers, but today, there is no substitute for experience based intuition. Reproducibility of cavity fill from casting to casting is still an issue. Luckily, the areas where cavity fill is incomplete can usually be repaired by adding material through fusion welding. Weld repair is labor intensive because it is done manually and because the casting surface must be cleaned prior to welding. In fact, weld repair is a significant cost element in the production of complex structural castings.

Investment cast titanium parts have the major advantage of being essentially net shape. After hot isostatic pressing, chemical milling, weld repair, and, if necessary, a stress relief heat treatment, these castings can be given a final simple chemical milling operation and are ready to use. Chemical milling after hot isostatic pressing is done to remove any reaction zone that exists between the mold and the molten titanium alloy. Hot isostatic pressing is done to close internal shrinkage. The introduction of HIP treatment of titanium casting had led to as much as a twofold improvement in fatigue properties and is one of the major factors responsible for the growth in casting use in fatigue limited applications. More complex castings, such as the ones shown in Fig. 3.29, often contain surface connected defects due to shrinkage, hot tearing, or incomplete mold filling that cannot be healed by hot isostatic pressing. In such cases, weld repair is used to close these asperities, usually after the HIP cycle. Weld repair of high value castings allows these parts to be used instead of being scrapped, but the weld repair is costly. Weld repair is typically done using the tungsten inert gas (TIG) welding process. Depending on the size of the casting defect, metal in the form of weld filler wire may be used to repair the defects. It is expensive to make weld wire from all the high performance alloys, including Ti-6Al-4V. For alloys available as sheet, narrow strips can be slit and used as a filler. As mentioned above, weld repair also is very labor intensive. These two factors add significantly to the cost of structural castings. Improved casting processes, therefore, would reduce the cost of titanium castings and could lead to an increased market share.

One such improved casting process currently under active development is permanent mold casting of titanium. Permanent molds have been used for casting lower melting point materials, e.g. Al and Mg, and less reactive high melting point materials such as steel. The challenge for permanent mold casting of titanium is to minimize the interaction of the molten metal with the mold while obtaining adequate superheat to permit good fill. Currently, castings about 200 mm in maximum dimension represent the size limit that can be produced by permanent mold casting of titanium.

3.5.2 Powder Metallurgy

Powder metallurgy (PM) has been used as a means of making net shape parts from many metals and alloys. The traditional low cost, net shape powder metallurgy method for making parts has been press and sinter. This method is used to make steel and copper based alloy parts for a wide range of machinery. In contrast, powder cost limits the range of potential applications for titanium PM parts. This is because the reactivity of titanium essentially precludes production of gas atomized powder and sintering without expensive measures to prevent atmospheric contamination. Currently, direct gas atomization of molten titanium is not possible on a large scale using an atomization process analogous to that currently being used for Fe, Ni, and Cu base alloys. Several organizations have developed small, laboratory scale atomization units and have successfully atomized titanium alloys. The quantity of powder produced in these units in a single run is typically less than a hundred kilograms. This gas atomized powder looks like Ni and Fe alloy powder made by gas atomization. That is, the powder particles are relatively small, nominally spherical, and exhibit smaller secondary spheres, called satellites, that are attached to the larger powder particles during atomization. An example of gas atomized Ti-6Al-4V powder is shown in Fig. 3.33a.

The most common and only viable method of making pre-alloyed titanium powder is the so-called rotating electrode process (REP). In this process, a titanium bar is spun at about 18 000 rpm in an inert gas filled chamber, similar to that shown schematically in Fig. 3.34. A heat source (either an electric arc or a plasma torch) melts the surface of the end of the spinning bar. When the heat source is an electric arc, the powder is known as REP powder and when it is a plasma torch, it is known as PREP powder. The molten titanium alloy at the surface is separated from the rotating electrode and impelled outward by the centrifugal force. Under the influence of surface tension spherical droplets are formed and these solidify in flight. The solid spherical pre-alloyed powder particles formed in this way have a size that depends on the rotational velocity of the electrode. Typical average particle size for powder made from 18 000 rpm operation is about 300 to 500 μm in diameter. Figure 3.33b shows these spherical powder particles. By comparison to the gas atomized powder shown in Fig. 3.33a, this spherical powder is relatively free from satellites that are commonly seen in gas atomized spherical powder. The REP powder also has a larger average size than the gas atomized powder.

a b

Fig. 3.33. Spherical, pre-alloyed titanium powder made (a) by the gas atomization process and (b) by the rotating electrode process, SEM

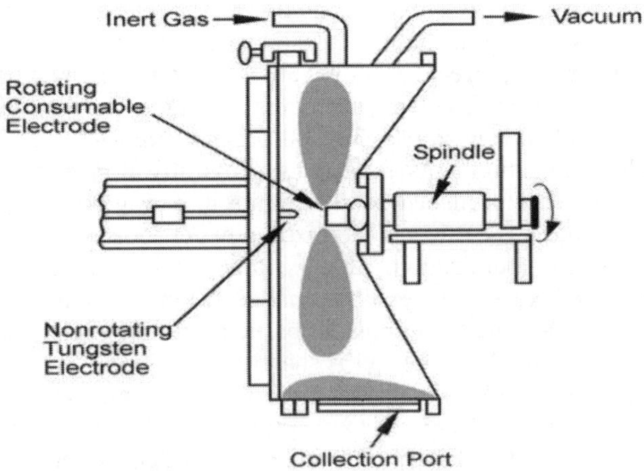

Fig. 3.34. Schematic of REP unit for making centrifugally atomized titanium powder (courtesy F. H. Froes)

In case of REP powder, some of the tungsten electrode can be eroded during atomization and this is incorporated into the powder as reactive inclusions. PREP powder eliminates the source of tungsten inclusions, but the plasma torch creates a significant quantity of very fine particles (almost like fume) and these must be filtered out before the powder is consolidated, which adds some cost. After sieving to obtain a uniform size distribution, this REP powder is easy to handle and has a good tap density (about 70%). Still, all powder handling must be performed in vacuum or inert gas to retain powder cleanliness, which also adds cost. Consolidation of this powder is done by canning and hot isostatic pressing (HIP) to form a fully dense body. Achieving as-HIP'd densities of 99.99% of the theoretical alloy density is relatively easy because, unlike Ni and Fe gas atomized powders, very few powder particles are hollow and filled with inert gas. This is an advantage of the rotating electrode atomization process.

The issues that must be overcome before PM titanium alloy parts gain broader acceptance include powder cost and the possibility of defects in the as-HIP'd body. For high performance applications where the attraction of near net shape processing is greatest, the presence of reactive inclusions, many of which are tungsten, limits the fatigue properties. The advent of PREP powder has reduced the incidence of inclusions but has not totally eliminated these defects. For static parts, the properties of PM titanium alloys are adequate and some of these have sufficiently complex geometry that PM could be attractive, if the cost were more attractive. There does not appear to be any obvious solution to the powder cost issue. Hence the future of PM titanium parts made from pre-alloyed powder seems uncertain.

It also is possible to make titanium alloy parts from elemental blended powders and this method has much more attractive economics. The process for making blended elemental powder metallurgy parts is shown in Fig. 3.35. From this it can be seen that finely divided unalloyed titanium is mixed with alloying elements (typically finely divided master alloy) in the correct proportions and blended in a twin cone blender. This mixture is then cold isostatically pressed (CIP'd) in an elastomeric mold to create a solid shape that is > 80% dense. It is then vacuum sintered to create a near net shape that is > 94% dense and that contains no surface connected porosity. This sintered shape can be HIP'd without a can which makes this process much less expensive. This is because the can is expensive and the removal of it after processing by hot isostatic pressing adds significant cost. Alternatively, sintered shapes can be extruded or forged in a closed die set to create other shapes as shown in Fig. 3.35.

The unalloyed fine titanium particulate is either the smallest particulate that comes directly out of the reduction reactor (called sponge fines) or fine particles created by crushing sponge. In the latter case, the hydrogenation/dehydrogenation (HDH) process is used to embrittle the sponge to make it more efficient to crush. In either case, these input materials are relatively inexpensive compared to pre-alloyed REP or PREP powder. The elemental titanium particulate typically has higher oxygen content. This restricts use of this material to applications that permit higher oxygen levels. An example of the morphology of sponge fines is shown in Fig. 3.36. Regarding sponge fines, the Na reduction reaction in the Hunter process proceeds at lower temperatures and less sintering of the sponge cake occurs as compared to the Kroll process. Earlier, up until about 1990, when there was still a Na reduction sponge plant in operation, sponge fines also were more plentiful and readily available as a by-product at an even lower cost.

Elemental blended PM parts rely on solid state diffusion to achieve uniform composition and microstructure. This dictates the input particle size, but this is not a significant limitation. This process is being used to make limited lot size parts for static or limited life aerospace components and sporting goods applications. Experimental automotive parts also are being made and tested. In this latter case, the parts contain a ceramic second phase that is added for increased stiffness and retention of adequate fracture resistance. In such cases, the ceramic additive is typically TiC or TiB_2. The elemental blend process is ideally suited for adding such particulate reinforcements since it already involves blending powders as a process step. A photo of a particulate reinforced automotive connecting rod is shown in Fig. 3.37. This part is cold isostatically pressed, then sintered, and then forged to create a fully dense part.

The foregoing discussion has briefly covered most aspects of titanium powder production and consolidation into components. The most promising role for titanium powder metallurgy appears to be for certain niche applications. There are numerous methods for making titanium powder and for consolidating it into shapes. These have been described in greater detail elsewhere [87, 88].

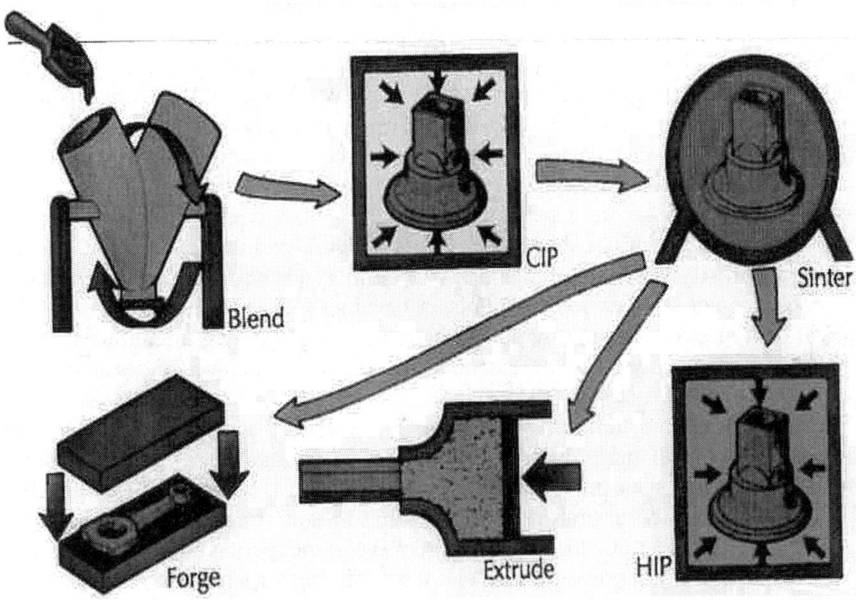

Fig. 3.35. Flow diagram for making titanium PM parts from blended elemental powders (courtesy S. Abkowitz, Dynamet Technologies, Inc.)

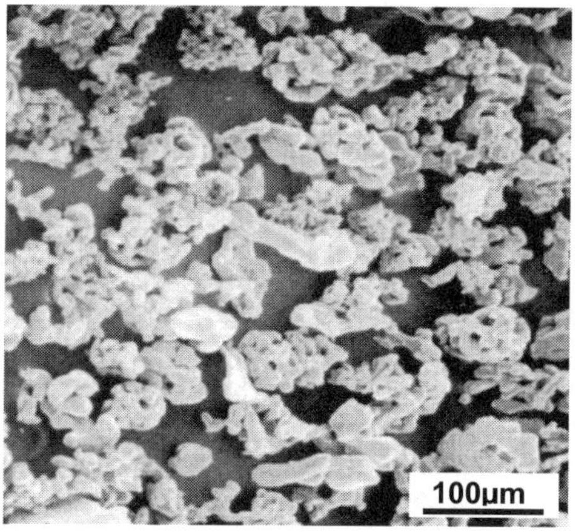

Fig. 3.36. Sponge fines used in making elemental blended titanium PM parts, SEM (courtesy S. Abkowitz, Dynamet Technologies, Inc.)

Fig. 3.37. Particulate reinforced titanium connecting rod made by PM method (courtesy S. Abkowitz, Dynamet Technologies, Inc.)

3.5.3 Laser Forming

Laser forming (also called laser deposition) is a relatively new net shape technology for fabricating titanium components. This method uses a focused laser beam to fuse (deposit) titanium powder at a desired location. The deposition point is controlled by placing the component being made on a numerically controlled table. By continuously manipulating the X, Y, and Z coordinates of the location of the laser focal point, which also coincides with the point of introduction of the titanium powder, a three dimensional shape can be generated. This technology allows the use of geometry descriptions consisting as the digital files that are now routinely generated during mechanical component design, to directly generate a near net shape titanium component. This is schematically illustrated in Fig. 3.38. In principle, laser deposition has the advantage of making a complex shape that is much closer to the final dimensions as compared to forging. An example of such a component is shown in Fig. 3.39. A realistic comparison shows that the cost of machining away much of the forged weight is offset by the cost of titanium powder as has been discussed previously (Sect. 3.5.2). Further, because the powder is fused during deposition, the as-deposited material has a fully lamellar microstructure, but with a finer structure than is typically seen in a casting because of the higher cooling rate. In anticipation of at least limited niche market production of laser deposited components, several large machines have been built. One of these is shown in Fig. 3.40.

While these machines are complex and expensive, their cost is small compared to a forging press or an investment casting foundry. Therefore, new component making capability in geographic regions where no casting or forging capability exists can be created through the acquisition of a laser deposition apparatus. This, of course, assumes that there is access to a reliable supply of high quality, affordable pure or pre-alloyed titanium powder. There is currently experimentation underway with elemental powders (sponge fines and granulated master alloys), but it is too soon to decide how effective this approach will be.

The longer view of the future of laser deposition of titanium is not entirely clear, but it seems realistic to suggest that this technique will gain acceptance at least for niche markets where rapid turn time and low part quantities are involved. In such circumstances, the ability to create parts directly from a CAD file without a die or a tool forging or casting production should be a competitive advantage.

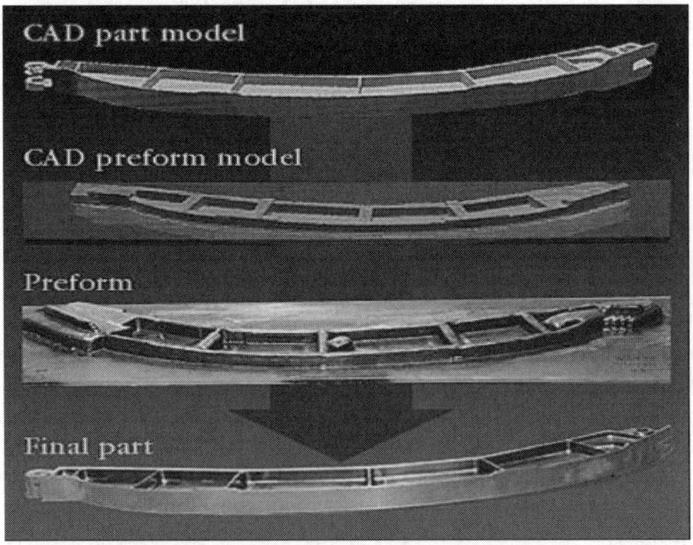

Fig. 3.38. Schematic sequence used to generate a near net shape titanium component

Fig. 3.39. As-deposited Ti-6Al-4V part (courtesy F. Arcella, Aeromet Corp.)

Fig. 3.40. Photo of large laser deposition apparatus (courtesy F. Arcella, Aeromet Corp.)

3.5.4 Conventional Sheet Forming

Titanium alloys are broadly considered difficult to form. This is in part because of their high yield stress and comparatively low elastic modulus. In combination these properties result in large elastic strains which are recovered after the load is released during a forming operation. This elastic strain is known as "spring back". Spring back creates difficulty in achieving the desired final contour or shape after a forming operation. This is typically overcome by a second hot sizing operation where the formed sheet is forced into a shaped die at an elevated temperature and allowed to "relax" to the desired shape by time dependent deformation (creep forming). This takes considerable time and, consequently, adds considerable cost to the final part. Further, if a significant volume of parts is required, more than one set of hot sizing dies may be needed and this also adds to the cost. There has been an effort to standardize the formability characteristics of titanium sheet using a dimensionless index called the minimum bend radius, T_R. This is defined as follows:

$$T_R = R/H$$

where H is the sheet thickness and R is the die radius. Small values of T_R correspond to improved formability. The values of this parameter vary between alloys and range from 1.5-2 for CP titanium to 4.5 for Ti-6Al-4V. Sheet formable β alloys such as Ti-15-3 have T_R values much closer to CP titanium than Ti-6Al-4V, typically around 2. Elevated temperature forming improves the formability of Ti-6Al-4V as shown in the graph in Fig. 3.41. Here the benefit of hot forming is clearly demonstrated, but there also is additional cost associated with this choice. Elements of this added cost include the need for heat resistant dies or tools, the requirement to heat these dies, and the long time on station necessary to complete the forming operation. In addition, there is the possibility of having to pickle the part (clean the surface chemically) after forming if there is any α-case or oxygen enrichment of the surface. Historically, these considerations have resulted in a reluctance to use titanium sheet for applications that require forming. The alternative is to fabricate sheet structures, but this is expensive, requires some type of joining, and adds to the part count in any system, which increases the cost. More recently, a third alternative, superplastic forming of titanium sheet, has been developed and this process will be discussed in the next section.

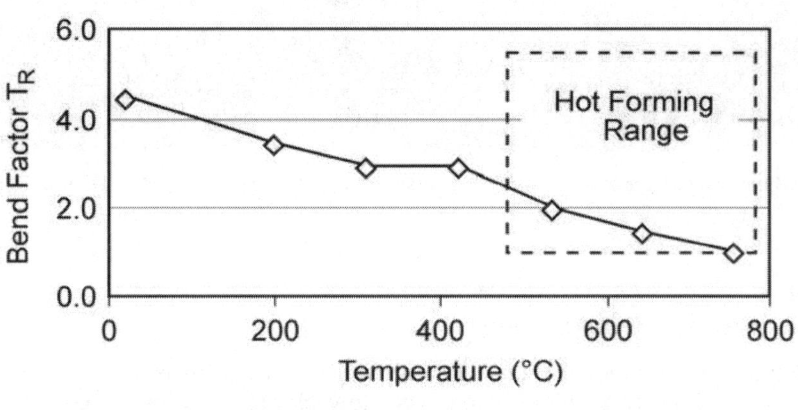

Fig. 3.41. Sheet formability index vs. temperature for Ti-6Al-4V (courtesy RMI)

3.5.5 Superplastic Forming and Diffusion Bonding

Titanium exhibits superplastic behavior and the capability for diffusion bonding. These two characteristics can be used individually or jointly for fabricating titanium components. A collection of papers describing various aspects and applications of superplastic forming (SPF) and diffusion bonding (DB) can be found in a conference proceedings [89]. A more recent summary of both processes also is available elsewhere [90].

The difficulties associated with conventional sheet forming of titanium alloys, especially Ti-6Al-4V, resulted in the commercial development of superplastic forming [91]. It has been known for quite some time [92] that titanium alloys such as Ti-6Al-4V exhibit superplastic behavior. Superplastic behavior is operationally defined as having a strain rate sensitivity exponent m of 0.5 or greater in the following expression relating flow stress and strain rate sensitivity:

$$\sigma = \sigma_0 \, (d\varepsilon/dt)^m$$

where σ_0 is the threshold flow stress at very low strain rate. As the strain rate sensitivity of the flow stress increases, the material becomes more resistant to formation of necks or other forms of localized deformation. The flow stress and m values are microstructure, strain rate and temperature dependent. Figure 3.42 shows the temperature dependence of the flow stress in the superplastic forming temperature range [93]. In alloys such as Ti-6Al-4V, the maximum value of m is about 0.7 and typically occurs at about 875°C at a strain rate of $10^{-4}\ s^{-1}$ or less. Detailed observations of the microstructure of materials after superplastic forming showed no evidence of an increased dislocation density. This has led to the postulation of new mechanisms of plastic flow in these alloys at the relevant temperature and strain rate ranges. One mechanism was postulated by Ashby and Verrall [94] who have described the mechanism of superplastic flow as diffusion induced "grain switching". Consequently, the microstructural length scale must be consistent with the times required for diffusion over these distances to occur. At temperatures of at about 875°C and at a strain rate of $10^{-4}\ s^{-1}$ this length is about 20 μm, which is about the normal microstructural size scale in sheet products of this alloy after normal processing (pack rolling). The flow stresses at these temperatures and strain rates are very low. Consequently, SPF permits the sheet to be deformed into a single die cavity under 0.2 MPa gas pressure, which is typically argon to prevent surface oxidation of the titanium sheet. This is schematically illustrated in Fig. 3.43. Because the flow stress is so low, there is no spring back after the superplastic forming process is completed. Even though the strain rates are quite low and the forming times are long, the SPF setup is simple and inexpensive. Consequently, several units can be run concurrently at manageable expense. Figure 3.44a shows a typical die used to make a deep pocket section by SPF. Figure 3.44b shows the actual part that is made using this die. Initially, the cost of SPF parts was considered high, but this cost has decreased with experience. There are now suppliers that produce SPF components which the original equipment manufacturers (OEMs) such as Boeing and Airbus can procure in the same way they buy forgings, castings, or other purchased components.

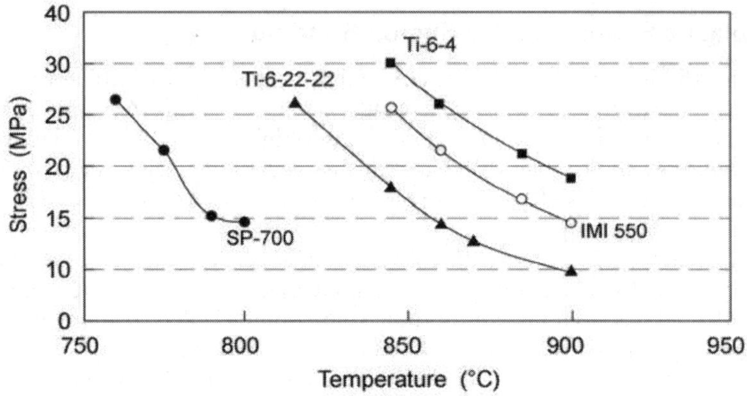

Fig. 3.42. Flow stress as a function of temperature for four titanium alloys at a strain rate of $5 \times 10^{-4}\ s^{-1}$ [93]

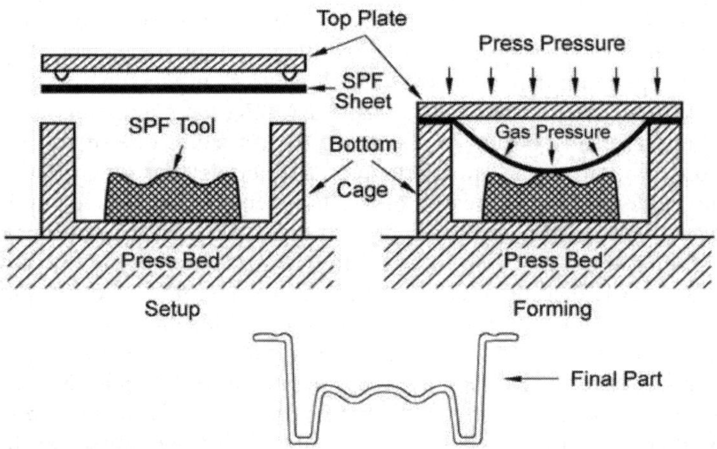

Fig. 3.43. Schematic drawing of superplastic forming of titanium sheet (courtesy RMI)

The capability of titanium for diffusion bonding is another intrinsic characteristic that can be exploited to enable a novel manufacturing process. This is the propensity of titanium to dissolve its own surface oxide when heated to ≥ 550°C in vacuum or high purity inert gas (i.e. a nonoxidizing environment). Consider two sheets of titanium alloy that are placed in intimate contact and a modest pressure is applied to create a high degree of contact. If this "sandwich" is heated to about 600°C or higher, the TiO_2 on the surface of the sheets dissolves, leaving pure surfaces in contact. The result is that two sheets bond together by inter-diffusion, resulting in a bond that is imperceptible even during metallographic examination. This method of joining is called diffusion bonding (DB). Diffusion bonding typically is performed at temperatures in the (α+β) phase field and, therefore, this process does not perturb the equiaxed microstructure of the pieces being joined. This is in contrast to fusion or friction welding, as will be described later in this chapter. Figure 3.45a shows the macrostructure of a diffusion bonded joint in Ti-6Al-4V. From this photo it can be seen that there is no evidence of the prior bond line between the two sheets. Figure 3.45b shows the microstructure of this area, also with no evidence of the prior bond line. This micrograph clearly illustrates the retention of the equiaxed α+β microstructure in the bond region. This can be important for component performance since the properties of a β processed material are different from those of α+β processed material as will be discussed in detail in Sect. 5.2.

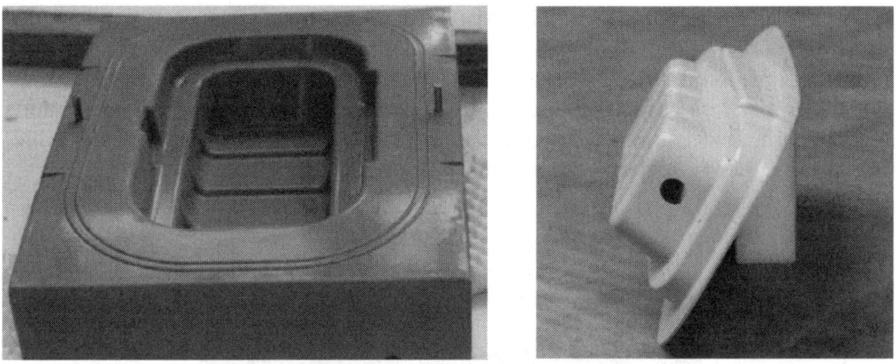

a b

Fig. 3.44. Superplastic forming of titanium sheet components: (a) One piece die (b) Resulting sheet metal part (courtesy W. Beck, FormTech)

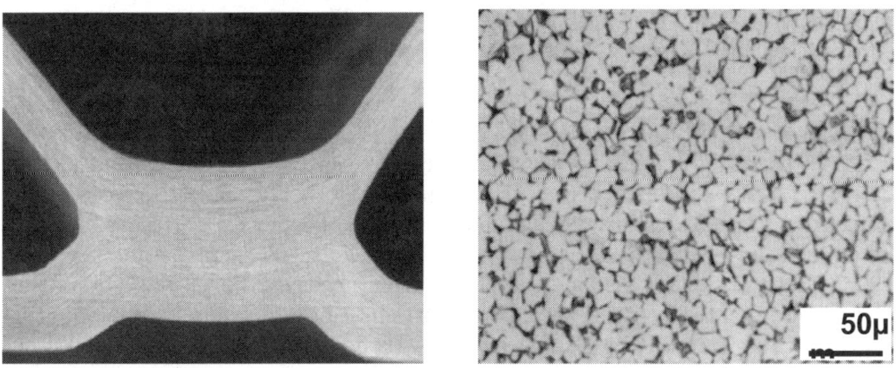

a b

Fig. 3.45. Diffusion bonded (DB) joint of two sheets in a SPF/DB component, LM: (a) Low magnification showing complete bonding (b) High magnification showing no evidence of the prior bond location (courtesy W. Beck, FormTech)

In principle, the DB process can be used to create joints between pieces of large sections, such as the wingbox for a large military aircraft as shown in Fig. 3.46. In practice, the joint preparation requirements necessary to ensure a high quality bond are quite exacting in terms of surface flatness and surface finish in the areas where the actual bonds occur. Further, it is necessary to inspect the DB joints and this inspection is both time-consuming and technically challenging. Consequently, the economics of this process become less attractive as described later. A sub-scale model wingbox of the component shown in Fig. 3.46 was fabricated early in the stages of DB process development. When this sub-scale article was fatigue tested, it was found that the cyclic life was less than expected. During postmortem evaluation several unbonded areas were found within the DB joints and these defects served as crack initiation sites. Detailed analysis led to the conclusion that the surface flatness of the bond areas was not sufficient to permit complete contact across the entire joint surface. The cost of implementing adequate joint preparation procedures is very high and increases with component size. When this constraint is considered, the cost of using the DB process for fabricating large components becomes less attractive, and the technical risk of unbonded regions remains as a concern. Because of these concerns, DB has never been adopted for production of large, heavy section components.

Fig. 3.46. Wingbox for a large military aircraft, considered as a candidate for fabrication by diffusion bonding (DB) methods (courtesy R. G. Broadwell, Wyman Gordon)

In contrast, the diffusion bonding of sheets is a more attractive possibility because the starting material is flat and because inspection of these joints, if required, is easier to do. Moreover, the combination of SPF and DB into a single process that enables creation of complex shapes with integral stiffeners becomes extremely attractive. In part, this is possible because the temperature ranges at which SPF and DB can be done coincide. Further, because the entire section being formed and bonded is superplastic, the constraint of the surrounding material that prevents complete bonding in heavy sections is not an issue. The combined processes have been successfully demonstrated and are commonly referred to as superplastic forming / diffusion bonding (SPF/DB). The use of a stop-off agent permits the selective bonding of regions of the sheet. If this bonding is followed by pressurization to separate the unbonded regions of the sheet (also called selective inflation) a complex honeycomb structure can be created from a three piece pack as shown schematically in Fig. 3.47. An additional benefit of the inflation is that it provides an in situ test of the bond strength. Such structures have a high section modulus and are structurally efficient in terms of low weight, high stiffness configurations. One example of a SPF/DB component is shown in Fig. 3.48. This is a manifold used to distribute cooling air for the variable clearance control system of the low pressure turbine case of a turbofan aircraft engine. This single part replaces an array of brazed tubes. Another important example is the use of SPF/DB processing in the manufacture of large, wide chord, hollow titanium fan blades for high thrust, high bypass ratio turbofan aircraft engines.

In the hands of a creative design engineer, SPF/DB enables the creation of complex, integrally stiffened, lightweight components. Such components are economically attractive because a single part can replace hundreds of parts, when fasteners such as rivets are included in the part count. Moreover, elimination of fasteners also eliminates the associated stress concentrations. This also improves structural efficiency. The exception to this is in failsafe structures where SPF/DB components have a continuous crack path that must be accounted for in the structural integrity analysis. Fortunately, structures that are SPF/DB candidates seldom are critical load bearing structures. This minimizes any penalty that safe life design might impose on the use of SPF/DB components. Today, the acceptance of SPF/DB is growing slowly. Components fabricated by this manufacturing method are still considered expensive and niche applications are required for them to be cost competitive. Analysis of the cost of a typical SPF/DB component shows that principal cost elements are the Ti-6Al-4V sheet and the tooling that will withstand prolonged times at 900-950°C. Therefore, an alloy that was less expensive to fabricate into sheet and could be formed at lower temperatures has the potential to reduce the cost of SPF/DB components and to increase the use of this forming method.

Nippon Kokan (NKK) in Japan has developed a sheet alloy expressly for use in SPF/DB applications. This alloy known as SP-700 (see Table 2.6) is easier to fabricate into sheet and has high m values at 775°C. This alloy also can be heat treated to high strengths after SPF/DB, so it is very attractive from a structural efficiency standpoint. The SPF response of SP-700 at 775°C compared to 900°C for Ti-6Al-4V results in an order of magnitude lower oxidation during the forming process that reduces post-forming pickling cost also. SP-700 is now being produced under license in the US and this should increase the acceptance of it as a viable lower cost replacement for Ti-6Al-4V for SPF/DB applications.

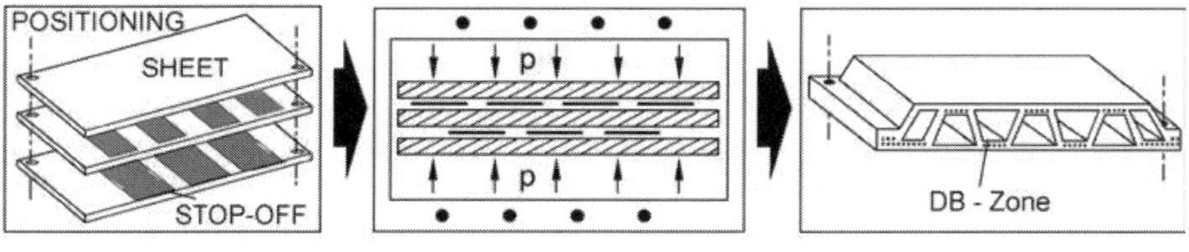

Fig. 3.47. Schematic of SPF/DB process to make honeycomb structure (courtesy W. Beck, FormTech)

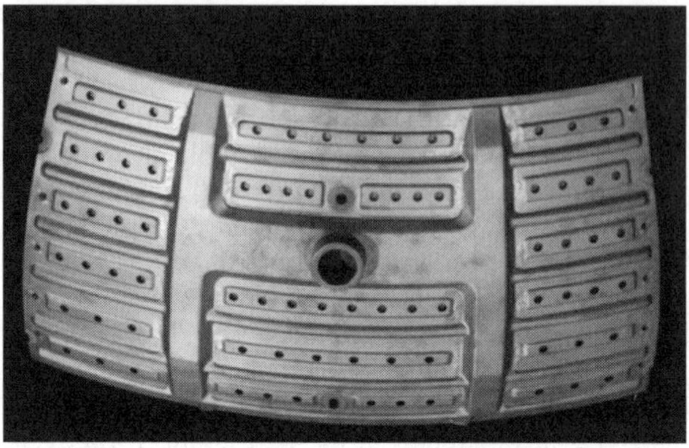

Fig. 3.48. Superplastically formed and diffusion bonded (SPF/DB) manifold (courtesy GE Aircraft Engines)

3.6 Conventional Joining Methods

Conventional joining as used here means all joining methods except diffusion bonding which has been described for reasons of logic in conjunction with superplastic forming in the previous section. The most common joining methods are fusion welding and friction welding. Brazing of titanium also is practiced to limited degree. The reactivity of titanium alloys poses special process requirements in terms of protection of the molten pool or even solid titanium if it is exposed to the atmosphere for an extended time.

3.6.1 Fusion Welding

A number of titanium alloys are generally considered weldable, if the definition is used that fusion welds can be made that are sound and free of cracks. Typically, it becomes more difficult to produce welded structures or products as the alloy strength increases. This is because the properties of the weld do not match those of the base metal and because some of the high strength alloys contain eutectoid alloying elements that impair the solidification, integrity, and properties of the welds. An additional issue is the availability of filler wire that matches the composition of the base metal. This point will be discussed later. Fusion welding of titanium alloys is more often used for joining sheet structures. Large pressure vessels and other heavily loaded structures also are fabricated by welding because of their size. For example, about twenty years ago, the US Navy became interested in welded pressure hulls for submarines in response to the Soviet Alpha Class titanium hull submarines. This interest lasted about five years and during this period there was some excellent welding research done on heavy sections of medium strength (700-750 MPa yield strength), high toughness alloys. This effort ceased in the early 1980's when it became clear that a titanium hull submarine was not affordable. Today, the majority of heavy section welding is performed on CP titanium during the fabrication of vessels for the chemical and petrochemical industry. An example of one such vessel is shown in Fig. 3.49.

Fig. 3.49. Example of a large pressure vessel being fabricated from CP titanium by manual fusion welding, welder shown by arrow (courtesy J. A. Hall)

Fusion welding of titanium (or any other metallic material) involves melting and resolidification of the base metal, and the part of the weld that has been melted is known as the fusion zone. Fusion welds solidify from the edge of the fusion zone and the solid/liquid interface grows in toward the weld centerline. Because the concentration of impurities is quite low in titanium alloys, cracking due to the enrichment of impurities, as is the case in ferrous welds, poses no real problem. However, if the base metal contains eutectoid alloying additions such as Fe or Cr, these elements segregate to the last liquid to freeze. This solute enriched liquid can exhibit shrinkage porosity along the weld centerline or can exhibit liquation cracking due to the thermal stresses. These defects can and must be avoided because they particularly affect properties such as fatigue life. Preheating the base metal reduces the resolidification rate and helps reduce or eliminate this porosity and cracking.

There also is a zone adjacent to the weld fusion zone that experiences a solid state thermal cycle that alters its microstructure. This region of the weld is known as the heat affected zone. The thermal excursion exceeds the β transus temperature in a portion of the heat affected zone. Thus, both the fusion zone and some of the heat affected zone exhibit β transformed microstructures and properties. Consequently, welding of α+β processed material creates a microstructural and mechanical behavior discontinuity. The microstructural changes induced by fusion welding are summarized in the diagram in Fig. 3.50 which shows a plot of local temperature as a function of distance from the weld centerline. Also shown in this diagram are the temperatures corresponding to melting and to the β transus, and the positions of the fusion zone (FZ) and heat affected zone (HAZ). The boundary of the fusion zone is the position of the solidus temperature (T_S) which has a clear meaning. The extent of the heat affected zone is more difficult to define but within the heat affected zone the position of the β transus is well defined. Figure 3.51 is a low magnification composite photo showing all of these zones and the base metal. In this example it can be seen that the β grains are larger and columnar shaped in the fusion zone. Figures 3.52 and 3.53 are higher magnification views of the microstructure at various positions in the heat affected zone. Figure 3.52 is from the position of the heat affected zone that has exceeded the β transus. In this region the fully lamellar microstructure is the same as the fusion zone but the β grain size is smaller. Figure 3.53 is from the heat affected zone that has not exceeded the β transus but where the temperature has increased sufficiently to increase the volume fraction of lamellar regions. The base material is shown for comparison in Fig. 3.54 and it is a typical mill-annealed microstructure for hot worked plate that is not fully recrystallized. In thin sections, the cooling rate can be high enough to produce a high strength microstructure in α+β alloys or lead to a retained β structure in β alloys. It is common to stress relieve welds at about 700°C. For stress relief of components with a finished surface, the temperature is typically limited to about 550°C due to the risk of surface contamination, even if vacuum heat treated. It is important to understand the effect that this thermal treatment has on the structure of the fusion and heat affected zones. For example, decomposition reactions that result in high strength, low ductility structures should be avoided. These concerns are less or nonexistent in CP titanium and lower strength titanium alloys such as Ti-3Al-2.5V. This is an important reason why these grades are considered weldable whereas the higher strength alloys are considered difficult to weld. Retention of parent metal properties, particularly ductility, in the heat affected and fusion zones becomes more difficult as the maximum strength of the alloys increases.

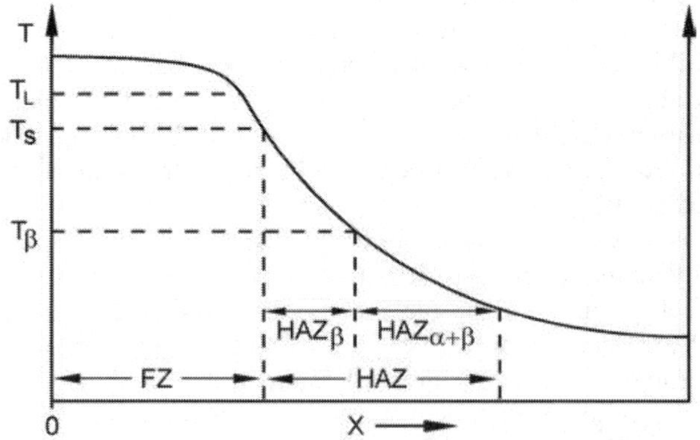

Fig. 3.50. Schematic drawing of temperature as a function of distance from the fusion weld centerline

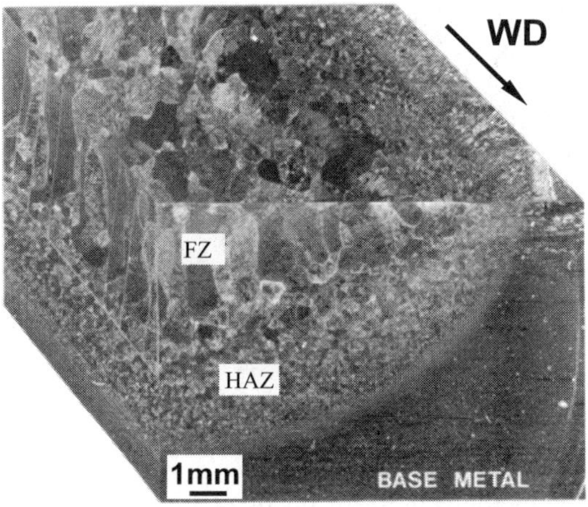

Fig. 3.51. Low magnification composite photo of a fusion weld, LM

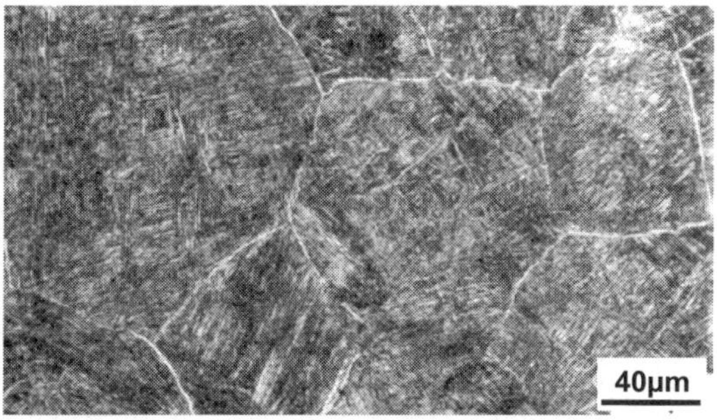

Fig. 3.52. Fully lamellar microstructure in a fusion weld heat affected zone (HAZ) that has exceeded the β transus, LM

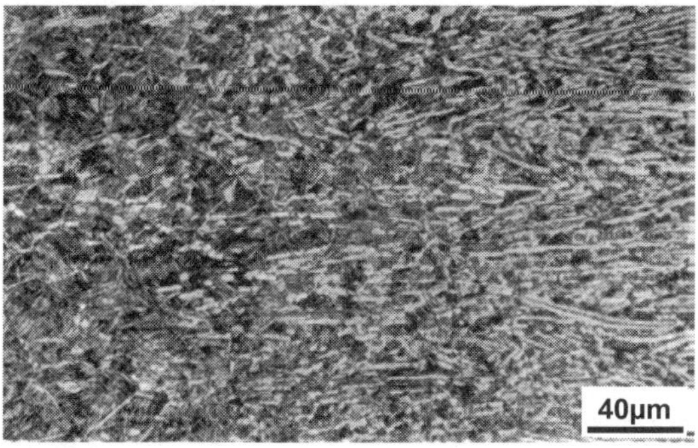

Fig. 3.53. Bi-modal microstructure in a fusion weld heat affected zone (HAZ) that has not exceeded the β transus, LM

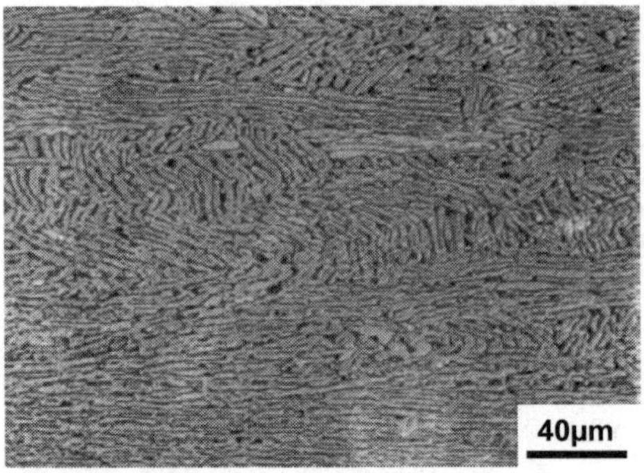

Fig. 3.54. Mill-annealed microstructure of base material, LM

There are five basic methods used for fusion welding of titanium. These are gas tungsten arc welding (GTAW), gas metal arc welding (GMAW), electron beam welding (EBW), plasma arc welding (PAW), and laser welding (LW). Attempts also have been made to use electroslag welding of titanium for heavy section applications, but this has only gained limited success and has never been reduced to a commercial joining process, at least in the Western world. Each of these methods has advantages for specific applications and these will be described in the following paragraphs of this section.

Gas tungsten arc welding (GTAW), also known as tungsten inert gas (TIG) welding, is perhaps the most common welding method used for titanium and its alloys. In part, this is because the majority of welded structures is associated with the fabrication of large components and equipment for the chemical and petrochemical industries. Many of these applications involve relatively thin sections for which the gas tungsten arc welding method is ideally suited. Any method of fusion welding of titanium and its alloys requires the protection of the molten weld from the atmosphere. The principle contaminants are oxygen and nitrogen from the atmosphere and hydrogen from any moisture (humidity) in the air. In gas tungsten arc welding, an inert gas shield around the torch is used to prevent contamination of the molten metal. Automated welding at high speeds also requires a trailing gas shield to minimize the exposure of the hot weld to air as it cools after the torch has passed. In through-thickness welds, a gas shield for the underside of the work piece (called the underbead) also is required to prevent contamination of the hot weld on the bottom surface of the work piece. This underbead protection is often supplied by inert gas that flows from holes in the welding fixture used to hold the work piece. This description of the precautions needed to prevent weld contamination suggests that welding titanium alloys is more complicated (and costly) than welding many other structural alloys. For repetitive welding operations, it often is easier to construct a welding chamber containing inert gas in which the welding is done. One example of this is the dry box used to weld first melt electrodes (Fig. 3.6) in a titanium melt shop. This is the approach taken, for example, in the production of welded CP titanium tubes. Here a tunnel filled with flowing argon is used to contain the tube, the torch, and the run out zone where the tube travels while it is cooling. The torch is stationary, the strip is formed into a tube, and the joint is fused as the tube passes under the torch during its travel along the length of the gas filled tunnel. The process of making welded titanium tube has become highly automated. CP titanium strip is uncoiled and fed into a series of dies that roll it into a cylinder. This cylinder is welded and emerges at the other end as finished welded tubing. As the tubing emerges, it is cut to length and is ready to ship or to be installed. There are many other less routine applications for gas tungsten arc welding and these involve manual welding where the welder manipulates the torch around a static work piece. In some cases the work piece is axisymmetric and can be rotated as the welder uses the torch to weld it.

Most titanium alloys are not easy to draw into wire because of their high strength and somewhat limited ductility. Consequently, production of titanium alloy filler wire is at best expensive and at worst prohibitively expensive or impossible to make at all. The production of filler wire from CP titanium grades 1 and 2 and the alloy Ti-3Al-2.5V is relatively straight forward, Ti-6Al-4V is available, but expensive. Filler wire made from other alloys has a very limited availability. This issue and possible solutions have been discussed earlier in Sect. 3.5.1 in connection with weld repair of castings. One solution to welding other alloys is to use under-matched filler and build up the weld joint cross section to compensate for lower yield strength in the weld. Another solution is to prepare the joints with sufficient precision that no filler is required and the mating surfaces of the base metal are fused together. Welds that use no filler metal are called autogeneous welds. Autogenous welding is more common for the original manufacture of components that are welded with an automated process. Welding with filler metal is common for manual welding of both new components such as fabrications or for repair of casting defects in structural castings or for the repair and maintenance of existing articles. In either case, gas tungsten arc welding of titanium is widely used and is a highly reliable process provided the changes in properties are taken into consideration and the necessary care is exercised to prevent contamination of the welds.

Gas metal arc welding (GMAW) of titanium and its alloys uses a torch where the weld filler wire is the electrode instead of the tungsten tip in the gas tungsten arc welding process. This process also is referred to as metal inert gas (MIG) welding. Gas metal arc welding is particularly well suited for automated welding where the weld gap is large, usually in thicker sections. This is often the case for large structures where preparation of the joint is more difficult. The precautions for protecting the weld from contamination described in the previous paragraphs for gas tungsten arc welding also pertain to gas metal arc welding. In principle, gas metal arc welding requires less precise joint preparation and for alloys where filler wire is not expensive, can be a cost effective process. This is particularly true where the use of this process allows the weld to be accomplished in a single or in fewer weld passes than for gas tungsten arc welding. Had the titanium hull submarine project gone forward, there is a very good prospect that it would have used gas metal arc welding to join the hull plates. Readers interested in more detailed descriptions of these welding processes including fixturing are referred to articles devoted to welding [95, 96].

Electron beam welding (EBW) is well suited for joining of titanium alloys, in part because it is typically conducted in a vacuum chamber, which naturally provides the necessary protection of the weld. Electron beam welding also has the intrinsic capability for making deep penetration (heavy section) welds in single pass because the electron beam is narrow and its energy density is very high. It is common to use electron beam welding for through-thickness welds of titanium plates that are 15 cm thick. Because the beam is narrow, the ability to align the beam with the joint becomes critical because the lateral dimension that is fused is small. This alignment is accomplished with elaborate fixturing that contact or connect with reference or datum points on the work piece. Another setup precaution is the use of "witness lines" to ensure that the fusion zone coincides with the joint position. Therefore, the initial setup for electron beam welding is time-consuming and exacting. Once the setup is successfully completed, high quality, deep penetration welds can be produced rapidly. This means that electron beam welding is more cost effective if a large number of parts are welded in one campaign to dilute the setup costs. Electron beam welds are usually autogeneous. This eliminates the cost of filler wire, but increases the cost of joint preparation. Because the weld fusion zone is narrow, it has small molten mass and the focussed electron beam creates a high ratio of fusion zone to heat affected zone. This factor minimizes the freezing segregation issues outlined earlier. Consequently, electron beam welding is a promising method of joining high strength and high temperature titanium alloys. For example, high temperature titanium alloys such as Ti-6242 and IMI 834 or a combination of both can be electron beam welded with sufficiently high quality, that the process can be used to join several stages of a rotor in an aero-engine. Historically, electron beam welding also has been used to fabricate high performance structures of difficult to weld alloys such as Ti-6Al-6V-2Sn (Ti-662), which also contains about 1% (Cu+Fe) in approximately equal proportions. The center wing carry-through structure of the F-14 military aircraft was forged from Ti-662 in two halves, machined and joined at its center using electron beam welding. The decision to fabricate this component was driven by the size of the forgings involved and the need to make them from relatively small billet. This, in turn, was due to the propensity of this alloy to develop nonuniform microstructures (beta flecks) if produced in larger ingot sizes. This aircraft is still flying and there have been no reported issues of weld deficiencies with the performance of this critical welded structure. Electron beam welding thus has the advantage of producing deep penetration welds in difficult to weld alloys without the need for filler wire. The requirement to weld under vacuum, typically in a chamber is a constraint, but sliding seal "out of vacuum" electron beam welded parts have been demonstrated to be possible, using a sliding seal to isolate the electron gun.

Plasma arc welding (PAW) of titanium alloys is a useful alternative mainly for gas metal arc welding, since this process creates wide fusion and heat affected zones. This process uses a plasma torch for the heat source in place of a conventional electric arc. Plasma arc welding typically imparts a higher level of energy to the heat source than electric arc welding and much more than electron beam welding. Because of the wide fusion zone, the plasma arc welding process is useful for through penetration welds in situations where a plate or sheet is joined to a hidden upstanding rib to create a T-shaped section. The width of the fusion zone compensates for misalignment of the torch with the underlying rib, making the process more robust. In the final production configuration of the wingbox shown in Fig. 3.46, the top cover plates were joined to the ribs using plasma arc welding. This aircraft also has been flying for many years without any issues associated with this critical welded structure.

Laser welding (LW) is similar in many ways to electron beam welding, because the laser beam is narrow and has a high energy density. Laser welding can be done without a vacuum using inert gas shielding to prevent weld contamination. The availability of high energy lasers is relatively new and there have been fewer opportunities to introduce laser welding for joining large structures until now. Laser welding can be thought of as a substitute for electron beam welding, but the penetrating power is yet to be demonstrated in the same way. In principle, laser welding has a level of flexibility that electron beam welding does not have because it does not require a vacuum. Laser welding works fine for titanium alloys but does not work for all metals, e.g. aluminum alloys, because the surface reflectivity of aluminum prevents coupling of the beam to the work piece. For welds on small components where the limited heat input is essential laser welding has been shown to work very well. For example, laser welding is used to close the CP titanium case for heart pace makers and to weld the electrode connectors in place. This allows a high quality weld to be created without any danger of heating the electronic module inside. This joining operation could not be performed with conventional heat sources such as a gas tungsten arc welding torch. It also is much better to use laser welding than electron beam welding for this application because there is no ionizing field with laser welding that could damage the electronic components inside the CP titanium case.

3.6.2 Friction Welding

Friction welding of titanium alloys is widely practiced as a means of obtaining a high integrity joint of materials with $\alpha+\beta$ processed (equiaxed) or β processed (lamellar) microstructures. There are three processes, two currently used in production and a third that is under development. For axisymmetric shapes, rotational inertia welding is the most commonly used process. In rotational inertia welding, one of the work pieces is held in a static fixture while the other piece is placed in a rotating holder that is attached to a flywheel of predetermined mass. This flywheel assembly is then spun to a predetermined rotational speed and the work piece is moved forward to engage the mating halves. These halves are pushed together with a predetermined axial force that heats the joint area by friction and expels some of the work piece from what previously were the faying surfaces of the two work pieces. This process of expelling materials from the joint is crucial to obtaining a clean joint. Modeling of this inertia process has been done and remarkable agreement is obtained between the geometry calculated in the simulation model and the actual part after inertia welding. The simulation result is shown in Fig. 3.55 whereas the cross section of the actual weld in the part is shown in

Fig. 3.56. The agreement is striking. This modeling capability permits more rapid specification of the process parameters for inertia welding and allows faster convergence on a set of welding parameters that can be used to produce welded parts with reproducible properties. Multistage compressor rotors for large gas turbine engines now are routinely inertia welded. The rotors are lighter weight than mechanically fastened rotors and have better life characteristics because they contain no bolt holes that are used to connect adjacent stages. An example of a rotor that has been prepared for inertia welding is shown in Fig. 3.57. The same rotor after welding and machining is shown in Fig. 3.58.

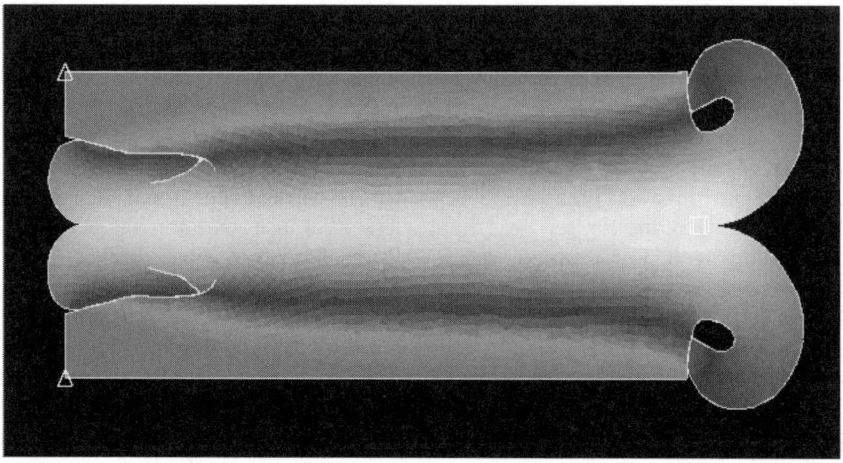

Fig. 3.55. Computer simulation of inertia weld showing the flash and cracks that develop in the flash (courtesy S. Srivatsa, GE Aircraft Engines)

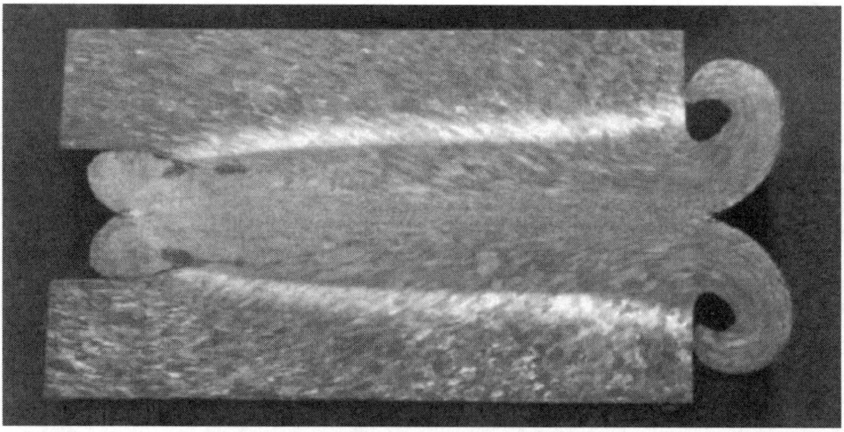

Fig. 3.56. Photo of macro etched cross section of inertia weld, compare to Fig. 3.55 (courtesy GE Aircraft Engines)

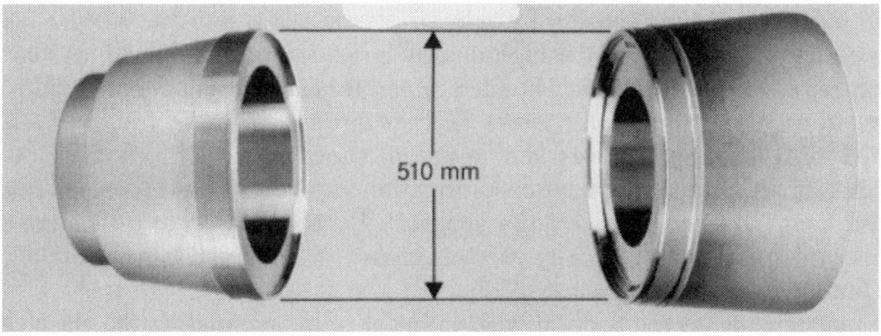

Fig. 3.57. Photo of the two work pieces before inertia welding to create a multistage compressor rotor for an aircraft engine (courtesy MTU)

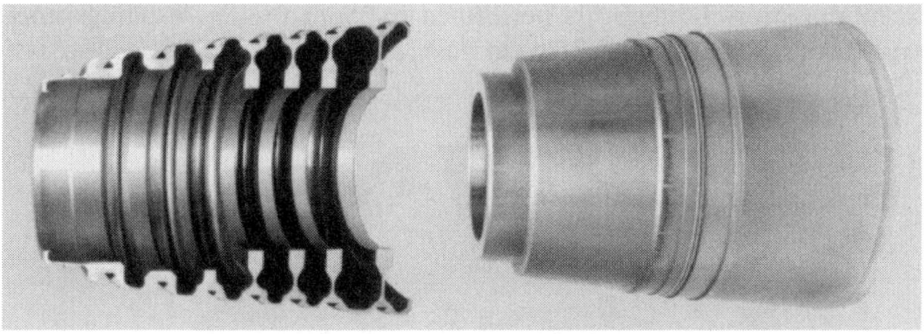

Fig. 3.58. Photo of inertia welded part (at right) made from the two pieces shown in Fig. 3.57, cross section of part (at left) showing the seven disk stages (courtesy MTU)

An alternate method of friction welding is linear friction welding [18]. This is useful for making friction welded joints of non-axisymmetric geometries. In this method, one part is oscillated back and forth in a straight line with a force applied normal to the plane of translation. As in the case of inertia welding, this oscillatory motion coupled with the pressure creates enough heat to soften the metal and allow the original surface material to be expelled into a flash, leaving a clean, high integrity joint. Linear friction welding is useful for attaching air foils to a hub for an aircraft engine rotor. Such a rotor is called an integrally bladed rotor. An example of the rotor during fabrication by linear friction welding is shown in Fig. 3.59. In this figure several blades can be seen that are already welded in place and the locations on the hub where two others will be attached. The benefit of using this technique is the cost saving of making a near net shape part and the improved properties of the air foils, since they can be individually forged and heat treated before attaching them. The alternate method of making a component such as this is to make a large forging with an envelope that encompasses the air foils and the hub, then machining the rotor from the forging. Clearly this latter method leads to a loss of much of the forging. Not only does this mean inefficient use of the forging, but the cost of machining the material away to make a final part with complex geometry also is substantial.

The friction welding process introduces plastic work into the weld zone and this is partly responsible for the high quality bond that occurs. There is clear metallographic evidence that the as-bonded material has undergone extension plastic deformation. A metallographic cross section through a linear friction weld between Ti-6Al-4V in the bi-modal and mill-annealed conditions is shown in Fig. 3.60. In this micrograph, the deformation of the primary α grains and the microstructural refinement is readily apparent. By comparison to the parent metal this finer microstructure typically exhibits higher strength and better fatigue properties but lower fracture toughness [97].

No matter whether the friction weld is linear or axisymmetric, the flash shown in Figs. 3.55 and 3.56 must be machined away before the part is usable. This is an exacting task because this is a high stress region and it is important to avoid any unintended notches or other stress concentrations. This operation is still considerably easier and less costly than machining the entire part from a forging. Without exception, stress relieving is performed for both friction welding processes, either prior or after removal of the weld flash.

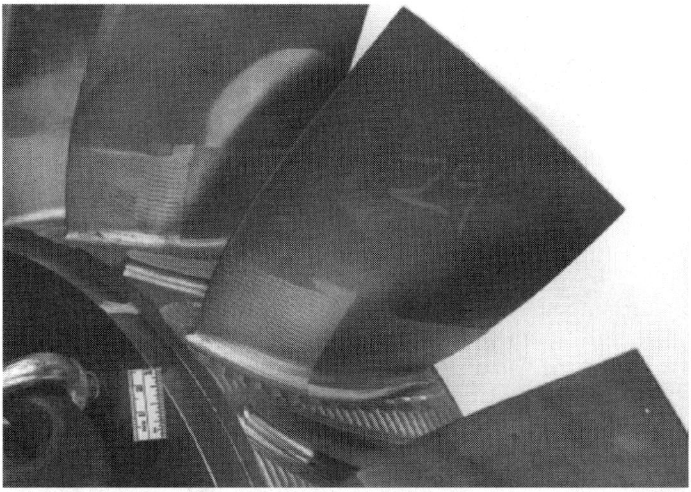

Fig. 3.59. Close-up photo of the airfoil joint area of an integrally bladed aircraft engine rotor being made by linear friction welding (courtesy MTU)

The third, currently experimental, friction welding process is friction stir welding (FSW). Friction stir welding has been under development for aluminum alloys for a number of years and is showing real promise as a means of joining sheets of high strength aluminum alloys such as 7475 that cannot be fusion welded. The FSW process uses an inert, rotating mandrel or tool and a force on the mandrel normal to the plane of the sheets to generate the frictional heat. The heat and the stirring action of the mandrel create a bond between the two sheets that is metallurgically sound and is created without melting the base metal. The apparatus used for friction stir welding is schematically shown in Fig. 3.61. The higher melting temperature and higher strength of titanium alloys poses a greater challenge for successfully developing a FSW process. The benefits are clear, however, and the payoff is great enough that there is significant activity on this subject in several laboratories around the world [98]. As with many new processes, other issues such as reliable inspection methods will require attention before this technique gains broad acceptance.

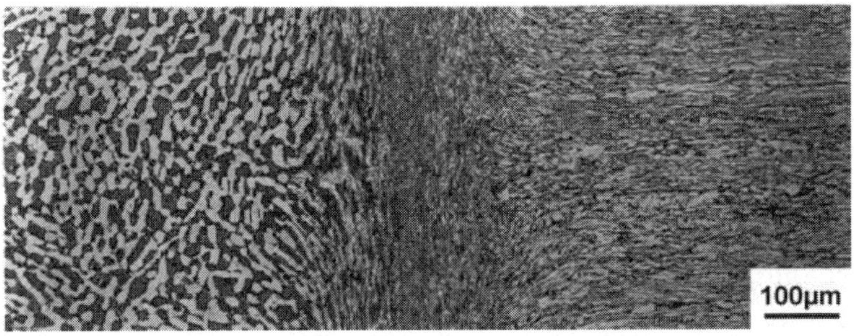

Fig. 3.60. Section through a friction weld, LM [97]

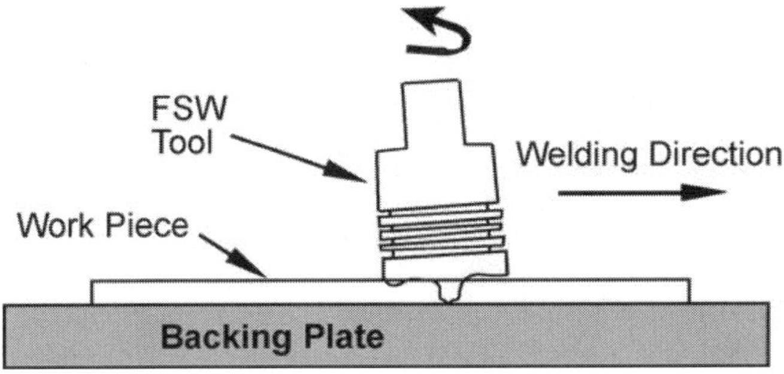

Fig. 3.61. Schematic of friction stir welding (FSW) apparatus (courtesy M. Juhas, The Ohio State University)

3.7 Surface Treatment

The use and proper execution of surface treatments such as shot peening and chemical milling are individually or jointly critical to the technological success of titanium and its alloys. This is because many fracture related events, particularly fatigue cracks, initiate at the surface of components. Shot peening has been used for many years on steel to create a compressive stress at the surface thereby enhancing the resistance of steel to fatigue crack initiation. The application of this process to titanium alloy components is not new, but is more recent than in the case of steel. Quite recently, the use of a laser to create a surface compressive stress has been shown to be highly beneficial and this technique also will be described.

Chemical milling is an important means of selectively removing material from the surface of titanium parts as a manufacturing method. It also is an important means of removing material that has become contaminated, for example by oxygen, during processing. In this case, removal of a hard, brittle surface layer intrinsically improves the resistance to crack initiation and fracture. The proper use of chemical milling is economical and efficient, but improper use can negatively affect the cracking resistance. Chemically milled surfaces are often shot peened to create or restore surface residual compressive stress.

This section discusses the practice and benefits of these surface treatment methods. The consequences of improper practice also are described and means of avoiding these unwanted effects are outlined.

3.7.1 Shot Peening

The resistance of titanium specimens or components with stress-free surfaces to fatigue failure is quite low, as will be mentioned in Chap. 5. Expressed as a fraction of the yield stress, the fatigue strength at 10^7 cycles to failure is typically 0.4-0.5. Further, the introduction of damage from manufacturing processes, such as machining, reduces this value to even lower levels. The introduction of a compressive stress in the near surface region of the material by local plastic deformation during shot peening provides protection against this further reduction in fatigue capability. A plot of local stress as a function of distance beneath the shot peened surface is shown in Fig. 3.62 [99]. This protects the integrity of the property values used in design. That is, the fatigue curves (S-N curves and Goodman diagrams for mean stress correction) are developed using specimens that are finished by low stress grinding followed by longitudinal mechanical polishing. Fatigue tests are run to obtain data that permit construction of so-called "95-99" curves. Such curves define the cyclic life at a given mean stress as a function of alternating stress for which there is a 95% confidence level that 99% of the data points will lie above the curve. Clearly, these are conservatively drawn curves, but such conservatism is appropriate for fatigue design. Inadvertent introduction of surface damage (nicks, scratches, gouges or abusive machining) can affect the conservatism of these curves creating unforeseen risk of fatigue failure. The beneficial effects of shot peening more than compensate for the presence of surface damage and, therefore, can be thought as introducing a "safety net" with respect to fatigue, provided the damage is not too severe or too deeply embedded. This is consistent with the general observation that shot peened fatigue specimens exhibit subsurface crack initiation sites. This suggests that compressive stress due to shot peening reduces the sensitivity of the material to surface crack initiation. Thus it would be expected that shot peening provides protection against crack initiation at manufacturing induced surface discontinuities. An example of a subsurface fatigue crack initiation site is shown in the fractograph in Fig. 3.63.

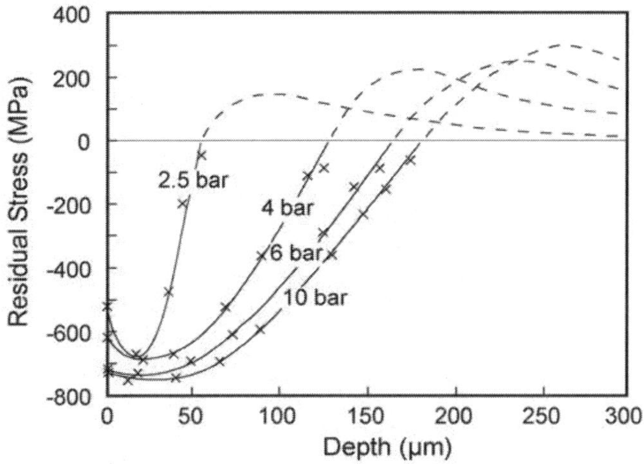

Fig. 3.62. Residual stress distributions after shot peening Ti-6Al-4V for 4 min at four different peening pressures [99]

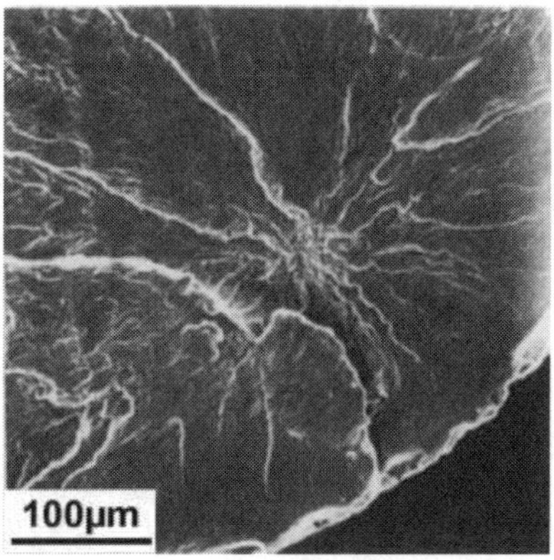

Fig. 3.63. Fractograph showing a subsurface fatigue crack initiation site in a shot peened specimen of Ti-6Al-4V, SEM [99]

The benefits of shot peening on fatigue performance depend on the alternating stress level and on the service temperature. The benefit diminishes as either of these operating parameters is increased. At low stress amplitudes ($N_F \geq 10^6$ cycles) at room temperature peening can extend the life by 30-50%. This improvement decreases at higher stresses and essentially disappears at $N_F \approx 10^3$ cycles to failure. The effect of temperature and time at temperature is less well quantified, but the benefit of residual stress clearly diminishes after long times at service temperatures where time dependent stress relaxation (stress relief) occurs. For most titanium alloys this stress relief becomes significant above 300°C. In such cases, the magnitude of the effect also is time dependent, but most design practice would only utilize the limiting value.

There are several external parameters that affect the magnitude of the shot peening effect. Included are type, hardness and size of shot, shot velocity and peening time. There are several good general references that describe practice and the effects of shot peening [100]. Early shot peening was done with cast, spherical steel shot, but more recently, there has been a consistent trend toward the use of conditioned cut wire shot instead of cast shot. This is because broken cast shot has sharp edges and can damage the material locally. The depth of the residual compressive field increases with shot velocity and with shot diameter. However, deeper compressive layers can cause macroscopic distortion in components with thin sections and this must be considered. One approach to obtaining some of the benefits of peening in these sections is to use alternate media such as glass beads or water droplets containing fine solids (vapor honing). Further, if shot peened components have features with dimensions of the order of the shot size, there is reasonable concern about the uniformity of the compressive layer due to simple geometry considerations. Other features, for example, where holes intersect a free surface, can get too much peening and suffer damage (overpeening). The recent trends toward increasing use of robotic peening equipment has improved the ability to compensate for this and also has reduced the amount of scatter in peening intensity. An example of a jet engine component and a robotic peening device is shown in Fig. 3.64. This device is housed in a six axes CNC automated peening machine such as the one shown in Fig. 3.65. As mentioned above, there also is a phenomenon known as overpeening. This refers to peening to such intensities that localized surface damage is introduced with a corresponding sharp reduction in fatigue capability. The peening parameters that have the greatest effect on intensity are shot size and velocity.

Fig. 3.64. Photo of jet engine component and robotic peening device (courtesy J. Whelan, Progressive Technologies, Inc.)

Fig. 3.65. Photo of a six axes CNC shot peening machine (courtesy J. Whelan, Progressive Technologies, Inc.)

The response of the material to shot peening also varies with microstructure and heat treatment condition. During low cycle fatigue testing, some titanium alloys can exhibit either cyclic softening or cyclic hardening, depending on the details of their microstructural condition. For example, it has been shown [90] that aging Ti-6Al-4V at 500°C for 24 hours causes it to cyclically soften and this alters the response of the material to shot peening (Fig. 3.66). For example, aging causes overpeening effects to be much more pronounced. Figure 3.66 clearly shows that long peening times are detrimental to fatigue life of a material that cyclically softens. The beneficial effects of shot peening as well as the effect of overpeening are shown in Fig. 3.67. From this figure it is clear that shot peening can be beneficial for improving the fatigue capability of titanium alloys, but it also is clear that judicious use of this technique is required to avoid unwanted side effects such as overpeening. Peening intensity is typically measured by peening metallic coupons called Almen strips and measuring the bowing or deflection due to the residual stress on one side of the strip. A peening requirement is usually stated in terms of the desired Almen strip deflection.

It is common to specify peening practices that ensure that the area to be peened is covered multiple times. This ensures that there are no regions that have not received shot peening. For example, specifying 200-400% coverage on fatigue critical areas is typical. Because of the penalty of overpeening just discussed, the appropriate intensity (pressure) must be selected to avoid this undesirable consequence.

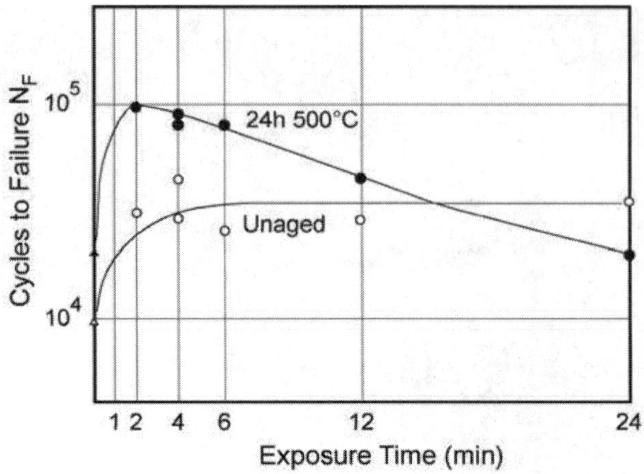

Fig. 3.66. Fatigue life vs. peening time in aged and unaged Ti-6Al-4V [99]

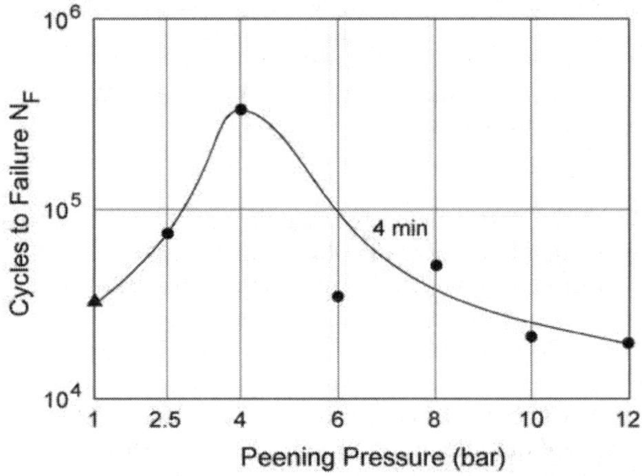

Fig. 3.67. Fatigue life vs. peening pressure showing the onset of fatigue life reduction resulting from overpeening [99]

3.7.2 Laser Shock Processing

There are alternate means of introducing a compressive stress at the surface of a material. One of these that has received recent attention is laser shock processing (LSP), more commonly referred to as laser shot peening by analogy to mechanical shot peening. This technique uses a high energy pulsed laser to create a localized pressure at the surface of the material. This pressure exceeds the yield stress and creates a residual compressive stress [101].

Currently a neodymium-glass (1.06 μm wavelength) laser with a pulse length of 15 to 30 nanoseconds and an energy per pulse of 50 joules or more is used. The typical laser impingement spot size ranges from 6 to 9 mm and depends on the optics and other details of the laser peening apparatus.

The laser is coupled to the work piece by two coupling layers, a transparent one (typically water) and an opaque one (typically paint). The actual coupling mechanism has been described as follows. The laser beam vaporizes some of the opaque layer. This thin vapor layer absorbs more of the laser energy and is rapidly heated to form a local plasma. This plasma expands rapidly but is confined at the metal surface by the transparent overlay, imparting a normal force to the work piece. This force can be as high as 100 kbar which greatly exceeds the local dynamic yield strength and creates a surface residual compressive stress, much like that created by a solid object impacting the surface (e.g. shot peening). The interaction between the laser and work piece is shown schematically in Fig. 3.68. The residual compressive stress layer from laser shock processing penetrates deeper into the part but has a lower maximum compressive stress level when compared to shot peening. This is shown in the diagram of Fig. 3.69 which compares the residual compressive stress profiles due to shot peening to that due to laser shock processing. The laser impingement site makes a shallow, gently curved impression on the surface, but this effect of laser shock processing on surface finish is less severe than that of shot peening. The additional benefit of the opaque (paint) surface layer on the component is that it also serves as a "witness" to permit easy inspection for complete coverage by the laser. With laser shock processing the incidence of surface damage that is the equivalent to overpeening is less common. Also, the surface roughening due to laser shock processing is minimal and, consequently, selected regions of a component can be treated. This reduces the time and cost required to treat a component. Still, today, the cost of laser shock processing is higher than for shot peening and it is only being used in cases where the benefits warrant the additional cost. Effort is being directed to using higher power lasers and larger spot sizes to reduce the time required to process one part. This is motivated by the desire to realize the significant benefits of laser shock processing on fatigue performance that have been recently demonstrated [102].

The principal attractiveness of laser shock processing compared to conventional shot peening is the increased depth of the compressive stress layer with about the same magnitude of the stress. This has been demonstrated in the following way. A deep notch such as a saw cut is introduced into a titanium alloy specimen and the perimeter of this notch is laser shock processed. During subsequent fatigue testing, no cracks were initiated from the saw cut and the specimen exhibited a normal (un-notched) life with the failure initiation site being well away from the saw cut. Fatigue testing of a comparable specimen that has not been laser processed shows a life of only a few thousand cycles with cracks being initiated at the saw cut. This is a significant result from a practical standpoint, because it demonstrates the potential to use laser shock processing as a means of refurbishing titanium alloy fan and compressor air foils that have sustained foreign object damage (FOD) during service. Since the incidence of such damage is near the front of the engine, where the air foils are made out of titanium alloys, this has the potential to save engine owners a great deal of money over the life of an engine. In military engines, the airfoils often are an integral part of the rotor. Two examples are impellers for centrifugal compressors and fan blisks. In these cases, the ability to refurbish damaged airfoils is even more important since they are permanently attached. A centrifugal compressor impeller that is being laser shock processed is shown in Fig. 3.70.

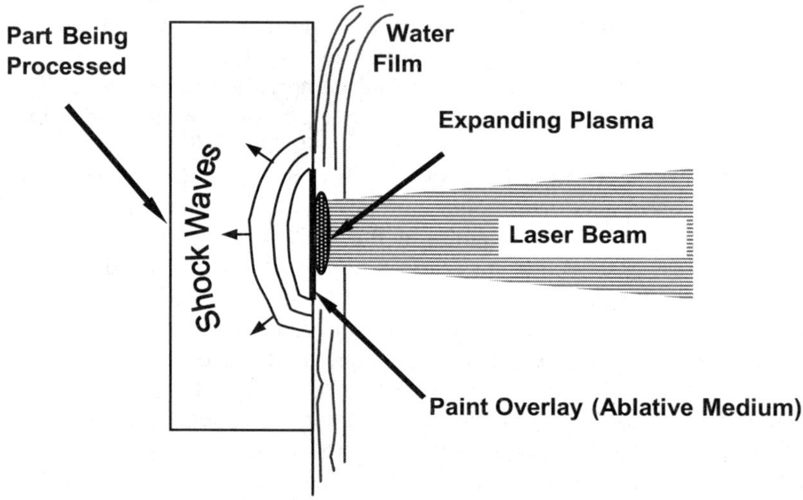

Fig. 3.68. Schematic of the coupling process between the laser beam and the work piece (courtesy LSPT, Inc.)

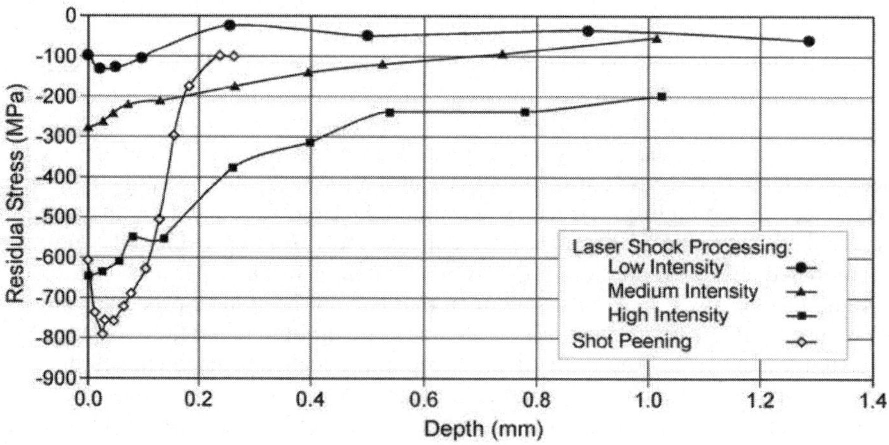

Fig. 3.69. Diagram comparing the compressive stress profiles due to shot peening and laser shock processing in Ti-6Al-4V (courtesy LSPT, Inc.)

Fig. 3.70. Aircraft engine compressor impeller in the process of being laser shock processed (courtesy LSPT, Inc.)

3.7.3 Chemical Milling

Chemical milling is a common way to selectively remove material from the surface of a component to create an array of features. Chemical removal of material from the surface that has been contaminated by oxygen during processing uses the same chemical reactions, but is usually called pickling by analogy to the terminology used in the steel industry. An example of a compressor casing for an aircraft engine is shown in Fig. 3.71. The waffle pattern shown on the surface of the casing is the result of chemical milling and is present to impart stiffeners at lower weight. This pattern is created by applying a masking agent to the cylindrical duct, often made from rolled rings, to form the pattern of the desired ribs of the waffle pattern. The masked cylinder is immersed in the chemical milling bath and the material in the "windows" between the ribs is dissolved until these become pockets of the desired depth. The purpose of this is to create an integrally stiffened, but lighter weight article. There is a good deal of art required in such an operation to avoid attack under the mask or undercutting of the ribs. Practitioners who fabricate titanium alloy components such as the casing shown in Fig. 3.68 perform chemical milling routinely. Although this is routinely possible now, it is the result of years of experience.

Chemical milling is done using a mixture of HF and HNO_3 acids in aqueous solution. The acid concentration and the bath temperature control the rate of material removal. As discussed in Chap. 2, Sect. 2.9.2, HF removes the surface oxide from titanium, allowing the metal to be dissolved. HNO_3 is an oxidizing acid and repassivates the titanium surface. This allows better control of the rate of metal removal during chemical milling. The metal dissolution reaction is exothermic, consequently, provision for water cooling of the bath must be made for large parts or continuous use. If the rate of metal removal is too high, gas evolution occurs at the metal/bath interface and the gas bubbles can cause uneven material removal. If the HNO_3/HF ratio is not maintained at ≥ 5, and is not controlled, excessive hydrogen liberation occurs and the titanium absorbs hydrogen from the bath during processing. The titanium surface during chemical milling is free of oxide, and hydrogen entry is easy if the hydrogen potential in the bath favors this. The chemical milling reaction depletes HNO_3 and thus the bath chemistry must be monitored and the acid ratio adjusted regularly. Hydrogen pickup during processing is harmful to the properties of the material and must be avoided. In alloys such as Ti-6Al-4V, hydrogen concentrations in excess of about 100 ppm can cause embrittlement either instantaneously or by delayed cracking. The latter is observed if there are residual stresses in the component from prior fabrication operations or if there is a microstructure or stress gradient, for example in the vicinity of a fusion weld or a notch. Delayed cracking results from the time dependent diffusion of hydrogen into regions where it has increased solubility (higher β phase volume fraction) or lower chemical potential (hydrostatic tension). At some point, for example during yielding at the notch, the critical conditions for TiH_2 formation are reached and cracking is initiated, because TiH_2 is brittle. The energy required to extend the crack in the presence of TiH_2 is relatively low and cracking or even failure can occur. In those instances where there is inadequate control of the chemical milling process, hydrogen content is increased during chemical milling. This hydrogen can be removed by vacuum annealing at temperatures of 600°C or higher and at pressures of 10^{-4} torr or lower. For large parts this creates some obvious issues regarding vacuum furnace size and availability and component distortion during annealing, but the capability of salvaging large, high value parts contaminated by hydrogen pick up is possible. Better management of bath temperature and composition makes this unnecessary.

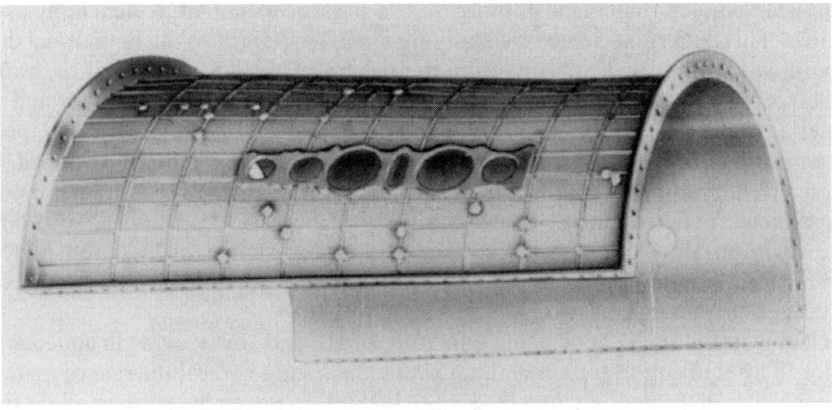

Fig. 3.71. Ti-6Al-4V compressor casing (upper half) with features created by chemical milling (courtesy GE Aircraft Engines)

3.7.4 Electrochemical Machining

Electrochemical machining (ECM) is a commonly used technique for making shapes that have a repeating geometric feature. For example, it is often used to create the airfoil shapes on titanium alloy bladed rotors for small gas turbines. The alternate approach is to make these parts by machining on an expensive five axis NC machine. The use of ECM provides a lower cost, more consistent quality product. This method uses a thin shaped electrode that has the desired geometry. This electrode is usually made of brass. The work piece, usually a forged block, is placed in a bath of electrolyte such as a saline solution. The electrode is attached to a mobile head and a DC voltage is applied between the electrode and the work piece, where the electrode is the cathode in the local electrolytic cell that is established. Because the electrode is quite thin, localized dissolution of the work piece occurs. The ECM machine head drive mechanism often has an automatic servomechanism that controls the distance between the electrode and work piece to maintain a high current density at the electrode tip. This assures a constant rate of local electrochemical dissolution. The result of this process is a deep, narrow cut that defines a shape that mimics the electrode. This creates a high degree of geometric flexibility that can be realized through skillful electrode design. When the cut reaches the desired depth, the voltage is shut off, the electrode is retracted and repositioned to make the next cut. The machining process is then repeated. Electrochemical machining is an efficient way to make complex shapes where, for example, there may not be enough clearance between adjacent features to permit access for a conventional machining tool. Small integrally bladed rotors are a good example of this. The disadvantages of ECM include the danger of local arcing between the tool and the work piece, leaving local damage and the accumulation of heavy metallic residue that is becoming increasingly expensive to dispose of properly. Great progress has been made in automating control of the tool/work piece distance, which minimizes that danger of arc damage.

The electrochemically machined surface can have a good finish and is free of any mechanical damage that traditional machining can introduce as discussed earlier in Sect. 3.4.3. Because the surface is stress-free, shot peening is required if fatigue is a concern.

3.8 Inspection Methods

The use of titanium alloys for many critical structural applications also has resulted in the development and use of a variety of inspection methods. These methods can be conveniently divided into volume methods that allow interrogation of the interior of the material and surface methods that permit detection of surface connected anomalies. These methods are complementary in nature and concurrent use of several of these methods will be described in this section. This permits achieving a high confidence level that no undesired conditions that could compromise properties exist in the material or component. Table 3.2 summarizes these methods. From Table 3.2 it can be seen that there are quite a lot of inspection methods and that these methods have a wide range of application, including inspection of field parts. Field part inspections are done during routine maintenance, or in some cases, where there is a concern for the integrity of a component after unexpected (premature) service failures. Failures from internal defects are obviously of greater concern compared to surface defects. Surface defects are more common, but also are easier to detect and, therefore, catastrophic failure due to surface defects is less likely.

The ability to consistently find small defects has improved the reliability of high performance structures and has led to reductions of unexpected service failures. The structural efficiency of these components also has increased because of the ability to design to higher operating stresses without increasing the risk of unexpected failure.

Table 3.2. Inspection methods for titanium alloy products and components

Method	Applications	Surface or Volume
Ultrasonic	Forging billet, forgings, bar, rolled rings, extrusions, field parts	Volume
Radiography	Castings, welds	Volume
Surface Etching	Finished forgings	Surface
Eddy Current	Finished machined features, field parts	Surface
Penetrant	Finished parts, welds, field parts	Surface
Surface Replication	Finished parts, field parts	Surface

3.8.1 Ultrasonic Inspection

Ultrasonic inspection of titanium and titanium alloys is the most common inspection method used when the material is intended for use in high performance applications [103, 104]. In this inspection method, ultrasonic waves are induced in the material using a piezoelectric transducer. These waves are typically longitudinal waves although shear waves can be used in special circumstances. The transducer operates at a characteristic frequency, typically 5 MHz. The transducer is coupled by water or other coupling media to the piece being inspected. Most factory inspections occur by immersing the part in a water tank. The detection of subsurface defects is based on the reflection of some of the incident ultrasonic waves from regions lying along their path. This reflection occurs whenever there is a region that has different acoustic impedance or resistance to the transmission of the ultrasonic waves. During operation the transducer sends waves, stops sending and waits to detect the reflected waves. There always is a reflection from the front and rear faces of the piece being inspected. These reflections are useful length markers to help physically locate sources of other reflections along the ultrasonic pathway. The reflected waves are displayed on an oscilloscope or, in the case of digital signal processing, on the screen of a computer monitor. A typical scan trace for one location is shown schematically in Fig. 3.72. The entire volume to be inspected is covered by either moving the part under the transducer or by moving the transducer over the part. For regular shapes, such as billet and bar, the transducer is usually fixed. For irregular shapes, such as forgings, the transducer is typically moved over the stationary part in a systematic pattern. This pattern is designed to ensure complete coverage of the piece being inspected and is called a "scan plan". Depending on the criticality of the application, the traces are recorded or the digital scans are stored. For applications other than rotor grade, an operator typically watches the oscilloscope and marks the materials when an ultrasonic indication is observed. These areas are typically reinspected and the defective portions cut out and discarded.

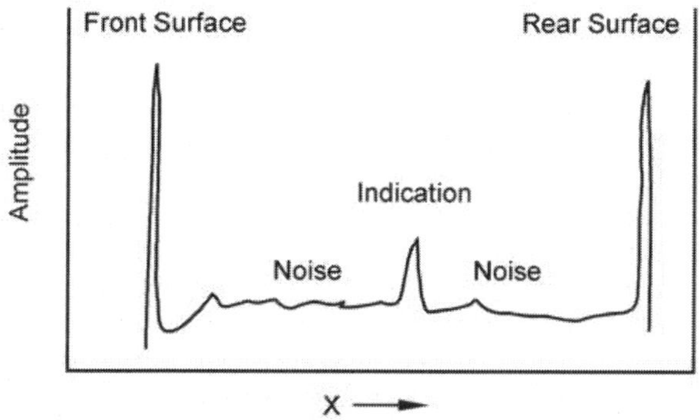

Fig. 3.72. Schematic ultrasonic scan of reflected amplitude vs. distance

In principle, ultrasonic inspection is a simple concept. In practice, modern ultrasonic inspection has become increasingly sophisticated and, therefore, more complex. In terms of its detection capability and the ability to obtain, interpret and understand the indications observed during inspection, a modern ultrasonic inspection apparatus bears little relationship to the original simple ones. Automated, computer controlled scanning is commonplace and digital images and image storage are more the norm than the exception. The detection capability is calibrated using standards. These standards are titanium blocks with small drilled holes in them that vary in size starting with 0.4 mm diameter (a no. 1 flat bottom hole which is 1/64 inch in diameter) and increasing in size. The calibration blocks are scanned to determine the smallest hole that can be detected. There is a detection requirement imposed on the facility by the buyer of the material or product (forging billet, forging, bar or plate) and this detection capability must be demonstrated. The standards also are used to assess the relative size of discontinuities that cause ultrasonic indications during inspection. This guides the subsequent analysis of the source and nature of these indications. In premium grade material used for gas turbine rotors, each ultrasonic indication must be cut out and analyzed by metallography and other means such as electron microprobe for local chemistry. This is a time-consuming practice, but one that has continuously improved the quality of rotor grade material by providing additional information regarding the defect nature and source. This information has proven to be valuable in subsequent process improvement. In addition, much effort has been expended in correlating ultrasonic indications with the indication source, its nature, location, morphology and size. This also has improved the ability to interpret false calls discussed earlier (Sect. 3.2.3).

Because there is not a one-to-one correlation between the size of the holes in the ultrasonic standards and the source of the ultrasonic indication in the material, considerable skill and experience are required to interpret the ultrasonic indications. It is very important to find defects at the forging billet stage to minimize the chance of defects being incorporated into forgings and to avoid having to scrap forged pieces with added value content. Further, both billet surface finish and larger section size pose a particular challenge for high resolution ultrasonic inspection. It is imperative that the maximum defect detection capability is used, but it also is costly and disruptive to the material production process to analyze "false positives". Thus there is a delicate balance that must be struck to ensure that all possible real defects are detected and that anomalous ultrasonic indications are recognized as such.

False positives are more commonplace in larger section sizes, especially in billet, where the structure is somewhat coarser. There still is an incomplete understanding of the origin of these indications. The elastic anisotropy of α titanium creates local changes in ultrasonic impedance. Since the elastic constants of α titanium, as outlined in Chap. 2, Sect. 2.3, differ by about 30% when measured parallel and perpendicular to the c-axis, microstructures that have α constituents with perpendicular c-axis can result in ultrasonic indications. Since the wavelength of longitudinal acoustic waves from a 5 MHz transducer are on the order of several hundred micrometers, coarse structures with average microstructural unit sizes comparable to this will scatter the acoustic waves resulting in spurious ultrasonic indications. Material with this condition is often said to be "noisy", meaning that frequent, low amplitude reflections are observed during ultrasonic inspection. In the extreme, this noise gives rise to false positives or insufficient inspection sensitivity necessary to meet the current requirements. The most effective solution to this situation is to process the billet material to create a fine microstructure. Alternate working and reheating to achieve microstructural refinement from recrystallization has been used to accomplish this objective with good success. The other characteristic of titanium alloys that affects ultrasonic inspection is the high attenuation of the acoustic waves. This is caused by the β phase when it is highly metastable. The damping of this metastable β phase can be several orders of magnitude greater than for other metals [105]. By comparison to Ti-6Al-4V and Ti-6242, the more heavily beta stabilized alloys, such as Ti-17 or Ti-6246, exhibit higher attenuation or damping. This can make it very difficult to penetrate thick sections with an adequate return signal that permits unambiguous defect detection. Aging the material to increase the β phase stability makes a significant improvement in the "sonicability" of the material. Consequently, aging before ultrasonic inspection has become a fairly common practice among material producers, particularly for inspection of β alloys, such as Ti-17 or Ti-6246.

The preceding discussion makes it clear that ultrasonic inspection of titanium alloy products becomes more difficult as the section size increases. In the past several years a new ultrasonic inspection method that allows inspection of heavy sections with comparable accuracy to that possible for smaller sections has been introduced. This method uses an array of transducers, each of which is focused at a different depth range on cylindrically shaped sub-elements of the whole volume of the round billet. This is illustrated in Fig. 3.73 showing, for example, an array of four transducers to inspect four zones. Essentially, each transducer is inspecting the equivalent thickness of one annular region as opposed to the entire section. This method has been called multi-zone because the billet is divided into a series of zones [106]. In reality the transducers are concentrically arranged in a single assembly. They have been shown separately in Fig. 3.73 to make the description of multi-zone inspection clearer. It has been conclusively demonstrated that multi-zone ultrasonic inspection is capable of finding defects that cannot be detected with the conventional, single transducer method. For example, Fig. 3.74 shows an example of strain induced porosity that was not detected in a titanium billet by conventional ultrasonic inspection. Figure 3.75 shows a similar ultrasonic image of a "hard alpha" (type I defect) in a billet. Here, also, this defect was not detectable using conventional ultrasonic inspection. Multi-zone ultrasonic inspection is capable of detecting defects that are smaller than those which can be detected using conventional ultrasonic inspection. This technique has allowed the requirement for minimum detectable discontinuity size to be reduced from a size corresponding to a 0.8 mm standard hole to a size of 0.4 mm. This change is highly beneficial with regard to improved reliability and improved precision in the calculated lifetimes for critical turbine engine components, both military and commercial.

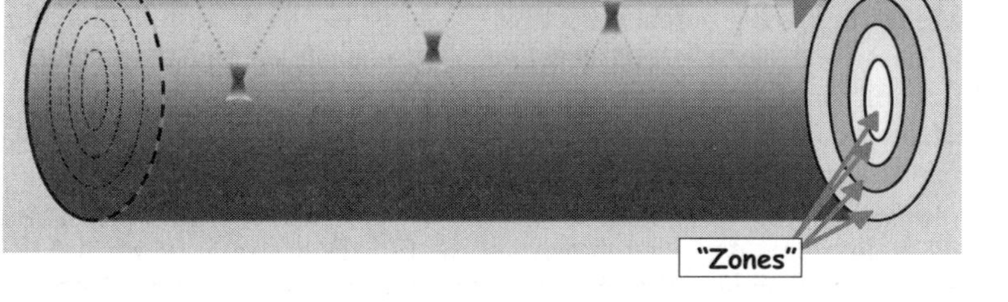

Fig. 3.73. Schematic of the multi-zone ultrasonic inspection method (courtesy GE Aircraft Engines)

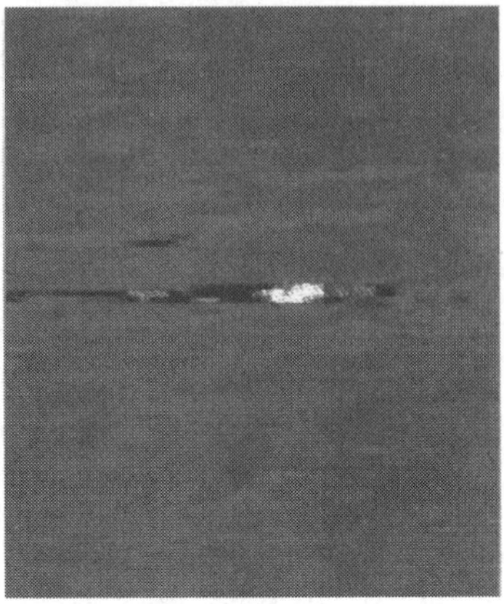

Fig. 3.74. Image from a digital multi-zone scan of a titanium alloy billet containing strain induced porosity (courtesy GE Aircraft Engines)

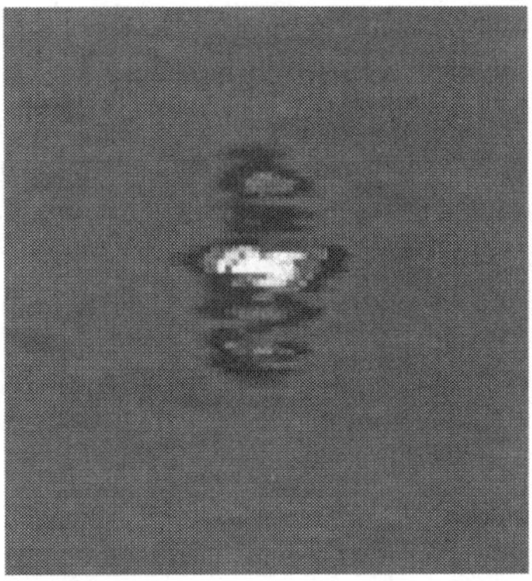

Fig. 3.75. Image from a digital multi-zone scan of a titanium alloy billet showing the indication that corresponds to a "hard alpha" (courtesy GE Aircraft Engines)

Today, it is common practice to inspect both the raw material (billet, bar or plate) and the semifinished component after forging. Given the progress that has been made in improving defect detection by ultrasonic inspection, this redundancy could be described as very conservative. The use of forging modeling and the availability of multi-zone inspected billet has resulted in essentially no defects being detected during the ultrasonic inspection of the forging. Consequently, this practice adds a great deal of cost with very little benefit. If the aerospace industry were to adopt "billet only sonic", a number of important changes in forging practice could be implemented. The most important of these would be the use of much nearer net shape forging practice. The discussion of "sonic shape" in Sect. 3.4.1 outlined the cost associated with this practice. The perceived risk and the fear of liability in the unlikely event of a serious failure has deterred the aero-engine community from adopting this "billet only sonic" practice until now. It appears that these obstacles remain as a major concern and barrier to taking this significant cost reduction step. Assuming the continuing absence of defects detected during ultrasonic inspection of forgings, it is conceivable that enough data may, at some point, permit essentially risk-free elimination of this redundant inspection requirement.

3.8.2 Radiographic Inspection

Radiography is a tried and proven method of finding internal discontinuities such as voids, pores and cracks. Radiography depends on the differences in mass for detection of these defects. Radiography is very effective in detecting W-rich defects in titanium. Early in the history of the titanium industry, this was useful when non-consumable melting was employed. The requirements for radiographic inspection of forgings remained for many years after the entire industry had adopted consumable electrode melting. Consequently, the effectiveness of radiography for the detection of cracks is strongly dependent on the relative orientations of the X-ray source and the crack. Symmetrical discontinuities such as pores and voids do not have this same limitation and radiography is commonly used for inspecting fusion welds. The majority of defects in welds are the result of solidification shrinkage or lack of fusion. Both of these defects are quite detectable using radiography. Many weldments also are on thin sections, such as sheet. The local mass difference from a void of given size is larger in thin sections, which also aids the detection capability of this inspection method. Generally speaking, radiographic inspection is less popular than ultrasonic inspection, largely because much of the required inspection is for heavy sections where the latter method is more sensitive. Today, radiography is used extensively for inspection of welds. It also is used for inspecting castings because many of the defects are due to shrinkage and are therefore readily detected by radiography. The irregular geometry and thin sections typical of castings are ideal for radiography and it is difficult to use ultrasonic inspection for these applications.

X-ray computer tomography also has grown in importance in recent years. This is because it is more sensitive due to the benefit from signal processing and because the resulting digital files can be directly converted for use by a computer aided design (CAD) system. Computed tomography machines are very expensive. Consequently, only large enterprises can justify the investment in these machines. This limits the extent to which X-ray computer tomography is and probably will be used as an inspection tool.

3.8.3 Surface Etching Inspection

Surface etching is an effective and complementary inspection method for critical titanium parts, such as rotors and air foils. The method consists of immersing a machined part with a good surface finish in a chemical bath that selectively attacks or decorates regions with different macrostructure and microstructure. Accordingly, etching is a useful means of detecting beta flecks (see for example Fig. 3.13), continuous α-case, the nitrogen stabilized alpha rich region near a "hard alpha" (type I defect, see Fig. 3.14a), type II Al-rich regions, and sometimes strain induced porosity (SIP). There are several etch bath compositions that are used for surface etching inspection. These are ammonium bifluoride (ABF), blue etch anodize (BEA), and dilute aqueous HF/HNO_3. These are described in Table 3.3. From this table, it can be seen that BEA is the most sensitive and this method has become the industry standard for large rotating parts and for fan blades used in large turbine engines, mostly for commercial aircraft. It is important that a robust etching process is selected because of the wide variation in experience of the shops that use these techniques during engine overhaul. In this regard, blue etch anodizing has been shown to be more robust than ammonium bifluoride etching, but the two methods are both adequate if properly employed. Blue etch anodize was originally developed to detect type II defects but has proven to be more broadly useful. An example of one such engine component after blue etch anodizing is shown in Fig. 3.76. In this figure, the nonuniform shades of gray are due to the presence of a beta fleck. In an actual etched part, this would be manifested as differences in shades of blue, which greatly enhances the ease of visual detection of such anomalies. By way of contrast, Fig. 3.77 shows the same portion of a titanium rotor that has been processed by ammonium bifluoride etching and by blue etch anodizing. This figure clearly shows the differences in sensitivity of detection that is characteristic of these two methods.

The airframe companies do not use surface etching inspections, partly because airframe parts are very large and partly because these parts generally are crack propagation rather than low cycle fatigue limited. The types of anomalies that surface etching reveals have the most detrimental impact on low cycle fatigue behavior where early crack initiation at defects is very harmful. Clearly the use of surface etching inspection for components such as rotors and large fan blades is an essential complement to the volume inspection done ultrasonically.

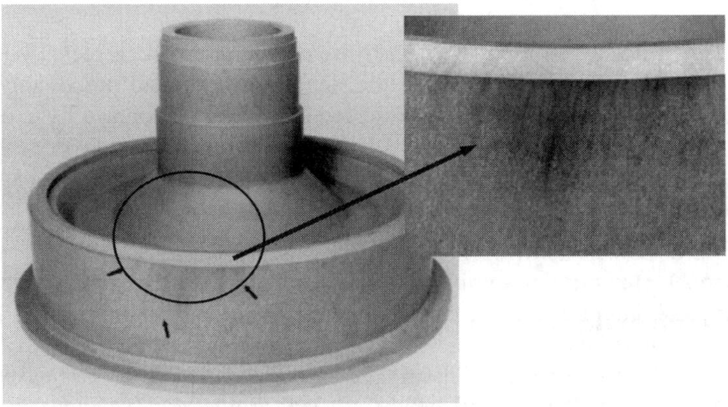

Fig. 3.76. Blue etch anodized part showing solute segregation (courtesy P. Wayte, GE Aircraft Engines)

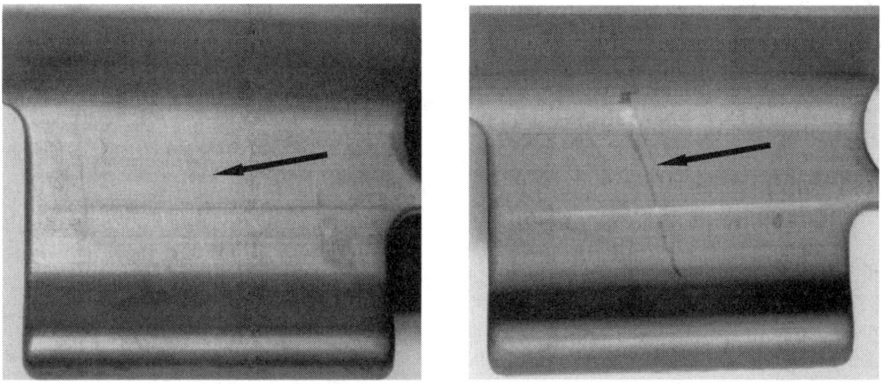

a b

Fig. 3.77. Same area etched with two different processes: (a) Ammonium Bifluoride (b) Blue Etch Anodize

Table 3.3. Surface etching inspection methods

Method	Bath Composition	Sensitivity
Blue Etch Anodize	Four step process: – Hot aqueous clean: 30 g/l Trisodium Phosphate, 3 g/l Boric Acid, wetting agent – Acid etch: Sodium Fluoride, Sulphuric Acid (concentrations vary between users) – Anodize: Trisodium Phosphate pH 8-9, anti-pit agent (part is anodic, control immersion time to achieve good contrast and sensitivity) – Back etch: Aqueous solution of 30% Nitric Acid, 1.5-3% Hydrofluoric Acid)	High
Ammonium Bifluoride	Two step process: – Acid etch: Aqueous solution of 15% Nitric Acid, 3% Hydrofluoric Acid – Ammonium Bilfluoride treat: Aqueous solution containing 18 g/l	Medium
Dilute HF/HNO_3	Single step process: – Aqueous solution of Nitric Acid (30%) and Hydrofluoric Acid (3%)	Low to Medium

3.8.4 Eddy Current Inspection

Eddy current inspection is a surface and near-surface inspection method. This inspection method uses a high frequency (10 kHz-2 MHz) electromagnetic probe to induce a current in the material near the surface of a component [107, 108]. Two factors affect electrical resistance measured by eddy current inspection, namely geometry and intrinsic electrical resistance. For inspecting titanium components geometry is more important. Discontinuities disrupt this induced current and cause a local increase in electrical impedance. This increase is readily detectable by the probe. By changing the probe frequency, the depth of the layer that is conducting the induced current changes as the inverse square root of the frequency. In CP titanium the skin depth is about 4500 μm at 10 kHz and decreases to about 450 μm at 1 MHz. For inspecting titanium alloy components, eddy current is commonly used to find geometric discontinuities that disrupt the conducting path (cracks, laps, etc.). This is in contrast to the use of eddy current evaluation of aluminum alloys, where the measured changes in electrical resistivity are used to differentiate various heat treatment conditions. This is because the variation in resistivity of titanium alloys due to changes in microstructural condition is much smaller. However, recent work in Europe has shown that eddy current methods can detect large beta flecks and the effects of abusive machining. This may be useful for inspecting deep drilled holes in fatigue critical parts.

In some titanium applications, for example disks in military aircraft engines, eddy current inspection is used as a complementary method to ultrasonic inspection (one for the volume of the component and one for the surface). The life calculation of these disks often is based on fatigue crack growth as opposed to crack initiation. Therefore, the life stringently depends on the initial discontinuity size that is used in the crack growth calculation. Smaller initial sizes result in longer lifetimes, so it is desirable to demonstrate the ability to detect the smallest possible discontinuity for use in the calculation. For engines in many military aircraft, the inspection time intervals are set according to a standard known as Engine Structural Integrity Program (ENSIP). Precision eddy current inspection has enabled the implementation of ENSIP. Eddy current inspection is ideal for this and dimensions as small as 25-50 μm can be reproducibly detected, even in the dovetail slots where the blades are attached. A photo of such a rotor is shown in Fig. 3.78. In this photo the dovetail slots can be clearly seen. It is obvious that inspection of these slots with any method other than a shaped eddy current probe that slides down the slot is problematic at best. Eddy current inspection is also used for field inspection of components in those cases where cracks are detected during engine overhaul. This would be impossible without custom shaped eddy current probes. This allows the remaining parts from the same production lot in the field to be either retired prior to failure or cleared for continued use.

To summarize, eddy current inspection is the most sensitive surface inspection method in use today. It is used to inspect local features where there is more than the usual level of concern about discontinuities. It is not used as a general surface inspection method because scanning the entire surface would be very time-consuming and therefore costly.

Fig. 3.78. Aircraft engine rotor showing the dovetail slots that are inspected by eddy current using shaped probes (courtesy GE Aircraft Engines)

3.8.5
Dye Penetrant Inspection

Dye penetrant inspection is among the oldest surface inspection methods, but is still widely used and provides useful information. Dye penetrant inspection is based on the adsorption and selective retention at surface connected discontinuities of a dye that is put on the surface and subsequently removed. The retention of the dye is caused by capillary action. Consequently, retention is improved with tighter discontinuities such as cracks. Such discontinuities can be very difficult to detect visually, even with the aid of magnification. Detection of the retained dye is aided either by the use of a developer or by observing the part under ultraviolet light after application and removal of the dye. In the former case the dye is typically red and the developer is typically white. Here, the dye is drawn out of the discontinuity by the developer, which turns red locally. In the latter case, the dye contains chemical species that fluoresce under ultraviolet light. This method often is called fluorescent penetrant inspection (FPI). Examination of the part is done in a darkened area (often a booth with curtains) and the dye fluoresces brightly making discontinuity detection easier.

In practice, the dye penetrant inspection process involves the following steps:

- Immerse the part in a bath containing the dye.
- Drain the part to allow the excess to run off back into the bath.
- Rinse the part either with solvent or water and drain dry.
- Apply the developer or examine in an ultraviolet light booth.
- Mark all indications for later detailed examination.

Dye penetrant inspection is conveniently performed in the field, kits are sold that contain dye and developer in aerosol cans that aid in the uniform application of these substances. Fluorescent penetrant inspection is not so easily done in the field because of the requirement of availability of electricity for the ultraviolet light and a darkened space to perform the examination. In general, the sensitivity of detection is greater when the fluorescent dye is used. Also, there is a dye penetrant inspection system that uses a developer film that can be removed intact and has quite sensitive detection capability. This developer is a polymeric film that is stripped off as a continuous film and can be retained as a permanent record of the inspection process. The use of fluorescent penetrant inspection is commonplace during the overhaul and repair of aircraft and aircraft engines. It is a reliable, quite sensitive method for detecting cracks in parts that have been in service. However, it is less sensitive than eddy current inspection but is very useful because it allows detection of surface-connected cracks. Cracks that are forced closed by residual stresses or that have very narrow surface openings are more difficult to detect using penetrant inspection. After disassembly, the individual parts can be inspected and those containing cracks can be retired or the cracks can be repaired, depending on the criticality of the component.

3.8.6 Surface Replication

An old, but useful technique for examining surfaces and surface connected defects is surface replication. This technique employs cellulose acetate film that has been softened by solvent. This softened film is pressed onto the area to be examined and allowed to dry. After it is dry, it is carefully stripped off and examined under a microscope. The softened film takes on the topography of the surface so any asperities such as cracks or machining laps are now contained in reverse relief to the actual features, but are nevertheless quite observable. There are several advantages to surface replication. First, it is easier to replicate difficult to access locations and then examine the replica than it is to observe these areas directly. Second, the replicas provide a permanent record of the inspection and can be stored and reviewed again at a later time. Third, replication is completely non-invasive and can be used in situations where legal or other considerations would preclude other means of examination that could alter the area in question. Surface replication is also frequently used on locally polished surfaces of a component, where an indication, either by blue etch anodizing or by ammonium bifluoride etching, was marked. The inspection of the replica by light microscopy or by scanning electron microscopy allows further evaluation of the indication.

3.9 Characterization Methods

This section is intended to describe the methods and techniques commonly used to characterize the constitution, microstructure and preferred orientation (texture) of titanium alloys. Titanium alloys are prone to local flow and smear during cutting and preparation. Consequently, they are subject to a variety of preparation artifacts. These will also be discussed since avoiding them is essential if the true constitution and microstructure are to be revealed. As in the preparation and examination of all materials, there are a variety of methods that work well when conducted by skilled practitioners. No attempt will be made to describe or even mention all of these methods. Instead, proven methods that have served the authors and other investigators of titanium alloys well over the years will be described. The methods described represent those regularly used by any investigator of titanium alloys. Therefore, this coverage of characterization methods seems consistent with the intended purpose of this book.

3.9.1 Light Microscopy

The polishing techniques used in preparation of titanium alloys for examination by light microscopy (LM) are also used in preparing samples for examination in the scanning electron microscope and for X-ray diffraction studies. This is because the preparation of titanium alloy specimens with a distortion free surface is central to all these methods. Accordingly, these preparation methods comprise the basis for much of the subsequent characterization efforts.

Titanium and its alloys can be polished either mechanically or electrolytically. The initial steps are the same: dry rough grinding followed by wet medium grinding followed by wet fine grinding using the normal SiC grinding papers.

Mechanical polishing is usually performed in two stages. The first stage is preliminary polishing using either coarse diamond paste with kerosene or coarse alumina in a water slurry. The second stage is fine polishing using either fine diamond paste or fine alumina (typically 0.05 μm diameter) in a slurry of water and very dilute HF. Titanium can smear during polishing and the HF removes any smeared metal on the surface. The HF requires the use of rubber gloves during polishing for hand protection. Smeared metal obscures the true microstructure and must be avoided. This is particularly true because the apparent structure of a specimen containing smeared metal can have a realistic and misleading appearance. Skillful metallographers are familiar with this matter and all have their own routine to avoid such preparation artifacts. Here, an awareness of this possibility is a sufficient caution for those inexperienced in the mechanical polishing of titanium and its alloys for metallographic examination. Mechanical polishing has the disadvantage of usually requiring the specimen to be mounted, either in a thermosetting compound such as Bakelite or Lucite or in an epoxy. This is time consuming and also is an issue during examination in the scanning electron microscope (SEM) as will be described later. Mechanical polishing also has the advantage of allowing excellent specimen edge retention if the mounting material is hard and adherent. This allows examination of the material surface in cross section for determining the presence of any undesirable surface oxygen contamination, both metallographically and by the use of microhardness. As described in Chap. 2, oxygen is an interstitial strengthener and an α stabilizer. Consequently, oxygen contamination of $\alpha+\beta$ alloys causes an increase in the α phase volume fraction near the surface. In extreme circumstances the surface can be entirely converted to α phase. This sometimes is called α-case. The hardness of the contaminated layer also is higher due to strengthening by the oxygen. In such cases, microhardness measurements every 5-10 μm traversing along a line perpendicular to the surface is used to define the extent of enrichment. The local hardness decreases with distance from the surface and approaches that of the interior (the "core hardness") at some point. This is the limit of the oxygen contamination and this dimension can be used to guide removal of material during chemical milling. Another common issue is the appearance of an apparent martensitic phase in mechanically polished metastable β alloys. Careful mechanical polishing or electrolytic polishing, as described next, eliminates this artifact as is illustrated in Fig. 3.79. As can be seen in this figure, properly prepared specimens show no transformation of the β phase (Fig. 3.79a). Figure 3.79b shows the transformation products that can be induced by mechanical polishing if care is not exercised.

Titanium and titanium alloys can readily be electrolytically polished and this is a useful method for quick examination. An electrolyte and the polishing conditions that work well are shown in Table 3.4. In addition to the ease of preparation, electrolytic polishing does not create any possibility of smearing. Electrolytically polished specimens generally are not as flat as mechanically polished ones and this loss of flatness can be a disadvantage during light microscopic examination at high magnifications. Today, much of the high magnification examination is done in the SEM, so this disadvantage is less serious. In fact, the depth of field of the SEM is one of its major advantages as a metallographic tool, as will be described in greater detail later in this section.

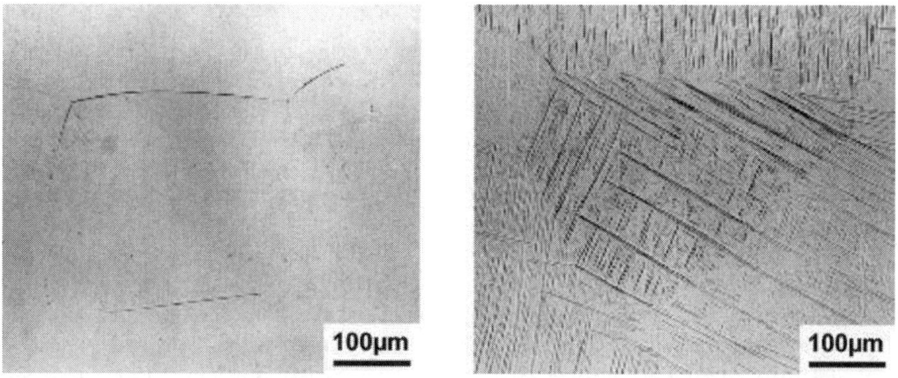

a b

Fig. 3.79. Illustration of mechanical polishing induced artifact in Ti-6246, LM: (a) Untransformed β phase in specimen that has been mechanically polished with proper technique (b) Stress induced martensite formed during mechanical polishing

After polishing is complete, etching can be done using Kroll's etch or an oxalic acid stain etch (Table 3.5). The latter is somewhat orientation dependent and helps elucidate α/α grain boundaries. Both of these etchants are widely used and both give excellent results. Kroll's etch is usually used with a swab but the oxalic acid etch is done by immersing the polished surface and watching until it appears cloudy. Re-etching for longer times is possible with both etchants, so short initial etching times followed by examination can help prevent overetching and the subsequent need for repolishing. Kroll's etch can be mixed in several strengths or acid concentrations. The etchants more concentrated in HF are particularly useful for deep etching for SEM examination. The compositions and application details of these two etchants are given in Table 3.5. There is a tremendous amount of "art" in the preparation of metallographic specimens. Consequently, there are numerous alternate preparation techniques in addition to those described above. However, the above methods have been shown to work well and should be sufficient for almost any metallographic examination need.

Alpha titanium, being hexagonal, is optically anisotropic and the use of polarized light provides useful qualitative information regarding relative orientation of α grains. The use of polarized light to estimate the degree of texture (preferred orientation) is quick and reliable. This method complements the orientation imaging microscopy (OIM) technique that will be described later in this chapter. An example of Ti-6Al-4V viewed under both bright field and polarized light illumination conditions is shown in Fig. 3.80. Examination under polarized light should be done in the as-polished (unetched) condition. The bright field image is obtained after etching. This pair of micrographs illustrates the additional information that can be gotten using polarized light.

Nomarski interference microscopy also is useful for examining titanium alloys because the rate of attack by the etchants is somewhat anisotropic. This provides a modest amount of surface relief that is accentuated when examined using a Nomarski objective. The Nomarski method gives images that are considerably more aesthetically pleasing. In reality these images contain little more information than normal bright field images. There is nothing wrong with aesthetics, but it also is important to keep the information point in perspective. An example of a Nomarski interference image is shown in Fig. 3.81.

Table 3.4. Electrolytic polishing of titanium (electrolytes and polishing details)

Electrolyte Composition	Polishing Conditions	Polishing Voltage	Comments
5% H_2SO_4, 1.25% HF, balance Methanol	Polish at ambient temperature in a recirculating cell (Struers, Buehler or other)	25-30 V (DC)	Recommended for LM preparation. Polish for about 30 seconds
300 ml Methanol, 175 ml Butanol, 30 ml Perchloric Acid (70-72%)	Polish cold at –30°C using a stainless steel cathode	12-20 V (DC)	Recommended for TEM preparation. Gentle bath agitation helps avoiding "bubble tracks"

Table 3.5. Metallographic etchants for titanium and titanium alloys

	Application Method	Etchant Composition
Kroll's Etch	Swab until surface becomes less reflective	95 ml H_2O, 3 ml HNO_3, 2 ml HF or 95 ml H_2O, 4 ml HNO_3, 1 ml HF
Oxalic Acid Stain Etch	Immerse and remove after polished surface appears "cloudy"	Equal parts of aqueous 10% Oxalic acid and 1% aqueous HF solutions

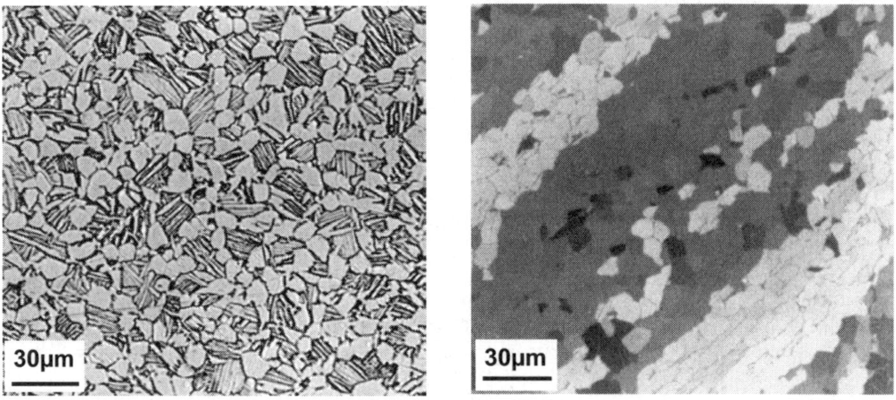

a b

Fig. 3.80. Presence of microtexture in Ti-6Al-4V, LM: (a) Bright field (b) Polarized light

Fig. 3.81. Nomarski interference image of a deformed Ti-Mo alloy showing the presence of twins, LM

3.9.2 Electron Microscopy

Examination of titanium or its alloys using electron optic devices such as the transmission electron microscope (TEM) and the scanning electron microscope (SEM) provides complementary information to that obtained from other characterization methods such as light microscopy and X-ray diffraction. This section will outline the use of the TEM and the SEM in the characterization of titanium and its alloys including a brief summary of the relevant experimental techniques.

3.9.2.1 Transmission Electron Microscopy

The ability to examine titanium and titanium alloys by thin foil transmission electron microscopy lagged behind similar capability for other metals such as Al and Cu alloys. This was largely because of the difficulty encountered in preparing high quality thin foils of titanium base materials. This was in part related to the reactivity of titanium including its propensity to absorb hydrogen during electrolytic polishing and chemical polishing. There are some early examples in the literature of hydrogen-related artifacts. Blackburn [51] was among the first investigators to demonstrate the routine preparation of artifact-free titanium thin foils by electrolytic polishing. He used the window method described elsewhere [109] along with the methanol, butanol and perchloric acid electrolyte and polishing conditions described earlier in Table 3.4. This polishing method is capable of preparing thin foils from a wide range of titanium alloy types and compositions and, if the polishing is done cold, minimizes the amount of hydrogen that is absorbed by the specimen. Today, jet polishing of 3 mm diameter disks using semi-automatic polishers explicitly designed to make TEM specimens is the preferred method of thin foil preparation. Jet polishing is fast, gives more reproducible results and lessens the risk of preparation-induced deformation in the thin sections being examined. The early commercial jet polishers were made by Fischione in USA, but Struers and other metallography equipment makers now sell jet polishers.

The biggest drawback of electrolytic polishing is the local rate of removal which is strongly influenced by the local electrochemical dissolution potential. In two phase alloys that have individual constituents of significantly different chemical composition, highly preferential attack of one constituent can occur. This can result in very nonuniform thinning or, in extreme cases, in only one constituent in the thin section. Thin foil preparation by ion milling eliminates this selective attack. Ion milling also has the advantage of not introducing hydrogen into the specimen, and may actually cause the hydrogen content to be reduced. This is because the ion mill exposes a pristine titanium surface in a vacuum under conditions of modest local ion beam heating. This method takes several hours minimum to produce a thin specimen, but it still has become popular because it is used for specialized situations, two of which will be described later (interface phase and spontaneous relaxation). Ion milling uses a beam of Ar ions accelerated by a high voltage (usually 10-20 kV) to sputter materials from the surfaces of a 3 mm diameter TEM disc. In materials with atomic numbers equal to or greater than titanium, there is minimal damage introduced into the specimen by the ion beam bombardment. Careful examination of titanium base specimens permits detection of some mottling due to the ions, but this does not interfere with forming good images and appears not to adversely affect the ability to examine the microstructure of the specimen.

The principle of thin foil preparation is the same no matter what material removal method is used. The material is gradually removed until the specimen is perforated. Along the edge of the perforation is a wedge of material that is thin enough to be electron transparent. The volume of material that is actually examined in a TEM specimen is extremely small, but the high magnifications used for the examination sometimes cause the investigator to overlook this point. The usefulness of TEM for characterizing uniformly distributed microstructural features is unsurpassed. For nonuniform features such as grain boundaries in large grain size materials, multiple foils typically must be prepared before the feature of interest is captured in the thin region. This is time consuming and can be very tedious. Higher accelerating voltage microscopes extend the amount of electron transparent material in a foil, but this still is of relatively limited help if the features of interest truly are inhomogeneously distributed.

Now a new device is available that allows thinning of TEM specimens at predetermined sites. This device is a focused ion beam device (known as a FIB). The FIB was originally developed for electronic materials studies where the defect density is very low ($<10^4/cm^3$). It is now clear that the FIB is a useful tool for metallurgical studies also. The FIB transforms the TEM from a powerful characterization tool for general microstructural features to a problem solving tool as well. This is because this device permits areas of particular interest to be selectively examined because the thin section can be deliberately located at any desired position within the material being examined. The use of FIB devices in metallurgy studies is just beginning, but this instrument promises to revolutionize microstructural studies in real structural materials of practical interest. This is especially true for materials containing inhomogeneously distributed microstructural features that may affect the performance of the material.

During the early TEM studies of titanium and titanium alloys, several thin foil artifacts were common, but these were not always recognized as such. In addition to the gross effects of hydrogen on obscuring the microstructure mentioned earlier, there is a second effect worth mentioning. This is the formation of an interface phase in $\alpha+\beta$ alloys and the complete distortion of α precipitates in β alloys. These effects have been described in several publications [110, 111]. The interface phase is illustrated in Fig. 3.82. It now is clear that this phase is a thin foil artifact. This clarification was demonstrated by jet polishing and ion thinning specimens from the same material sample. The jet polished specimens contained the interface phase whereas the ion thinned specimens did not. This result is convincing because it shows that the interface phase is a hydrogen-related artifact. In reality, once this is understood, the interface phase does not cause any particular confusion in the basic interpretation of $\alpha+\beta$ titanium alloys microstructures. However the interface phase does obscure the structure of the alpha/beta interface itself as can be seen in an ion-milled specimen, Fig. 3.83. In those cases where the interface is the object of study, ion milling is essential. If the electrolyte temperature is allowed to get above about –20°C, there is the risk of significant hydrogen pickup during polishing. This can lead to sufficiently high bulk concentrations that hydride (TiH_2) precipitation can occur. These hydrides can form as plates, giving rise to displacement fringe contrast that can be mistaken for stacking faults. An example of this fringe contrast due to hydrides is shown in Fig. 3.84. Heating specimens containing such features to a few hundred degree Celsius in the microscope causes these hydrogen related features to disappear. Hydrogen related effects are the only reasonable explanation for these changes that occur over such a small temperature range. Polishing under conditions where the electrolyte is too warm also can result in surface hydrides that are small and coherent and which cause strain contrast. An example is shown in Fig. 3.85. These surface hydrides also disappear during heating, which is consistent with their identification as hydrogen related features.

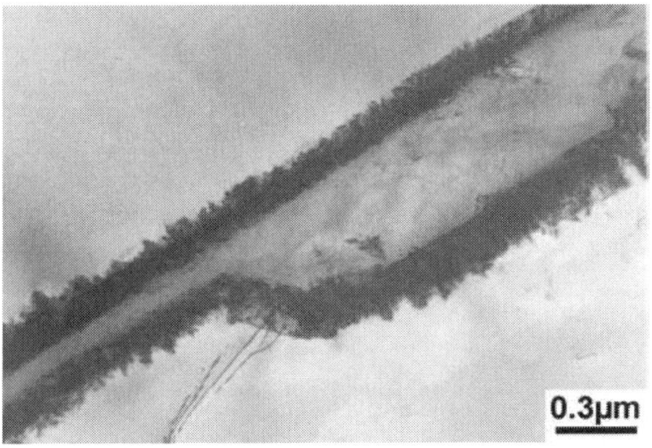

Fig. 3.82. Interface phase in an electropolished Ti-6Al-4V specimen, TEM

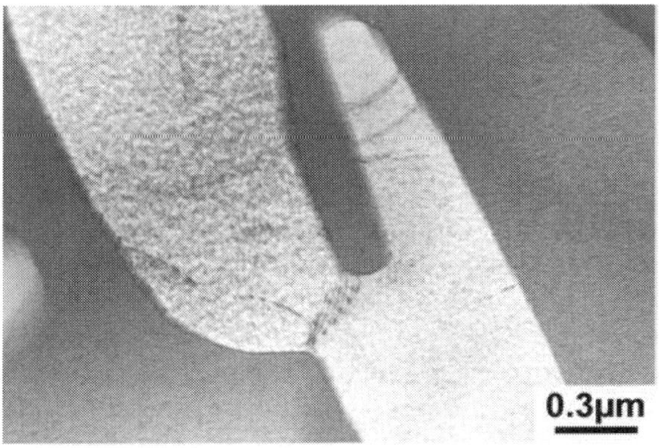

Fig. 3.83. Ion milled specimen showing absence of interface phase, TEM (courtesy D. Banerjee, DMRL)

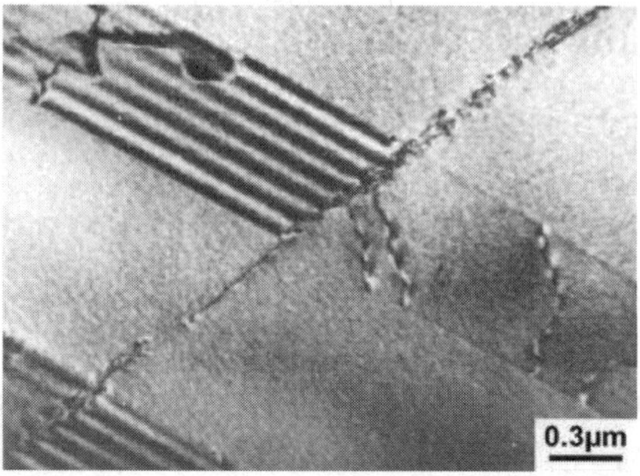

Fig. 3.84. Displacement fringe contrast due to the presence of thin hydrides in Ti-5Al-2.5Sn, TEM

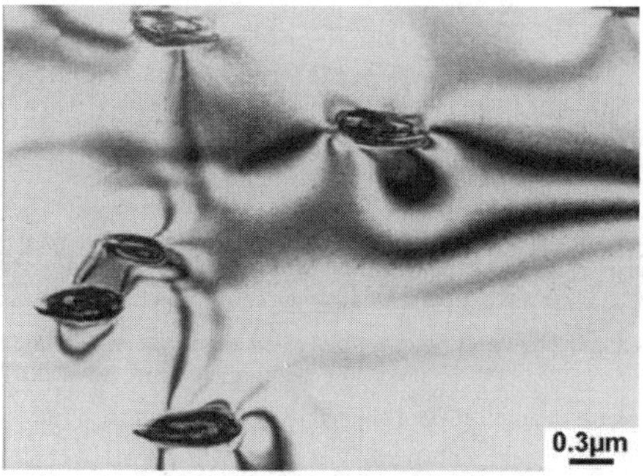

Fig. 3.85. Strain contrast around electrolytic polishing induced surface hydrides along thin edge of the specimen, TEM

Another characteristic of TEM examination of titanium alloys is the occurrence of "spontaneous relaxation" of metastable phases when the bulk constraint is removed during thinning. There are two examples worth noting. First is the transformation of orthorhombic martensite to a face-centered cubic structure in thin foils [112]. The second is the spontaneous shearing of metastable beta phase in the thin regions of the foil, leaving "martensite-like" features in the image that can be mistakenly thought to be part of the real microstructure [113], as shown in Fig. 3.86. Ion milled specimens containing orthorhombic martensite or metastable β phase, do not exhibit these two thin foil artifacts [114]. This article suggests that the "spontaneous relaxation" occurs by movement or creation of interfaces in the thin specimens and that a small amount of damage from the ion milling prevents this from happening. This is another advantage of ion milling.

Fig. 3.86. Example of thin foil artifact: "Martensite-like" features in metastable β phase, Ti-6246, TEM

3.9.2.2
Scanning Electron Microscopy

The intrinsically fine microstructural scale of titanium alloys requires examination at relatively high magnifications. This, coupled with some of the difficulties in mechanical polishing of titanium alloys, presents a challenge to prepare specimens with flatness of field suitable for obtaining high quality light micrographs at high magnifications. Consequently, the use of the SEM for metallographic examination at higher magnifications has become common. The high depth of field and good resolution of the SEM make examination of electrolytically polished samples easy and the results are very good. An example of a SEM micrograph is shown in Fig. 3.87 which gives a level of detail that would be very difficult to reveal by LM. In addition to electrolytic polishing the sample, etching with the more concentrated (95-3-2) Kroll's etch (see Table 3.5) provides good differentiation between constituents.

The excellent depth of field of the SEM also has allowed the development and use of another interesting technique for characterizing the relation between fracture topography and microstructure. In this technique, a portion of the fracture surface is preserved by masking using standard electroplating masking lacquer. The adjacent fracture surface is then electrolytically polished until it is smooth. The specimen is then etched using again the 95-3-2 version of Kroll's etch. The lacquer is then removed with solvent and the fracture surface and the adjacent polished surface are cleaned and dried. When the sample is placed in the SEM, the intersection of the microstructure and the preserved fracture surface can be viewed directly. Both the microstructure and the fracture surface are simultaneously in focus in the SEM image because of the depth of field of the SEM. This technique permits the effects of microstructure on fracture topography to be directly observed. Two examples of the application of this technique are shown in Figs. 3.88 and 3.89. This technique has proved to be very useful for advancing the understanding of the fracture topography of titanium alloys, including the relationship to the underlying microstructure.

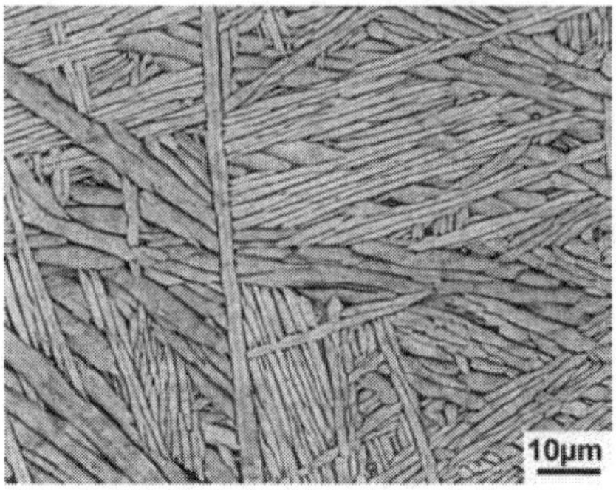

Fig. 3.87. Colony structure in Ti-6Al-4V resolving small secondary α phase plates and thin β phase "ribs" between the α phase plates, SEM (courtesy M. Juhas, The Ohio State University)

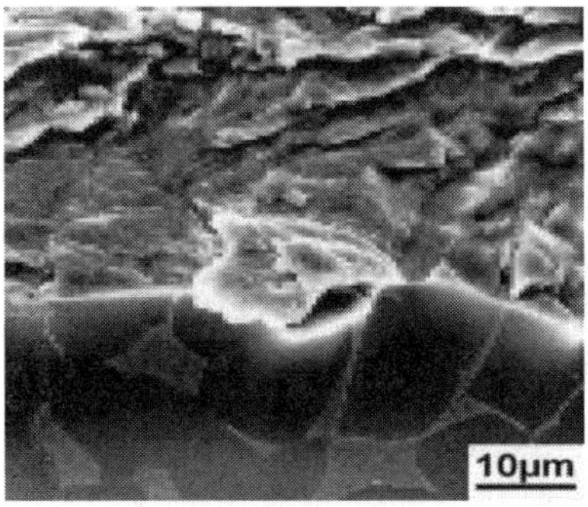

Fig. 3.88. Plateau etched specimen showing the relation of fracture topography to underlying microstructure, SEM

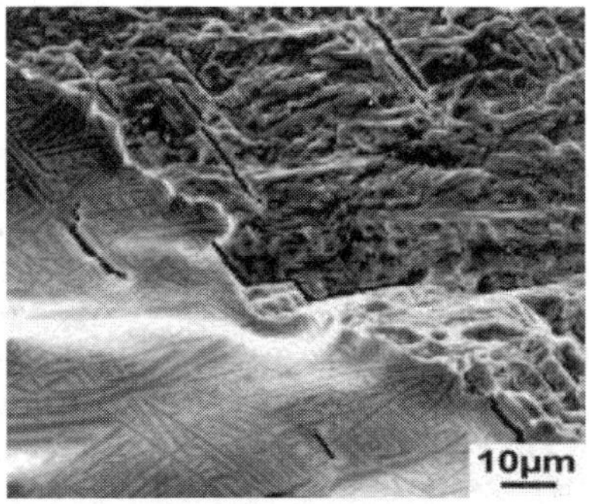

Fig. 3.89. Plateau etched fatigue fracture surface showing the relation between secondary cracks and α plates, SEM

Recently an improved capability to generate backscattered electron diffraction patterns and to automatically index them in real time with the aid of a computer has led to a new type of imaging [115]. This is called orientation image microscopy (OIM). This imaging method collects the crystallographic orientation of individual microstructural constituents as determined in the SEM by backscattered electron diffraction and generates an image that depicts the variations in crystallographic orientation. The image is generated with the use of an orientation polyhedron with surfaces that are mis-oriented from the specimen surface normal by preset amounts. For example, if the basal pole of an α grain is within 5° of the beam, the pixels belonging to this grain are presented in the image as one color. Grains that have a mis-orientation greater than this but less than, for example 10°, can be presented as another color. The resulting image is an orientation map, which gives complementary information to the actual image. The orientation polyhedron is shown in Fig. 3.90. Normally this polyhedron would have different colors for each face, but the different gray levels shown here provide a reasonable idea of this part of the technique. Figure 3.91 shows two orientation images generated using this imaging technique. From this figure the large regions of consistent gray level in Fig. 3.91a is consistent with the presence of microtexture, while the more random distribution of gray shades in Fig. 3.91b is consistent with very little microtexture. As illustrated earlier, the use of polarized light can qualitatively provide the same information in a much shorter time. The use of OIM should be viewed as a quantitative complement to the polarized light method. Because of the cost of OIM facility and the time required to generate OIM images, it is recommended that polarized light be used before this more elaborate technique. In cubic materials where there is little or no optical anisotropy, OIM is useful in understanding local orientation, for example where it is not clear whether boundaries are high angle grain boundaries or sub-boundaries.

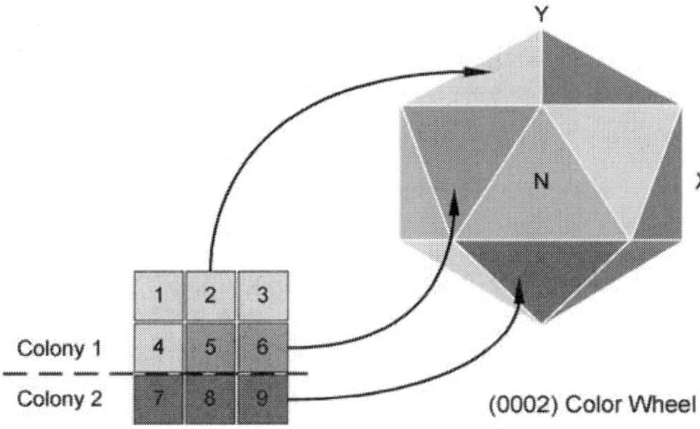

Fig. 3.90. Orientation polyhedron that allows interpretation of orientation images

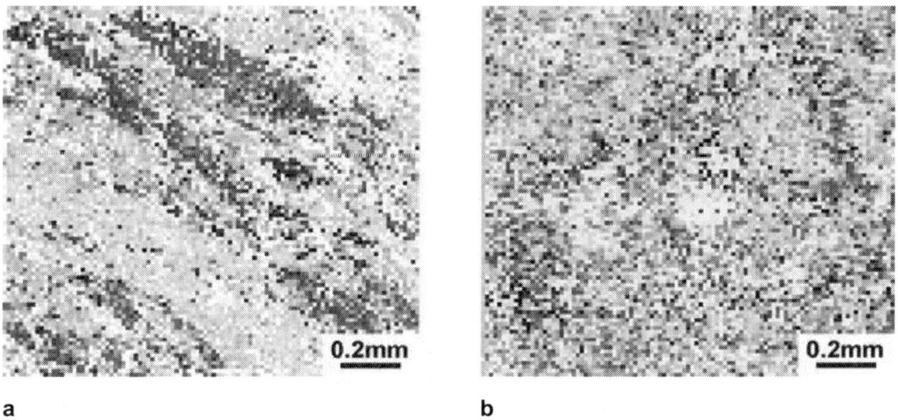

Fig. 3.91. Orientation images of Ti-6Al-4V: (a) Strong microtexture (b) Without significant microtexture (courtesy A. P. Woodfield, GE Aircraft Engines)

3.9.3 X-Ray Diffraction

X-ray diffraction has been used to determine the structure and amount of phases in crystalline materials for a long time [116]. There are comprehensive texts on this technique, so there is little need to provide a detailed description of this technique here. Instead, there is one characteristic of titanium that merits discussion in connection with X-ray diffraction studies. The most useful wavelength for conducting X-ray diffraction studies of metals is Cu Kα because it provides a reasonable balance between penetration and diffraction peak resolution. However, titanium fluoresces under Cu Kα radiation that increases the background intensity in diffraction patterns obtained using this radiation. This increased background can mask low intensity peaks. There are two approaches to eliminating the effects of sample fluorescence. One is the use of a diffracted beam monochromator, which only allows the diffracted Cu Kα radiation to enter the detector. In such a monochromator, the most efficient monochromator crystal to use is pyrolytic graphite. This method works very well and allows acquisition of high quality patterns with good peak to background ratios. The other method is to use an energy dispersive detector so only the Cu Kα radiation is accepted by the detector. If this option is selected a detector with about 125 eV resolution is required for good results. Everything else being equal, the diffracted beam monochromator is simpler and works very well. Because α titanium is hexagonal, solid titanium samples are seldom isotropic. Consequently, the relative intensities of the diffracted peaks seldom match the standard values for powder diffraction patterns. In the extreme, reflections can be missing altogether. The use of a specimen spinner helps for samples where the directionality lies in the plane of the sample. This is often the case for flat rolled products such as sheet and plate, but may not be the case for forgings, where the typically weak texture also has a different orientation. The point of mentioning this here is simply a caution to avoid drawing detailed conclusions from X-ray patterns obtained from solid specimens if these conclusions can be affected by the presence of texture, which is usually present in titanium mill products.

X-ray diffraction also is used to determine pole figures used in describing the nature and amount of texture present in titanium and titanium alloy products. X-ray pole figures provide information regarding average preferred orientations in a volume of material. If needed, pole figures can be complemented by orientation image microscopy (see Sect. 3.9.2.2). The effects of texture on the properties of titanium alloys depend on both the type of texture, including the degree or intensity, and on the specific alloy. A recent book by Kocks treats this matter in detail [117]. Texture effects will be discussed in connection with individual alloy classes later in this book.

Finally, X-ray diffraction is used to determine residual stresses resulting from processing of titanium alloys. This residual stress can shift the mean stress level in a component and affect fatigue behavior. It also can lead to stress assisted migration and concentration of hydrogen and can cause the occurrence of stress corrosion cracking where it might not otherwise be expected. Measurement of residual stress by X-ray diffraction has been performed for years on a routine basis for cubic metals such as aluminum alloys and steel. The elastic and plastic anisotropy of α titanium and the potential for crystallographic texture makes the reduction of X-ray diffraction data into residual stress components more difficult. Consequently, the reader interested in measuring residual stresses in α+β titanium alloys must be prepared to spend some time in planning the residual stress measurements and in analyzing the results.

3.9.4 Mechanical Testing

The purpose of this section is to call attention to aspects of mechanical testing that are peculiar to titanium. The section is not intended to be an exhaustive discussion of mechanical testing per se. Such discussions can be found in a variety of books on this subject, see for example [118, 119].

The preparation of titanium and titanium alloy fatigue specimens for endurance testing must be done very carefully, because, titanium is not a hard material like steel. The specimens require mechanical polishing after machining to obtain a reasonable surface finish that also is free of any residual stresses due to machining or subsequent polishing. Clearly, surface residual stresses will affect the initiation of fatigue cracks and interfere with obtaining true fatigue lives. Residual scratches also can behave as notches so mechanical polishing should be done in the longitudinal direction to avoid any scratches (notches) oriented normal to the loading axis. Laboratory fatigue specimens are often electrolytically polished to provide a maximum consistency in surface quality and minimize the scatter due to this factor. For research purposes, this is a useful approach. For the generation of fatigue data, mechanical polishing is a better approximation to the surfaces of titanium components that would be used in manufactured products. In these cases longitudinal mechanical polishing gives data that approaches the behavior of such components in service. The goal is to generate data that is neither conservative nor non-conservative. There are several companies that specialize in specimen preparation and testing which provide consistent results that meet this description. It is not surprising that these companies often are located in the proximity to aircraft or aero-engine companies. Figure 3.92 shows the variation in fatigue life that can be obtained due to different specimen preparation methods, such as rough mechanical polishing, electrolytic polishing, fine mechanical polishing, and shot peening [120]. If the electrolytically polished curve is taken as the "true" fatigue capability of the material, it becomes clear that specimen preparation methods and surface condition can lead to either better or poorer fatigue life.

The test environment also plays a significant role in determining the fatigue results, In this case, testing in inert gas, vacuum, and air of varying relative humidity, all affect the life. Examples on the influence of environment on mechanical properties will be shown in Sects. 5.2.6 and 6.2. It is important to be aware of this point when examining the literature and if self-consistent results are to be obtained.

Hydrogen that is present in the material as a residual impurity or that may have been introduced during specimen preparation also can have a negative effect on the measured fatigue life. Thus, the precautions outlined in Sect. 3.9.2.1 regarding hydrogen pickup should be carefully observed in order to avoid misleading results due to hydrogen contamination.

Titanium exhibits significant room temperature creep and also can undergo room temperature recovery after straining. These effects can affect the extent of primary creep measured in a creep test. Therefore, caution must be exercised in the alignment of the loading system and in the loading methods used, and the loading rates should be carefully monitored and controlled to ensure reproducible creep results. In creep testing titanium, considerably more care must be exercised to obtain reliable data compared to that required for Ni base alloys.

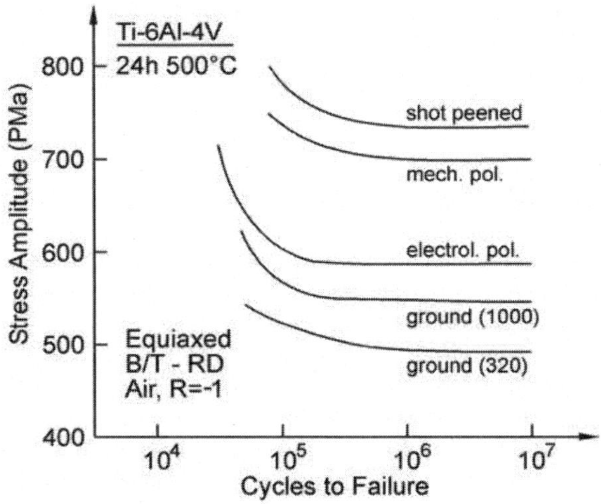

Fig. 3.92. S-N curves for Ti-6Al-4V showing the effect of specimen preparation and surface condition on fatigue life

4 Commercially Pure (CP) Titanium and Alpha Alloys

This chapter describes the processing, microstructure, and properties of α titanium alloys, with emphasis on the various grades of commercially pure (CP) titanium (often referred to as CP-Ti). A definition of α titanium alloys was included in Chap. 2, but it may be useful to include a more detailed one here. All α titanium alloys are based on the low temperature, hexagonal allotropic form of titanium. These alloys can contain substitutional alloying elements (Al or Sn) or interstitial elements (oxygen, carbon, or nitrogen) that are soluble in the hexagonal α phase. These alloys also contain limited quantities of elements that have limited solubility such as Fe, V, and Mo. Table 2.6 lists a representative group of α titanium alloys and grades of CP titanium, along with representative selections of alloys belonging to the α+β and β classes. The grade designations are taken from the American Society for Testing and Materials (ASTM). As the beneficial application of this class of titanium alloys has been recognized, the use has increased. Further, specific alloys have been formulated to improve to environmental resistance of CP titanium and α titanium alloys or to provide comparable performance at reduced cost where expensive additions such as palladium are involved. Consequently, there has been a proliferation of alloy grades. There are now at least 16 identifiable alloys or grades.

Table 4.1 lists these alloys and respective grade number, together with the composition limits or ranges of these alloys depending on the element involved and also minimum yield strength values. All alloys in this class derive their characteristics from the hexagonal α phase, as mentioned earlier. For some purposes the class should be subdivided to allow a clearer discussion of behavior trends. All CP titanium grades are grouped together because none of the grades derive strength from the substitutional alloying elements present including Fe, Pd, or Ru. The remaining alloys in Table 4.1 comprise the group called α titanium alloys here. In principle, all alloys in the table are α titanium alloys but subdividing this broad class as suggested above makes discussion of their properties and applications less confusing.

The excellent corrosion resistance of CP titanium compared, for example, to stainless steel, has made it an attractive material of construction for chemical and petrochemical processing equipment [61]. It also has become popular for heat exchangers and other piping applications because of its weldability and good general fabricability, both into tubing and the subsequent shaping of the tubing for specific applications [121]. Although CP titanium is more expensive at the outset than stainless steel, articles made from it often have lower life cycle cost because of the superior durability of CP titanium in service [122]. For the most part, the initial selection of CP titanium for a particular application is primarily related to properties such as corrosion resistance and fabricability. When one of the other α titanium alloys is selected, this usually is because CP titanium has insufficient strength for the intended application. The effects of composition on mechanical properties, especially strength, usually are a secondary consideration relative to the corrosion resistance. There are growing numbers of applications such as pressure vessels where mechanical properties are an equally important criterion for materials selection. For these applications it is critical for the design engineer to consider the variation in mechanical behavior of the different alloy grades when making the material selection. While the desired level of corrosion resistance always is necessary, it may not be sufficient for other applications, e.g. pump impellers, where vibrations and high mean stresses place strength related constraints on alloy selection. This subordinate importance of mechanical properties for many applications is a major distinction between this class of titanium alloys and the others described in Chaps. 5, 6, and 7.

The organization of this chapter is similar to that of Chaps. 5, 6, and 7, i.e., the relation between processing and microstructure is first described (Sect. 4.1), followed by a discussion of microstructure and properties in Sect. 4.2. The chapter concludes with illustrations of some applications for CP titanium and α alloys in Sect. 4.3.

Table 4.1. Chemical composition and minimum yield stress for CP titanium and α titanium alloys

Grade or Alloy	O (max.)	Fe (max.)	Other Additions	$\sigma_{0.2}$ (MPa)
CP Titanium				
CP Titanium Grade 1	0.18	0.20		170
CP Titanium Grade 2	0.25	0.30		275
CP Titanium Grade 3	0.35	0.30		380
CP Titanium Grade 4	0.40	0.50		480
Ti-0.2Pd (Grade 7)	0.25	0.30	0.12-0.25Pd	275
Ti-0.2Pd (Grade 11)	0.18	0.20	0.12-0.25Pd	170
Ti-0.05Pd (Grade 16)	0.25	0.30	0.04-0.08Pd	275
Ti-0.05Pd (Grade 17)	0.18	0.20	0.04-0.08Pd	170
Ti-0.1Ru (Grade 26)	0.25	0.30	0.08-0.14Ru	275
Ti-0.1Ru (Grade 27)	0.18	0.20	0.08-0.14Ru	170
α Titanium Alloys				
Ti-0.3Mo-0.9Ni (Grade 12)	0.25	0.30	0.2-0.4Mo, 0.6-0.9Ni	345
Ti-3Al-2.5V (Grade 9)	0.15	0.25	2.5-3.5Al, 2.0-3.0V	485
Ti-3Al-2.5V-0.05Pd (Grade 18)	0.15	0.25	2.5-3.5Al, 2.0-3.0V, (+Pd)	485
Ti-3Al-2.5V-0.1Ru (Grade 28)	0.15	0.25	2.5-3.5Al, 2.0-3.0V, (+Ru)	485
Ti-5Al-2.5Sn (Grade 6)	0.20	0.50	4.0-6.0Al, 2.0-3.0Sn	795
Ti-5Al-2.5Sn ELI	0.15	0.25	4.75-5.75Al, 2.0-3.0Sn	725

*For all grades shown, typical values of C and N are 0.08-0.10 and 0.03-0.05, respectively

4.1 Processing and Microstructure

Processing of α titanium alloys can be conveniently separated into two parts. The first is the material processing at the producer to create a product with the desired dimensions and with a microstructure (including texture) that ensures the proper response during component manufacturing operations. The second is the application of manufacturing processes (still processing, but of a different type) needed to create components and the effect of microstructure on the effectiveness of these processes.

4.1.1 Material Processing

From Tables 2.6 and 4.1, it is clear that there are a number of titanium alloys and grades of CP titanium that are classified as α titanium alloys. The constitution of this class of alloys is predominantly α phase with a small volume fraction (several per cent maximum) of β phase [123, 124]. The microstructure of α titanium alloys also is simpler than that of the higher strength α+β and β alloys discussed later in this book. Alpha alloys have sufficiently small total alloying additions that they essentially do not respond to heat treatment, but this characteristic contributes to the excellent weldability of alloys belonging to this class. Further, thermomechanical processing of these alloys is done to control the crystallographic texture and grain size, but is not used to manipulate the microstructure in a similar manner as done in α+β (Chaps. 5 and 6) and β alloys (Chap. 7).

CP titanium is largely produced as flat rolled products (sheet and plate). While some castings of CP titanium are used for pump impellers (see Fig. 3.30), and other complex shaped components where the net shape capability make casting an attractive option. Casting of titanium alloys has been discussed in Chap. 3, Sect. 3.5.1. Very few CP titanium forgings are made and used. This is, in part, because the titanium alloys belonging to this class are readily welded. Consequently, welded fabrications are typically more cost-competitive than forgings and the properties of the welds are nearly equivalent to the base metal which eliminates property variation of welds as a design concern. This section, therefore, will focus on processing of sheet and plate.

There are several exceptions to this generality about limited CP titanium forging use. One is the use of rolled cylinders such as shown in Fig. 3.28. Another are the ends of large pressure vessels such as the one shown in Fig. 3.49 where the plates are hot formed in what is essentially a forging operation. There also are forgings still being made from Ti-5Al-2.5Sn, but as will mentioned later, there are few new forging applications of this alloy because it is difficult to process and therefore expensive.

The processing of CP titanium starts with either a round VAR ingot (Fig. 3.7) or a rectangular cross section slab cast from a cold hearth furnace (Fig. 3.11). This ingot is rolled in a blooming mill or forged to create a large slab that has a smaller, uniform grain size. The trend is toward the use of larger ingots or cast slabs, which eliminates or minimizes the possibility for cross rolling. This slab is hot rolled to an intermediate product form called hot band. Hot band is coiled and, depending on the intended finish product gage, is either pickled and annealed or just annealed in preparation for final rolling to the final product. The processing flexibility is determined to a significant degree by the material grade. Grade 1, 2, and 12 are quite amenable to continuous cold rolling to gage. Grade 4 and some heats of grade 3 require elevated temperature pack rolling much as is employed to make Ti-6Al-4V sheet. This processing sequence is schematically captured in Fig. 4.1.

Much of the CP titanium sheet used today is produced as coil. During coil production, the material only receives a modest amount of cross rolling to achieve the desired product width. After this step, the gage is reduced to the desired dimension by unidirectional rolling and winding into coils such as those shown in Fig. 3.20. A typical processing sequence for α alloys consists of a series of three operations or steps (including homogenization) as is shown in Fig. 4.2. Table 4.2 summarizes the intent of steps II and III and the important parameters associated with these steps. In general, α titanium alloys require less homogenization time because the low solute concentration minimizes any concerns about freezing segregation in the ingot. Because of this reduced tendency toward solute segregation during freezing, large, homogeneous slabs can be directly cast from a cold hearth furnace (see, for example, Fig. 3.11), and homogenization of these slabs may be omitted completely. Direct rolling of these CP titanium slabs can be done with minimal conditioning and low material yield losses because the material has good ductility and low notch sensitivity. CP titanium made by continuously rolling as-cast slabs is the most cost competitive titanium mill product currently available.

Table 4.2. Important processing parameters and resulting microstructural features of CP titanium

Processing Step (see Fig. 4.2)	Important Parameters	Microstructural Features
II	Deformation Degree	– Texture Intensity – α Grain Size
III	Annealing Temperature	α Grain Size

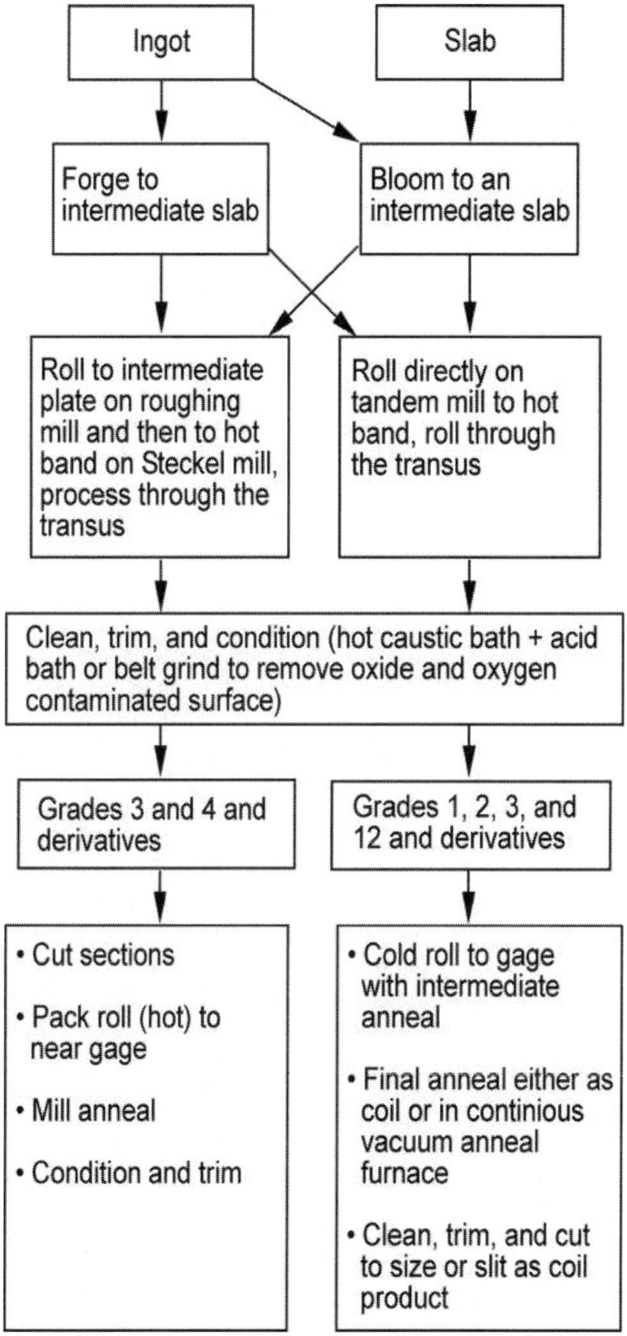

Fig. 4.1. Schematic processing sequence for making CP titanium sheet and strip (courtesy J. A. Hall)

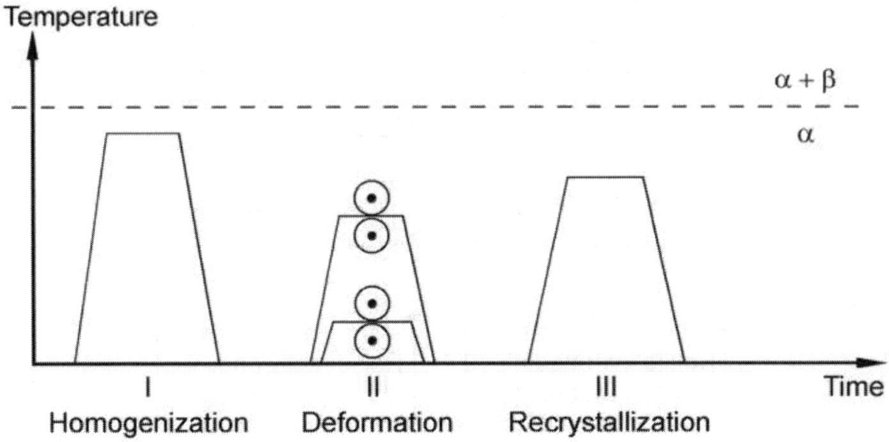

Fig. 4.2. Processing route for CP titanium (schematically)

Unlike most of the higher strength alloys, many α alloys also can be extensively cold rolled, but α alloys are seldom used in the unrecrystallized condition. This is because cold worked titanium has little strength advantage since it has very low work hardening and because the cold worked material has reduced ductility which constrains subsequent component manufacturing options. Therefore, the term processing, when applied to CP titanium, usually includes recrystallization annealing after the final rolling operation. Annealing can be done in a continuous annealing oven that is in the strip line between the final rolling stand and the coiler or by batch annealing of finished coils. The microstructure of α alloys after processing consists of recrystallized α grains with dispersed β phase. An example of this dispersed β phase is shown in Fig. 4.3. This β phase is present because there always is a small amount of Fe present in all grades of CP titanium. Fe has a very low solubility in the α phase and it therefore is rejected to form the β phase, either during solidification or subsequent cooling. This β phase remains stable down to room temperature. Table 2.6 shows the maximum Fe content in each grade of CP titanium. The typical Fe content depends on the grade of CP titanium and increases with increasing grade number. The Fe is present as a deliberate alloying addition because the β phase pins the α grain boundaries and helps control the grain size during recrystallization. Smaller grain sizes can be used to increase the yield strength as described by the Hall-Petch relationship when higher strength is desired [125, 126]. Intermediate grain sizes improve the formability by promoting more extensive twinning. The crystallography of deformation twinning of the α phase has been described in Chap. 2, Sect. 2.4.2, but an example of twins in CP titanium is shown here in Fig. 4.4. The final grain size in CP titanium (grades 1-4) principally is a function of both the amount of work put into the material during the final rolling operation and the recrystallization temperature (Table 4.2). The dependence of grain size on recrystallization annealing temperature after 3.5 h is plotted in Fig. 4.5 [127]. Grain growth can occur during continued annealing once the recrystallization is complete, but the β phase reduces the rate of grain growth and enhances the ability to process material with reproducible grain sizes. The effect of Fe content on grain size is illustrated in Fig. 4.6. In the as-received condition, the two production lots with different Fe levels (0.15% and 0.03%) had similar grain sizes. After an additional annealing treatment of 1h at 700°C the material with 0.15% Fe did not show any noticeable grain growth (Fig. 4.6a) whereas the material with the low Fe content of 0.03% exhibited secondary grain growth (Fig. 4.6b).

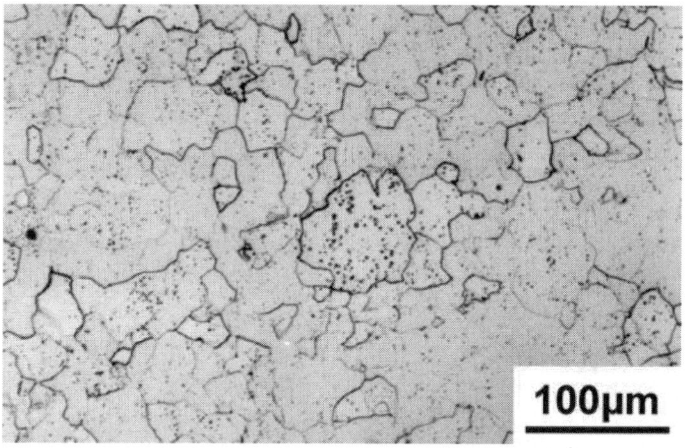

Fig. 4.3. Dispersion of Fe stabilized β phase in CP titanium grade 3 (0.15% Fe), LM

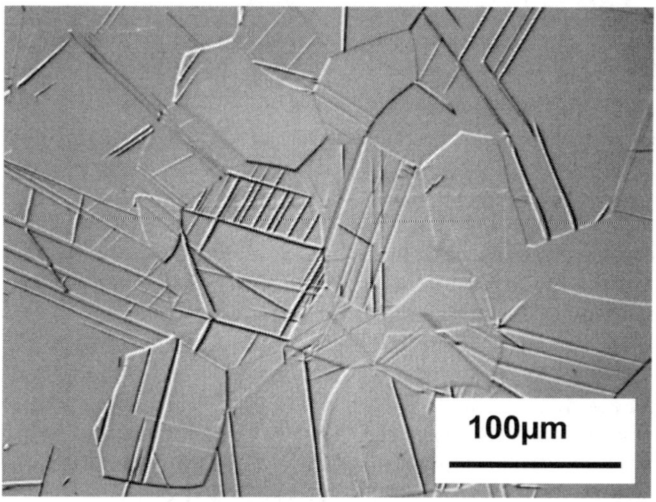

Fig. 4.4. Deformation twins in CP titanium, LM (courtesy J. A. Hall)

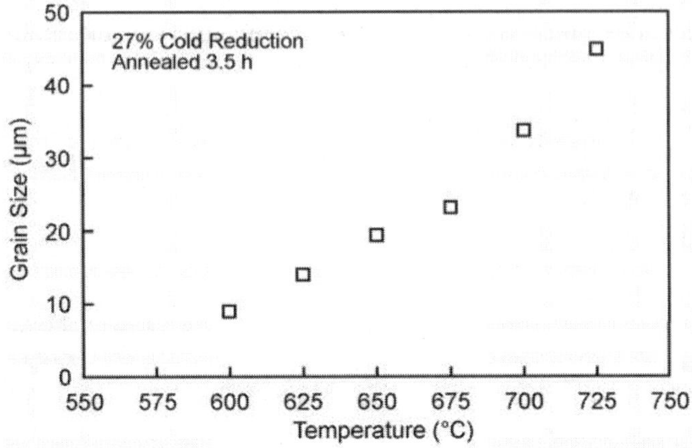

Fig. 4.5. Effect of annealing temperature on grain size of CP titanium (grade 3) [127]

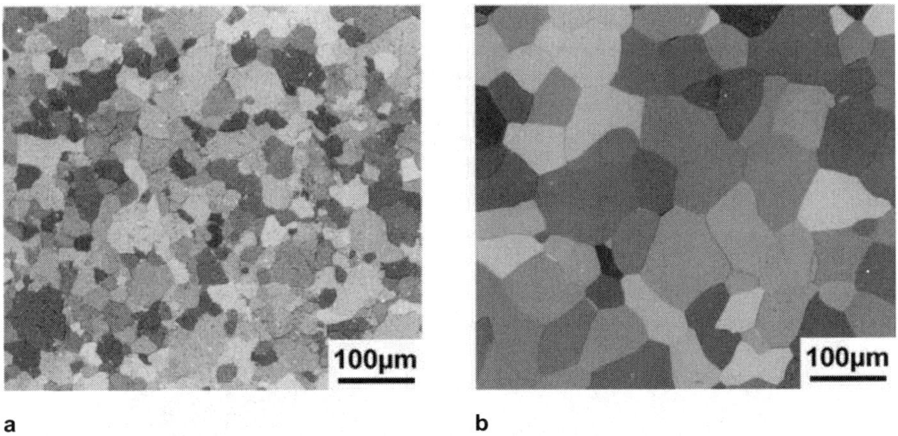

Fig. 4.6. Effect of Fe content on the grain structure of CP titanium, LM, polarized light: (a) 0.15% Fe (b) 0.03% Fe

The Fe stabilized β phase also has a higher solubility for hydrogen and its presence improves the hydrogen tolerance of this class of titanium alloys. This is important since many of the applications for this class of alloys is in corrosive environments where there is a relatively high possibility for hydrogen pickup during service. The presence of hydrogen can have major effects on properties, as will be discussed in Sect. 4.2.

The α alloy Ti-5Al-2.5Sn was originally developed in the early 1950's as an intermediate strength, higher temperature alloy. Given this historical fact, it should not be surprising that Ti-5Al-2.5Sn is considered a difficult alloy to produce. Many of the newer alloys have been formulated with a more balanced view of producibility and performance. Ti-5Al-2.5Sn is produced in two grades as shown in Table 4.1. The difference between the grades is the composition. The ELI grade (ELI stands for Extra Low Interstitials) has lower oxygen, lower Fe, and the maximum Al concentration is lower. All of these differences are directed to making the material more ductile at lower temperatures because a major use for this alloy is for cryogenic tanks. During hot working, Ti-5Al-2.5Sn is mostly α phase because it contains only 0.25-0.5% Fe and no other β stabilizing elements. It also has a high β transus (1040°C for ELI grade), and the (α+β) phase field exists only over a very narrow temperature range. Strain localization in the α phase that leads to shear cracking of varying degrees of severity is commonly encountered in Ti-5Al-2.5Sn during hot working to produce billet, bar, sheet, or plate. Severe shear cracking can cause the work piece to be scrapped. Less severe cracking can be manifested as shallow surface cracks which necessitates extensive conditioning before final processing. Both of these events add to the final product cost of this alloy. Because of the difficulty encountered in producing this alloy, the production sequence is comparatively uneconomical. For example, the hot working operation is typically performed at temperatures very high in the α phase field and with smaller strains per pass than would be used for processing the equivalent product in Ti-6Al-4V. At these high temperatures and at low strains recovery is rapid. Consequently, it is challenging to produce Ti-5Al-2.5Sn with grain sizes as small as commonly observed in Ti-6Al-4V. Consistent with this point, the yield stress of Ti-5Al-2.5Sn can vary by as much as 50 MPa between plate and sheet, reflecting the difference in grain size between these two product forms. Hot rolling this alloy near 1000°C, which is commonly done to minimize cracking, introduces a high degree of preferred orientation (texture) with the basal plane poles aligned in the transverse direction. This type of texture is called a transverse (T) texture, discussed later in Sect. 5.1.2 and shown in Fig. 5.8. Consequently, in Ti-5Al-2.5Sn sheet it is common to find a 70-100 MPa increase in the transverse yield stress compared to the longitudinal direction.

4.1.2 Processing into Components

Alpha alloys, especially CP titanium, are used for a variety of applications where corrosion resistance is important and where density corrected strength is competitive with other corrosion resistant alloys such as austenitic stainless steel or Ni-Cr-Mo alloys such as C-276. As mentioned earlier, formability and weldability of CP titanium are major considerations in selecting it for a specific application because these characteristics affect final product cost. Welded CP titanium tubing has become an important intermediate product that now is manufactured in large quantities using automatic tube machines. These machines use coiled CP titanium strip as input stock. This stock is slit to width, bent into a round cross section by passing it through a series of rolls, then the mating edges are welded together and the resulting tube is cut to length, all in a continuous process. The principal input material requirements for making high quality welded tube are uniform response to the bending applied to create the tube shape and good weldability. The bending strains are relatively low and for CP titanium, grades 1 and 2, ductility is not an issue. Uniformity of local yielding in the circumferential direction also is required to maintain roundness of the tube and this requirement can be readily met if the starting material has uniform grain size and texture. Many welded tubes also are put into heat exchangers by roller expansion. Roller expansion uses a tapered rotary mandrel that plastically deforms or expands the tube diameter. During expansion the tube diameter increases until it contacts the perimeter of the hole in the sheet that holds the tubes in place. This is called the tube sheet [121] and can be seen in Fig. 4.7. Roller expansion creates a seal between the heat exchanger tube and the tube sheet. Uniform response to roller expansion requires a uniform gage (including the weld) and a constant circumferential yield stress to ensure the same degree of expansion (plastic flow in the radial direction) at all points along the tube circumference. This includes the yield stress of the weld and the heat affected zone. There are several other issues associated with successful roller expansion. One is adequate ductility to support the plastic strain imparted during the operation. Another is the force required to complete the operation, which is determined by the flow stress. Since roller expansion is a manual operation the latter point becomes a significant limitation as the size of the tubes and the tube wall thickness increases. At some point the required forces exceed the physical strength capability of many roller expander operators. An example of a shell type heat exchanger containing welded CP titanium tubes is shown in Fig. 4.7.

Other CP titanium components are deep drawn. Deep drawing is more demanding than bending because it involves biaxial straining. A reproducible response to biaxial stretching requires material that has reproducible values of the plastic strain ratio, r. The parameter r is the ratio of plastic strain in the width and thickness directions of the sheet during plane strain deformation. High r values correspond to less thinning during biaxial stretching which allows deeper pockets to be drawn in a single operation without tearing. Higher r values correspond to higher effective yield strength in the thickness direction compared to the width or other in-plane directions. This elevation in effective yield stress minimizes through thickness strain and the attendant thinning during biaxial deformation. The highest r values are obtained in a material with a basal (B) texture (see Fig. 5.8) because biaxial stretching imposes a c-axis strain. Unfortunately, this texture is very difficult or inconvenient to produce, because it requires cross rolling. Increasing oxygen also causes the resolved shear stress values for the various α titanium slip modes (see Chap. 2) to diverge, with $\vec{c}+\vec{a}$ slip having the highest values. Therefore, the r values of textured CP titanium, grades 3 and 4, can be increased compared to those for grades 1 and 2, but the overall tensile ductility of the higher oxygen grades also is lower. One commonly used measure of drawability is the ratio of the starting blank size to the punch diameter used to deep draw a circular cup. This maximum value is called the Limiting Draw Ratio (LDR) and is material dependent [128]. CP titanium is considered to be reasonably formable, but the yield strength requires considerable press forces. In all metals, drawing pockets that are deeper than their diameter requires multiple drawing operations.

Sheet forming of titanium alloys has been discussed in Chap. 3, Sect. 3.5.4. As part of this discussion, the phenomenon of "spring back" and its consequences has been described. Spring back in CP titanium is a greater problem than for other, competing materials. One method of reducing spring back is to use warm forming. The flow stress of CP titanium is reduced at warm forming temperatures. Consequently, there is an attendant reduction of elastic strain. For any material, the elastic strain at the yield point is the quotient of the yield stress and modulus of elasticity. This elastic strain contributes to spring back during forming and causes dimensionally accurate net section shapes more difficult to make. Relative to austenitic stainless steel, α titanium alloys have a high ratio of yield stress to elastic modulus. The resultant elastic strains in CP titanium can range from about 1.5 to 4 times larger than in stainless steel. Warm forming at 200-300°C reduces the flow stress [129] and helps minimize spring back. At temperatures of a few hundred degrees Celsius, oxidation of the titanium is of little concern, making warm forming an attractive process.

Fig. 4.7. Shell type heat exchanger containing welded CP titanium tubes (courtesy J. A. Hall)

4.2 Microstructure, Composition, and Properties

The properties of α titanium alloys depend on composition (i.e. oxygen content for CP titanium) and on processing history, because this history controls the grain size and the preferred orientation (texture). By comparison to α+β and β alloys, the properties of α titanium alloys have a more direct composition dependence. In general, the sheet products show a continuous variation of the yield strength and the elastic modulus between the longitudinal and transverse directions [26]. This variation is due to the crystallographic texture always present in α titanium alloys. As shown in Chap. 2, the elastic constants of α titanium are higher when measured parallel to the c-axis compared to a perpendicular direction. This difference in elastic constants increases in Ti-Al alloys, presumably because Al is a substitutional element which contracts the α phase lattice. In contrast, oxygen is an interstitial element in the α phase and has essentially no effect on modulus of elasticity, as seen in Table 4.3. Thus, the composition dependence of texture effects on modulus is smaller in CP titanium than for alloys such as Ti-5Al-2.5Sn.

The options available for strengthening single phase titanium alloys such as the α phase alloys are relatively few and there are practical limits to the extent to which they can be used. The basic mechanisms for strengthening α phase alloys are solid solution strengthening, both by interstitials (oxygen, carbon, and nitrogen) and substitutional elements (aluminum, tin, and zirconium), grain size strengthening, texture strengthening, and precipitation hardening by α_2 phase formation. The strengthening of titanium alloys has been generally discussed in Chap. 2, Sect. 2.8. The strengthening mechanisms relevant to α titanium alloys are summarized here in Table 4.4. Among these possibilities, solid solution strengthening by interstitials (oxygen in particular) leads to strain localization as will be discussed later. Solid solution additions and very small grains both restrict the operation of deformation twinning and reduce the formability of the material. Texture strengthening is possible in principle, but is highly directional. Texture that provides strengthening in the plane of the sheet also reduces the possibility for the improvement in formability which can be achieved from a basal texture type, as also discussed earlier in this chapter. Precipitation hardening by α_2 phase also results in strain localization which can severely reduce the tensile ductility of the material.

Table 4.3. Typical mechanical properties of CP titanium

Material	E (GPa)	$\sigma_{0.2}$ (MPa)	UTS (MPa)	Elong. (%)	σ_{10^7} (MPa) (R = –1)	$\sigma_{10^7}/\sigma_{0.2}$
Grade 1	105	170	240	24	-	-
Grade 2	105	275	345	20	-	-
Grade 3	105	380	445	18	280	0.73
Grade 4	105	480	550	15	350	0.73

Table 4.4. Available strengthening mechanisms in α titanium alloys

Strengthening Mechanism	Dependence	Examples/Limitations
Grain Size	$d^{-1/2}$	Fine grains limit twinning
Interstitial Solid Solution	$c^{1/2}$	Strain localization >2500 ppm oxygen
Substitutional Solid Solution*	c	Strain localization >5% Al eq.
Texture	c-axis orientation	Max. strength when loaded along c-axis
Precipitation	$r^{1/2}$, $f^{1/2}$	Occurs >5.5% Al eq.

*[130, 131]

CP titanium, particularly grades 1 and 2, typically is produced as coil. During coil production, the rolling is largely unidirectional and the type and intensity of texture present is attributable to this working practice. The typical texture present in CP titanium has the maximum concentration of basal poles lying along the direction that connects the sheet normal and the sheet width (transverse direction). The point of maximum basal pole concentration is inclined about 30 degrees away from the sheet normal toward the transverse direction. A typical basal pole figure for CP titanium is shown in Fig. 4.8 [132]. The texture in CP titanium is different from the typical texture in flat rolled α+β alloys, such as Ti-6Al-4V. In Ti-6Al-4V, a strong basal/transverse (B/T) texture exists (see Chap. 5, Fig. 5.8). During biaxial stretching the texture in CP titanium causes reasonably high resolved stresses along the basal poles. Loading in biaxial tension is equivalent to through-thickness compression and is accommodated by through-thickness thinning of the material during forming. Under this type of loading, the operation of twin modes is beneficial to formability of CP titanium because twins allow deformation when the stress axis is parallel to the c-axis. This fact, coupled with the lower yield stress levels in this class of alloys, qualitatively explains why these alloys are preferred for sheet forming application in comparison to higher strength alloys, such as Ti-6Al-4V or Ti-5Al-2.5Sn, for which twinning is uncommon.

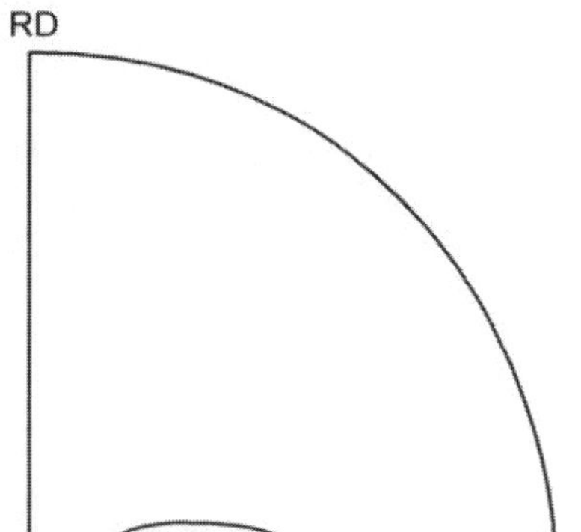

Fig. 4.8. Crystallographic texture (0002 pole figure) of unidirectionally rolled CP titanium sheet [132]

The maximum oxygen concentrations (in addition to Fe) contained in the various grades of CP titanium are listed in Tables 2.6 and 4.1. Oxygen is a powerful solid solution strengthener in the α phase and the variation in tensile properties of grades 1 through 4 of CP titanium reflects this point (Table 4.3). In this class of alloys the oxygen and Fe contents are carefully controlled during the alloy formulation since these factors have a major effect on the strength of the final product. At sufficiently high oxygen concentrations (≥ 0.25%), the slip character changes from wavy to planar (see Sect. 2.8.1) and the propensity for twinning declines rapidly [35, 52]. This slip mode transition is illustrated in Fig. 4.9. This change in deformation behavior is accompanied by a reduction in the formability of the higher strength grades of CP titanium. This is in part due to the occurrence of strain localization associated with planar slip and in part due to the reduction in the amount of twinning. The effect of oxygen on deformation mode persists only up to modest temperatures (about 300°C). This is reflected in the convergence of the allowable tensile stress versus temperature curves for several grades of CP titanium as shown in Fig. 4.10 [133]. Thus, warm forming of grades 3 and 4 is relatively common if the required forming strains are large.

Grain size strengthening is an important means of altering the strength in CP titanium. The effect of grain size on the flow stress at various strains of CP titanium grade 3 is shown in Fig. 4.11 [134]. The effects of grain size and oxygen on strength are essentially additive and, together, can lead to yield stress values as high as 480 MPa in CP titanium grade 4. As in many other metallic materials, the strength and ductility of CP titanium are inversely related for a constant grain size. This is illustrated by the data in Table 4.3, which also shows that the high cycle fatigue strength of CP titanium at 10^7 cycles is about 0.7 of the yield strength. Given the wavy to planar slip transition that occurs between grades 2 and 3, one might expect grain size effects to have a pronounced effect on fatigue crack initiation, but there is very little systematic fatigue data for these materials. The general lack of fatigue and fracture data for CP titanium reflects the earlier suggestion that selection of this class of materials is typically based on superior corrosion resistance and fabricability, rather than on mechanical properties, per se. Moreover, titanium applications that require good strength and fatigue behavior will lead in most cases to the selection of higher strength titanium alloys because these properties are significantly better and the cost is only incrementally higher, but the increase depends on product form. For example, in sheet applications, the cost of CP titanium sheet is enough lower than that of Ti-6Al-4V sheet that CP titanium will usually be preferred.

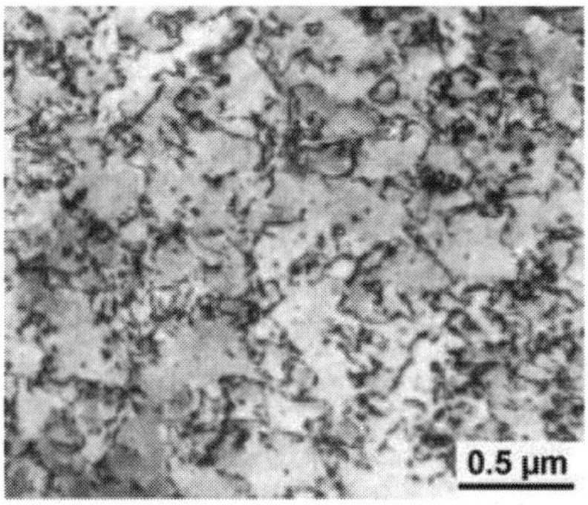

a

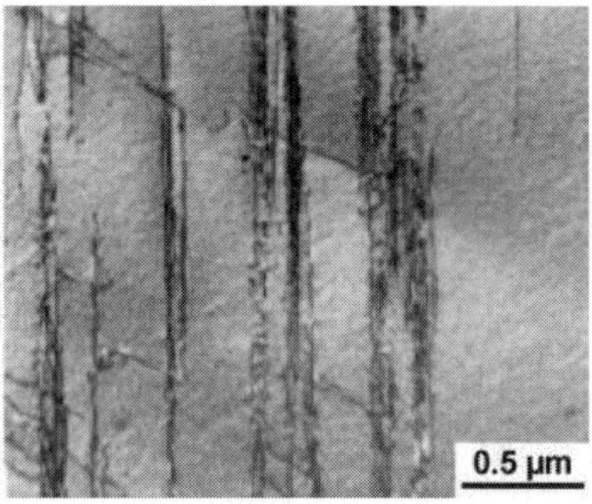

b

Fig. 4.9. Effect of oxygen on slip mode in CP titanium, TEM: (a) 0.15% oxygen (b) 0.50% oxygen [52]

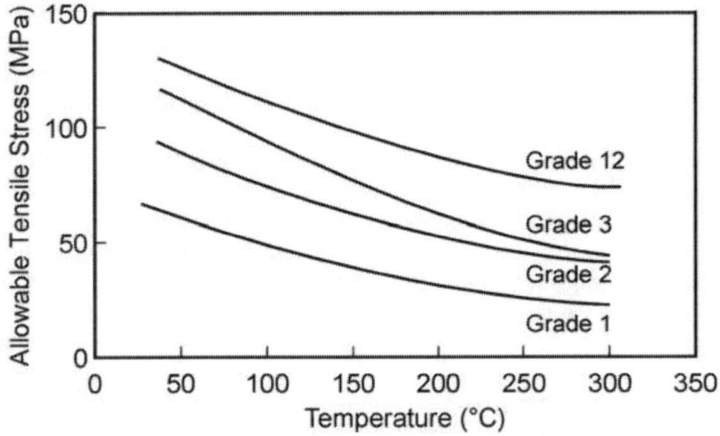

Fig. 4.10. Allowable tensile stress as a function of temperature for several CP titanium grades and Ti-0.3Mo-0.8Ni (grade 12) [133]

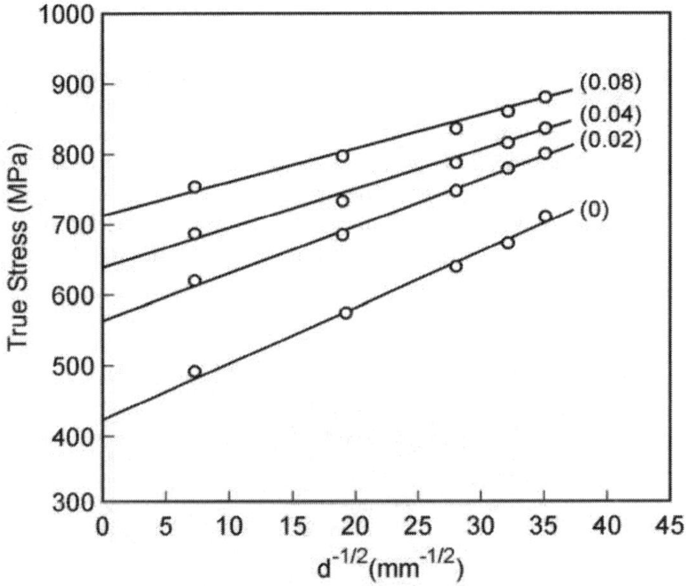

Fig. 4.11. Flow stress at room temperature for four plastic strain levels versus grain size of CP titanium [134]

The β phase, when present as a ductile inclusion, can act as a crack arrest site under most loading conditions. Processing, such as warm forming, or extended service at temperatures of 200-300°C can cause the Fe stabilized β to decompose into a brittle β+ω mixture [135]. When this occurs, the fracture resistance of CP titanium is reduced in fatigue and monotonic loading. Similarly, β decomposition can occur in grade 12 as discussed later. This point is important because there is a tendency to ignore the possibility of time dependent property changes in materials that are not heat treatable.

The fracture modes of CP titanium are somewhat different compared to the higher strength α+β or β alloys. The most apparent differences are the presence of large, elongated fracture features due to extensive plastic flow during crack extension and the much larger void sizes in CP titanium. An example of this ductile stretching and tearing fracture in CP titanium grade 3 is shown in Fig. 4.12 [136]. In part, this is because the unalloyed α phase is not very strong and quite ductile when it contains less than 2500 ppm oxygen, or when the α phase contains less than 3% Al and less than 2000 ppm oxygen. The other difference between α titanium alloys and the higher strength classes of titanium alloys is the relatively low number of boundaries that are barriers to slip transmission. In CP titanium, these boundaries are mostly α/α grain boundaries, whereas they are α/β boundaries in high strength alloys. In fact, the high strength alloys derive much of their strength from boundary strengthening, so these alloys have a high density of α/β boundaries, almost by definition. These α/β boundaries in the higher strength alloys are sites of strain accumulation and strain incompatibility and therefore often are sites for void nucleation during ductile fracture. In contrast, voids in CP titanium are mainly nucleated at grain boundaries. These voids grow by plastic stretching and ultimately coalesce by ductile tearing of the ligaments between the voids. Consequently, CP titanium normally exhibits very different fracture topography, compared to the high strength alloys. Grades 3 and 4 usually contain Fe stabilized β phase. Void nucleation also occurs at the associated α/β interfaces in these alloys. Among the alloys listed in Tables 2.6 and 4.1, the alloy Ti-5Al-2.5Sn is an exception to the earlier statement regarding Al and oxygen content. This alloy will be discussed separately at the end of Sect. 4.3.

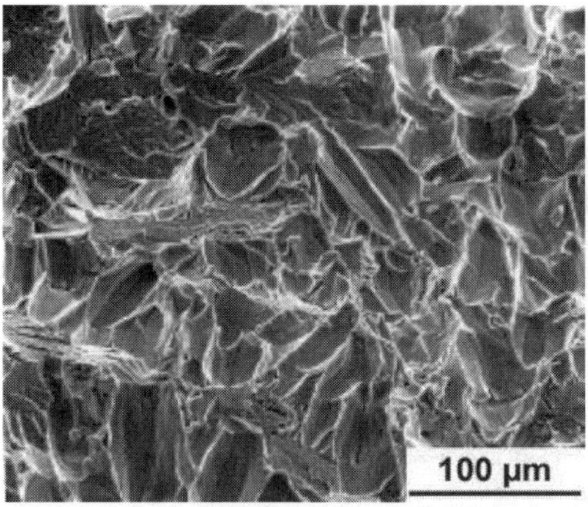

Fig. 4.12. Ductile fracture mode at room temperature, CP titanium grade 3, SEM [136]

At low temperatures and in combination with high loading rates the α phase becomes very strong, especially in grades 3 and 4 that have high oxygen content. Under such circumstances plastic flow is limited and the α phase can fracture by cleavage. The impact resistance of CP titanium grades 1-3 is compared to two other alloys in Fig. 4.13 [137]. Because of the potential for brittle cleavage fracture in CP titanium the use of grades 3 and 4 should be restricted if low service temperatures in combination with high loading rates are anticipated. The cleavage fracture mode is shown in Fig. 4.14. Cleavage fracture occurs along a crystallographic plane, which in α titanium is the basal plane. Cleavage fracture typically occurs when the local normal stress across the cleavage plane exceeds a critical value. This fracture criterion is in contrast to reaching a critical local plastic strain that causes microvoid nucleation to occur during ductile fracture. Therefore, in the hexagonal α phase, the role of texture directly affects the propensity for cleavage since the relative orientations of the basal plane and the loading axis affect the normal stress component across the cleavage plane. The propensity for cleavage fracture has been reported to decrease with decreasing grain size. This appears to be related to the increasing difficulty of nucleating cleavage cracks in small grains [138]. Thus, components intended for service at low temperatures are more suitably made from fine grained, lower oxygen grades of CP titanium. In these cases, the strengthening due the fine grain size partially offsets the strength reduction of the low oxygen material.

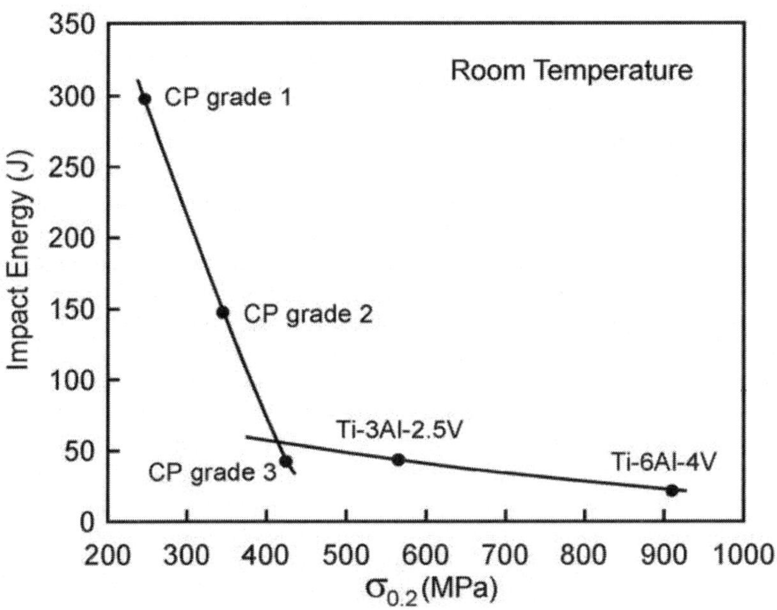

Fig. 4.13. Charpy impact toughness versus yield stress for severeal CP titanium grades and the alloys Ti-3Al-2.5V and Ti-6Al-4V [137]

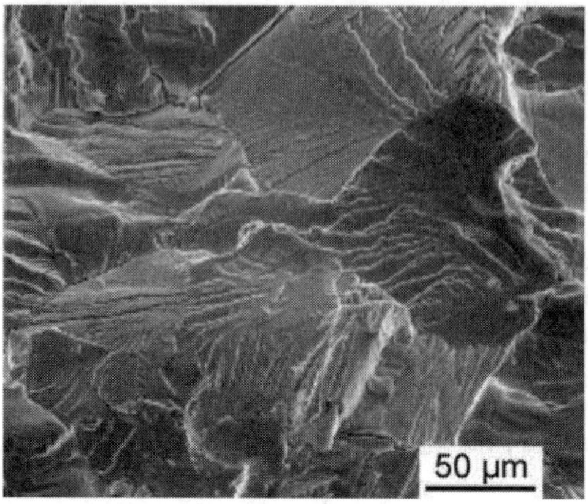

Fig. 4.14. Cleavage fracture at low temperatures and high loading rates, CP titanium grade 4, SEM

Alpha titanium alloys are among the most corrosion resistant structural materials available for use in a variety of aggressive environments. Examples of such environments are described in Table 4.5. As has been generally discussed in Chap. 2, titanium is very corrosion resistant in oxidizing environments where the protective oxide film remains intact and creates a passive surface. In reducing environments, such as sulfuric, phosphoric, and hydrochloric acids, this film breaks down unless inhibitors are present. The corrosion resistance of CP titanium in reducing environments is improved by the addition of small amounts (up to 0.2%) of platinum group metals (PGM). Palladium is the most common of these additions. The addition of 0.2% Pd renders α titanium very corrosion resistant in reducing environments without inhibitors. This is important for applications where inhibitor use is impractical or not possible. However, these grades are almost twice as expensive as the corresponding CP titanium grades without Pd additions. This cost difference is simply attributable to the high cost of Pd and therefore has limited the use of these alloys. Nevertheless, CP titanium grades 7 (0.25% oxygen maximum) and 11 (0.18% oxygen maximum) containing 0.2% Pd are used in numerous applications. These grades are comparable in cost to Ni-Cr-Mo alloys such as C-276, mentioned earlier. Recently, it has been shown that smaller Pd concentrations have nearly the same beneficial effect in stabilizing the surface oxide in reducing environments. Consequently, there also are now grades 16 and 17, which contain 0.05% Pd and which cost 30% less than grades 7 and 11. These low Pd grades have been successfully applied in all but the most aggressive reducing environments. Even more recently, it has been shown that another, less expensive element (Ru) can be added to CP titanium with nearly comparable improvement in the corrosion resistance in reducing environments [64]. In this case, it takes about 0.1% Ru to achieve an effect comparable to that of 0.05% Pd. Even so, because of the lower cost of Ru, the CP titanium containing 0.1% Ru costs 20% less than grades 16 and 17 at comparable oxygen levels. Consequently, there are now two new Ru containing grades of CP titanium available, grades 26 and 27. The strengths of CP titanium grades 7 (0.2% Pd) and 16 (0.05% Pd) are the same as the corresponding CP titanium grades that contain no Pd additions. However, the improved corrosion enhancement has been attributed to the presence of small Ti-PGM intermetallic precipitates, either Ti_2Pd or TiRu. These precipitates affect the local hydrogen (cathodic) overvoltage and therefore extend the range of conditions where the surface remains protective. These precipitates are small and occur in a low volume fraction. Consequently, the absence of any effect of these alloying additions on strength is anticipated for two reasons. First, the small volume fraction and large spacing of the precipitates would not cause dispersion strengthening. Second, the solid solubility of Pd and Ru in the α phase is low and any minor solid solution strengthening is overshadowed by the strong effect of oxygen. Therefore, the balance between cost sensitivity and the required corrosion resistance determines the alloy or grade selection.

Table 4.5. Industrial uses for titanium because of good corrosion resistance

Industry	Equipment	Environment
Power generation	Condensers, heat exchangers, flue gas scrubbers	Aqueous solutions of various purity, SO_2 containing gases
Water plants	Desalinization heat exchangers	Sea water
Petrochemical industry	Heat exchangers, well heads, pipe and down hole hardware	H_2S containing brines
Pulp and paper	Diffusion washers in bleaching section of process	Chlorides containing liquids
Chemical industry	Dimensionally stable electrodes	Cl_2 and Cl_2 compounds
Metal production	Cathodes for electrowinning Cu, Au, and Zn	Various aggressive aqueous solutions
Mineral dressing	Pressure vessels at high T and P	Various aggressive aqueous solutions
Biomedical devices	Orthopedic implants, surgical implants, surgical implements	Human body and autoclave sterilizers
Spacecraft	Cryogenic tanks	N_2O_4, liquid O_2, liquid H_2

Many applications for the various grades of CP titanium are for process equipment in the chemical and petrochemical industries. Much of this equipment operates at elevated temperatures (e.g. in the range 100-150°C) and this has a direct impact on the corrosion resistance of the materials of construction and on the mechanical strength of the materials, especially CP titanium. Therefore, the effects of elevated temperature service must be considered during the materials selection process, for example, for tube bundles in heat exchangers as shown in Fig. 4.7.

The severity of a given environment increases with increasing service temperature in part because the propensity for titanium to absorb hydrogen increases with temperature. The other factor is the increasing tendency for crevice corrosion to occur with increasing temperature. Crevice corrosion results from the formation of concentration cells at joints and other places where the corroding medium is confined by the local geometry. In situations such as this, the oxygen is locally depleted and the local corrosion environment becomes reducing, causing the surface oxide to break down. Alloying with Pd (e.g. grade 7) or even with Ni+Mo (e.g. grade 12) improves the resistance to crevice corrosion, especially at elevated temperatures as was shown earlier in Fig. 2.31 [61]. CP titanium also has good intrinsic resistance to crevice corrosion, especially at ambient temperatures, but at elevated temperatures, the grades containing PGM additions are superior for reasons already described.

The role of hydrogen in degrading the mechanical properties of CP titanium and of many high strength titanium alloys is well known. The use of titanium alloys under conditions where hydrogen pickup is anticipated should be avoided. There are circumstances where the potential for this is less obvious. Included would be galvanic couples where the area ratio of the anode to the cathode is large. Other examples are the cathodic regions in crevice corrosion and situations where elevated temperatures exacerbate hydrogen pickup. There are several views of the detailed mechanism responsible for the loss of ductility and reduced fracture resistance. It is clear that alloys containing enough hydrogen to cause hydride precipitation are brittle and it has been accepted for some time that the origin of this brittleness is related to the presence of the brittle TiH_2 phase. Consequently, a discussion of the details of TiH_2 formation is included below because of the importance of this reaction. However, there also is clear evidence for the loss of ductility at lower hydrogen concentrations, i.e. around the specification limit (150 ppm), where the presence of hydrides is not well documented. In both cases, the fracture occurs by a brittle, transgranular mode with the fracture plane lying on the basal plane of the α phase. Thus, there exists the possibility of two mechanisms for ductility loss at lower hydrogen concentrations. One possible mechanism is that the increased hydrogen concentration in slip bands on the basal planes lowers the fracture stress directly [139]. The other is that the increased hydrogen concentration in the slip bands forms a thin, difficult to detect or even transient hydride layer. It is not within the scope of this book to review all the information associated with hydrogen effects in titanium and its alloys. However, it is important to point out that loss of ductility in the absence of hydrides can still be related to hydrogen.

A principle mode of failure of CP titanium in aggressive environments, that is in regimes where the pH level and the temperature fall in the range of hydrogen pickup shown in Fig. 2.31, is hydrogen embrittlement. In such circumstances hydrogen concentrations high enough to cause hydride formation are common. In α titanium alloys, the cause of embrittlement is formation of titanium hydride (TiH_2), which has the CaF_2 structure and which is a brittle phase. Cracks can form in the brittle TiH_2 if the fracture stress of the hydride is exceeded. At high volume fractions of TiH_2 these cracks can grow into the α phase matrix, leading to fracture at low overall strains. The details of TiH_2 formation are thus important to summarize here because of the relationship between hydride formation and the mechanical behavior of CP titanium. The formation of TiH_2 has an attendant volume change (+18%) associated with it and this volume change must be accommodated by plastic flow in the matrix [140]. This accommodation effect has several consequences with regard to the kinetics of hydride formation. The first of these is a strong dependence on the flow stress of the matrix because this becomes the restraining force for the plastic flow required to accommodate formation of the precipitate. This, in turn, affects the apparent solubility of hydrogen in any particular alloy. Alloys in which the α phase is solid solution strengthened require greater supersaturations of hydrogen to create an adequate driving force for the formation of dislocation loops in the matrix. These loops are necessary to accommodate the increase in volume. An example of these loops around a hydride precipitate is shown in Fig. 4.15. Once these loops have begun to form, the hydrides grow rapidly. Externally applied stresses in excess of the yield stress create dislocations that also can provide the necessary accommodation. Consequently, the possibility of large hydrogen supersaturation exists in alloys in which the α phase is strengthened, either by oxygen or by solid solution elements such as Al or Sn. One result of this can be the rapid formation of a large volume fraction of hydrides if the local yield stress is exceeded. Another consequence of the volume change associated with hydride formation is that hydrogen is selectively concentrated in any region where a hydrostatic tensile stress field is present [141]. This becomes an important consideration from a design standpoint, especially if the article being designed will utilize higher strength alloys. Even when care is exercised to minimize the presence of stress concentrations at notches during design, it is essential that other sources of tensile stress concentrations are not introduced during manufacturing. Examples could include improperly machined radii or welds with incomplete fusion. From a materials selection standpoint, the use of grades 1 or 2 at room temperature is acceptable. For elevated temperature application, where crevice corrosion is of greater concern (Fig. 2.31), the use of grades 7, 11, or 12 becomes more important notwithstanding the increase in cost. These alloys generally have superior crevice corrosion resistance that is more often the motivation for their selection.

Fig. 4.15. Dislocation loops around hydride precipitates in α phase of titanium, TEM

Metals and alloys that draw their corrosion resistance from the presence of a passive oxide layer are often susceptible to crevice corrosion. This is because the region within the crevice becomes the site of a concentration cell and in the presence of chlorine ions the passive film breaks down, leaving the metal unprotected. The use of α titanium alloys is common in situations where there is a strong possibility of crevice corrosion because they generally exhibit better resistance to this type of attack than stainless steels and Ni alloys. It is relatively easy to test the resistance of a material to crevice corrosion and the results are straightforward to interpret. A common test for crevice corrosion susceptibility consists of making a layered stack of dissimilar materials with an insulating material separating them. The stack is compressed by a bolt that is electrically isolated from the specimen and is torqued to a predetermined value to achieve a reproducible crevice geometry. An example of such a stack is shown in Fig. 4.16. These stacks are immersed in the medium of interest with regard to crevice corrosion susceptibility. A commonly used medium that discriminates between the susceptibility of different candidate metals and alloys is saturated brine containing acid. The results of a typical test are shown in Fig. 4.17 where it can be seen that CP titanium, samples A and B, is susceptible to crevice corrosion whereas Ti-0.2Pd, samples C and D, is not.

One alloy of the α titanium class that has not been discussed until now is the Ti-0.3Mo-0.8Ni alloy, known as grade 12. This alloy has better corrosion resistance than CP titanium grade 2, but is not as good as grade 7 (Ti-0.2Pd), as was shown in Fig. 2.31. This figure shows that grade 12 has much better crevice corrosion resistance than grade 2. This makes it an economically preferred choice for many applications where crevice corrosion is a concern. Grade 12 is attractive for use where the improved corrosion resistance is adequate because it is considerably less expensive than grade 7. Grade 12 has comparable crevice corrosion resistance to grade 7 in some environments, but much poorer resistance in others. It also has been shown that the constitution of grade 12 can be quite complex, depending on the thermal history. For example, annealing at temperatures in the 650-700°C range can cause the Ni and Mo stabilized β phase to decompose into Ti_2Ni or ω or both. There is a complex interdependence between corrosion rate and alloy constitution as affected by heat treatment. This is shown in Fig. 4.18 [142]. It appears that use of grade 12 at elevated temperatures without a carefully selected prior heat treatment can cause time dependent reduction in corrosion resistance in service. Such nonconservative practices are always wise to avoid.

The remaining alloy usually associated with this class of alloys is Ti-3Al-2.5V (grade 9). It is shown in the α alloys section of Table 2.6 and in Table 4.1. This alloy is often called a "near α" alloy [32]. It frequently is substituted for α alloys containing less solute when higher strength is required together with good weldability and acceptable corrosion resistance. It is particularly useful because it can be readily fabricated into seamless tubing by extrusion or in a pilger mill. The corrosion resistance of this alloy is not as great as that of the various alloy grades discussed above, but the superior strength of grade 9 in combination with fabricability (including welding) and good resistance to a variety of harsh environments make it a good choice for many applications such as hydraulic tubing in aircraft and large pressure vessels where weight is critical from a transportation standpoint.

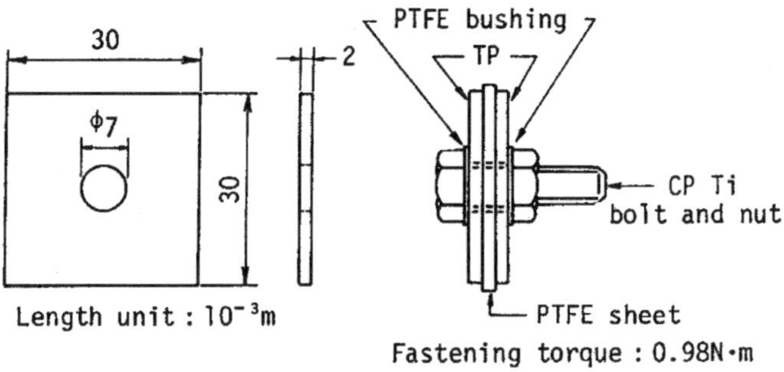

Fig. 4.16. Drawing of a crevice corrosion test specimen "stack" (courtesy J. A. Hall)

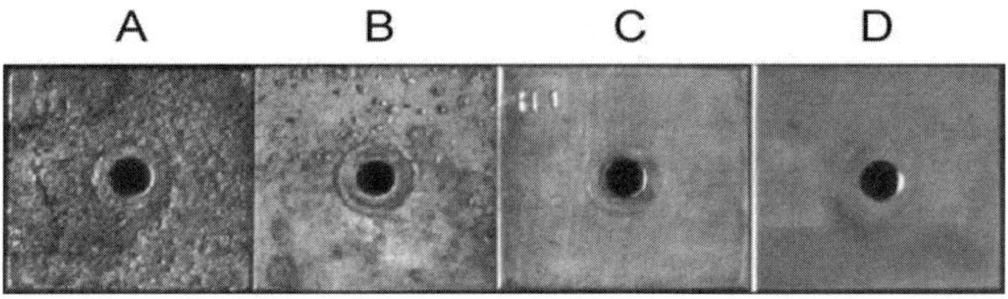

Fig. 4.17. Crevice corrosion specimens of CP titanium grade 2, specimens A and B, and Ti-0.2Pd (grade 7), specimens C and D (courtesy J. A. Hall)

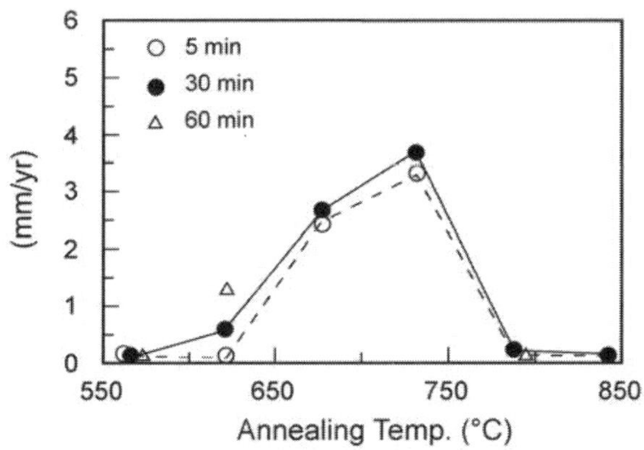

Fig. 4.18. Corrosion rate as a function of annealing temperature for three different times of the α titanium alloy Ti-0.3Mo-0.8Ni (grade 12) [142]

4.3 Properties and Applications

It already has been said that the major uses of CP titanium and other α alloys are for process equipment in the chemical and petrochemical industries. This is the case if the applications are ranked by quantity of material used annually. There are a number of applications for CP titanium in other industrial sectors. These applications, which also rely on the superior environmental resistance of this class of alloys, are listed in Table 4.5 including some of the operating environments.

Tube and shell heat exchangers are a common use of CP titanium and such a heat exchanger was already shown in Fig. 4.7. This type of heat exchanger consists of a large number of individual welded tubes. These tubes have been roller-expanded into a tube sheet at each end of the heat exchanger. This type of heat exchanger derives its principal advantages from two situations, one where there is the potential for fouling; the other is where the medium involved carries solids. These situations create the potential for blockages to occur in some of the individual tubes during service. In this type of heat exchanger, such blockages do not significantly impair the overall operation of the apparatus resulting in a "fail-safe" situation. In contrast to competing materials, CP titanium is immune to chlorine bearing compounds that are effective for removing fouling. The ease of cleaning fouled heat exchangers is another reason that CP titanium is an attractive material choice.

The wide range of environments that CP titanium can operate in and the ease of fabrication of the tubing and the heat exchanger are the main reason for the selection of CP titanium for tube and shell heat exchangers. The cost of these heat exchangers made from CP titanium ranks higher than those made from 70/30 Cu-Ni alloy or from 316 stainless steel but less than those made from zirconium alloys, Inconel 625, or C-276 (Ni-Cr-Mo). Very large tube type heat exchangers called tube and frame heat exchangers are made from CP titanium. An example of one of these is shown in Fig. 4.19. The size of this apparatus requires the tubes to be supported by a frame rather than using a tube sheet, but otherwise the operational principles and benefits are essentially the same.

Another commonly used design is called a plate and frame heat exchanger. An example of one of these devices is shown in Fig. 4.20. These devices use stacked panels to obtain maximum heat exchange area and achieve very large heat transfer surface areas per unit volume of device. Further, these devices are modular and additional modules can be added to accommodate additional heat extraction requirements. The panels are reasonably complex in shape. Thus the material of construction must have good formability. Some of the individual pieces that comprise these panels are shown in Fig. 4.21. CP titanium grades 1 or 2 meet these fabricability requirements and, when coupled with the intrinsic corrosion resistance of this class of material, the various grades of CP titanium have become a natural choice of material.

In all heat exchangers, the flow rate of the cooling water determines, to a significant extent, the efficiency of the device in terms of volume of process fluids in a given time. In those cases where seawater is used as the cooling water, CP titanium has a significant advantage over other less expensive alternatives such as the 70/30 Cu-Ni alloy. For example, CP titanium can be used in surface condensers at higher seawater flow rates than possible for the Cu-Ni alloys because the passive oxide film on CP titanium remains intact even at high flow rates. CP titanium surface condensers have been confidently sold with 40 year warranties for use in high seawater flow rate conditions. Situations such as this can improve the return on capital in a plant and make CP titanium a better choice in terms of life cycle cost.

Fig. 4.19. Tube and frame heat exchanger made from CP titanium (courtesy J. A. Hall)

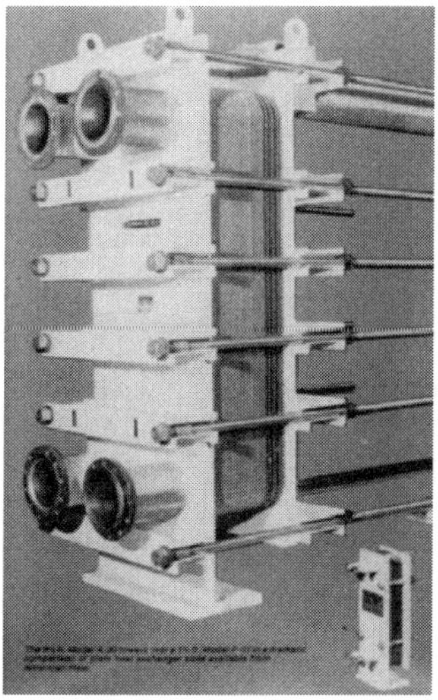

Fig. 4.20. Plate and frame heat exchanger made from CP titanium (courtesy J. A. Hall)

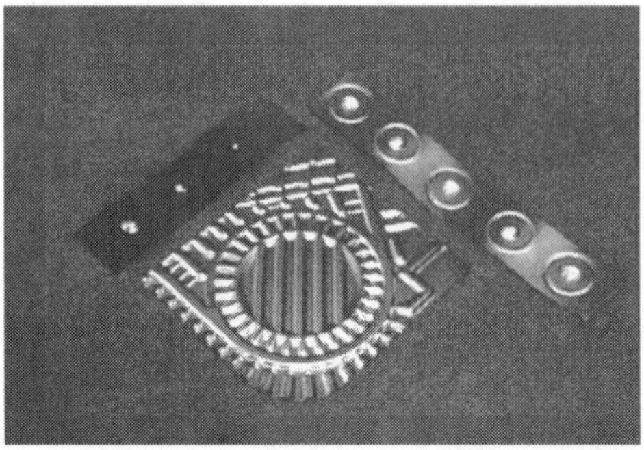

Fig. 4.21. Complex shaped pieces used in assembling the panels for a plate and frame heat exchanger, CP titanium (courtesy J. A. Hall)

CP titanium pressure vessels are also commonly used because of strength, fabricability, and corrosion resistance. An example of such a pressure vessel was shown in Chap. 3 (Fig. 3.49) and was discussed in Sect. 4.1.1 in connection with material processing. These large pressure vessels are fabricated by welding several large, seamless rolled cylinders together, an example of one of these cylinders was shown in Fig. 3.28.

Another component that operates in a very aggressive environment is the bleaching section of the pulp and paper production equipment. The operating environment is sufficiently aggressive that only CP titanium will work successfully over long periods. In addition, these structures are large and the specific strength of CP titanium alloys is helpful in transporting and siting these large components. An example of one of these structures is shown in Fig. 4.22.

CP titanium also is used in emission control systems for coal burning power generation plants. There is growing pressure to clean the effluent from these coal burning power generation plants. Doing this requires removal of sulfur containing compounds including SO_2. Sulfur compounds are much more compatible with CP titanium than with stainless steel or Ni alloys. Consequently, this application is growing, even though the cost of the CP titanium components is higher.

The higher strength alloy Ti-5Al-2.5Sn has been used for cryogenic tanks for spacecraft. Ti-5Al-2.5Sn also has been used for components in the liquid hydrogen fuel system of the Space Shuttle main engine. Both of these applications are driven by the good weldability and acceptable ductility at cryogenic temperatures of this alloy. Given the well-known susceptibility of titanium alloys to hydrogen uptake, this may be a surprising application. It has been found that contact between liquid hydrogen at 10 K and titanium alloys does not result in any interaction or uptake. Thus, when the engine has stopped running it is essential to purge any gaseous hydrogen from the system using either nitrogen or inert gas. This practice assures there is minimal contact between gaseous hydrogen and the titanium components. Hydrogen uptake and embrittlement possibilities are therefore minimized and this practice has proved successful.

There is another, quite old alloy that is similar to Ti-5Al-2.5Sn. It is Ti-8Al-1Mo-1V and this alloy does not have any significant heat treatment response, at least in the way that $\alpha+\beta$ alloys have. Consequently, it is reasonable to discuss this alloy in conjunction with Ti-5Al-2.5Sn even though, strictly speaking, it is an $\alpha+\beta$ alloy from an alloy constitution standpoint. Because of the high aluminum content Ti-8Al-1Mo-1V precipitation hardens by α_2 formation. Ti-8Al-1Mo-1V also has a higher modulus of elasticity than any of the $\alpha+\beta$ alloys because of its high aluminum content. As a sheet product it has a yield stress that is 70-140 MPa higher than mill annealed Ti-6Al-4V sheet. Ti-8Al-1Mo-1V was originally considered attractive because of this strength advantage. In fact, it was chosen as the principal alloy for the sheet structure of the American supersonic transport (SST) airplane in the 1960's. It was during this contemplated application that the significant susceptibility to stress corrosion cracking of this alloy was discovered. As a result of this discovery, this alloy has not been considered as attractive for any applications where this limitation could be an issue. Perhaps the most important remaining application for Ti-8Al-1Mo-1V is for fan blades for medium bypass ratio aircraft gas turbines engines. This is an application where high modulus of elasticity and good annealed strength are important characteristics. Today, the cost of this alloy, because of the difficulty experienced in producing it, and the stress corrosion cracking susceptibility have eliminated it from serious consideration for new applications. Nevertheless, it is a technologically interesting alloy and deserves to be mentioned here.

Fig. 4.22. Example of a very large structure used in the bleaching section of the pulp and paper production, CP titanium (courtesy J. A. Hall)

5 Alpha + Beta Alloys

In this chapter on $\alpha+\beta$ alloys the basic correlations between processing, microstructure, and properties of the whole group of $\alpha+\beta$ alloys listed in Table 2.6 will be outlined including the so-called "near α" alloys designated for applications at high temperatures. Only the special features of the latter group of alloys in the area of processing, microstructure, and properties will be described in Chap. 6.

The discussion of processing in this chapter refers to the creation of microstructures in the final product, e.g. forgings, plate, sheet, bar, and castings. The practical methods of shaping these products have been described in Chap. 3, but the fundamental parameters of processing that determine microstructure (and properties) are described here. Similar considerations can be important in intermediate products such as forging billet.

5.1 Processing and Microstructure

In the $\alpha+\beta$ alloys three distinctly different types of microstructures can be obtained by changing the thermo-mechanical processing route: fully lamellar structures, fully equiaxed structures, and so-called bi-modal (duplex) microstructures containing equiaxed primary α (α_p) in a lamellar $\alpha+\beta$ matrix. A common but less defined microstructural condition is the so-called mill-annealed structure which is discussed at the end of Sect. 5.1.3.

5.1.1 Fully Lamellar Microstructures

Lamellar microstructures can be obtained fairly easily in the final steps of the processing route by an annealing treatment in the β phase field (β recrystallization). For this reason, this microstructure is often also called "β annealed" structure. In the schematic drawing of the whole processing route (Fig. 5.1) this recrystallization treatment in the β phase field is step III. The deformation process (step II) can be done by forging or rolling, either in the β phase field or in the ($\alpha+\beta$) phase field. In industrial practice, the material is usually first deformed in the β phase field because of the lower flow stress and then deformed in the ($\alpha+\beta$) phase field to avoid large β grain sizes. Similarily, the recrystallization temperature in step III is usually kept within 30-50°C above the β transus to maintain control of the β grain size. As a result, the β grain size of fully lamellar microstructures is typically about 600 µm.

The most important parameter in the processing route is the cooling rate from the β phase field in step III, because the cooling rate determines the characteristic features of the lamellar microstructure, such as the size of the α lamellae (α plates), the α colony size, and the thickness of α layers at β grain boundaries, as can be seen from Table 5.1 in which the important processing parameters are listed together with their major influence on microstructure. An example of the variation in the lamellar microstructure as a function of cooling rate from the β phase field is shown in Fig. 5.2 (LM) and Fig. 5.3 (TEM) for the Ti-6242 alloy. The micrographs in this example illustrate that the microstructural features listed in Table 5.1 in the row for the processing step "Cooling Rate", namely width of α lamellae, α colony size, and width of α layers at β grain boundaries, all decrease with increasing cooling rate. The cooling rate in this example varies from slow cooling with 1°C/min, typical for furnace cooling, to fast cooling with 8000°C/min which can only be obtained by water quenching of relatively thin sections (≤ 10 mm thickness). For these two extreme cooling rates, examples for the Ti-6Al-4V alloy have been shown already in Sect. 2.6 covering the subject "Phase Transformations" (Figs. 2.15 and 2.17). The microstructure resulting from the medium cooling rate of 100°C/min (Figs. 5.2b and 5.3b) is typical for the majority of commercial cooling rates, for example fast cooling (water quenching or forced air cooling) of thicker section sizes (forgings, plates, etc.) or air cooling of thin sheets. The change from a colony or Widmanstätten type of microstructure to a martensitic structure, i.e. the critical cooling rate avoiding the nose in a CCT diagram, is naturally alloy dependent, but for the most common α+β alloys, such as Ti-6Al-4V or Ti-6242, this change occurs at cooling rates faster than 1000°C/min. Thus, martensitic microstructures are rarely present in structural components of α+β alloys.

It also is possible to obtain a large range of cooling rates for castings, just as in case of forgings. The recent development of permanent mold casting (Sect. 3.5.1) allows much higher cooling rates to be attained than in investment castings of comparable size. Accordingly, additional control over microstructural scale can be achieved.

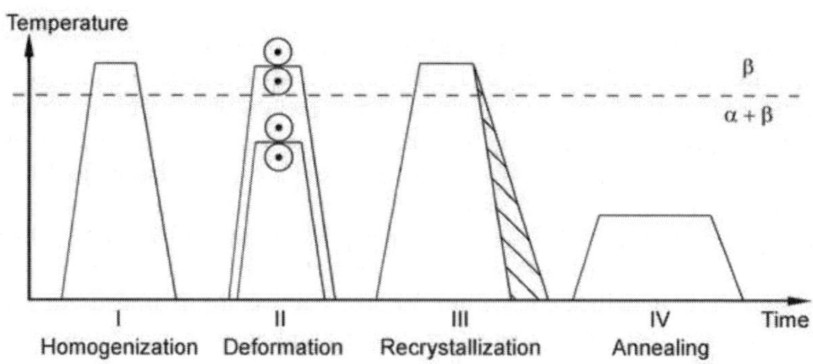

Fig. 5.1. Processing route for lamellar microstructures of α+β titanium alloys (schematically)

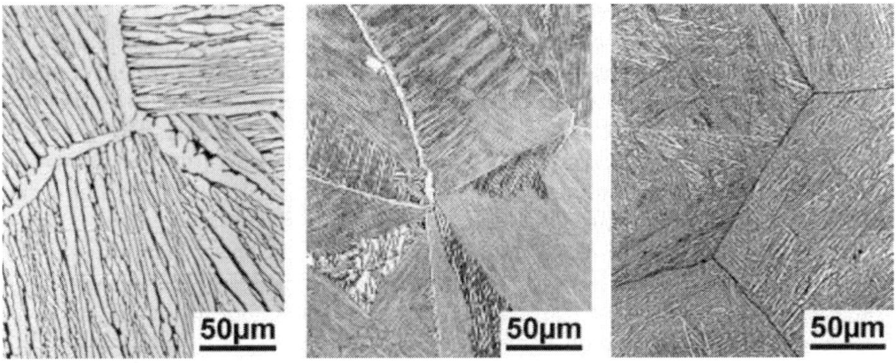

a b c

Fig. 5.2. Effect of cooling rate from the β phase field on lamellar microsturctures, Ti-6242, LM: (a) 1°C/min (b) 100°C/min (c) 8000°C/min

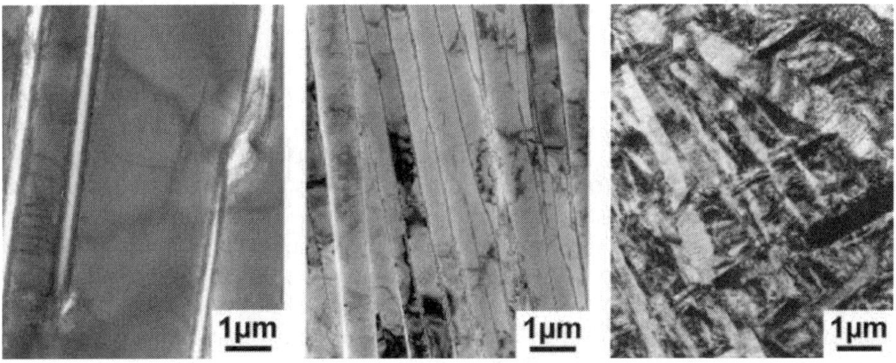

a b c

Fig. 5.3. Effect of cooling rate from the β phase field on lamellar microsturctures, Ti-6242, TEM: (a) 1°C/min (b) 100°C/min (c) 8000°C/min

Table 5.1. Important processing parameters and resulting microstructural features for lamellar microstructures

Processing Step (see Fig. 5.1)	Important Parameters	Microstructural Features
III	Cooling Rate	– Size of α Colonies – Width of α Lamellae – GB α Layer
IV	Annealing Temperature	– Ti_3Al in α – Secondary α in β

Although both microstructural parameters, the width of individual α plates and the size of α colonies, decrease with increasing cooling rate, their main size change occurs during different ranges in cooling rate. The width of the α plates decreases drastically from about 5 μm in slowly cooled material to about 0.5 μm for a cooling rate of 100°C/min (compare Figs. 5.3a and 5.3b), whereas a further increase in cooling rate leads only to a reduction to about 0.2 μm (average width of martensite plates) with a fair amount of thicker martensite plates present in the microstructure, see Fig. 5.3c. In contrast, the α colony size, which can be as large as half of the β grain size in slowly cooled material, e.g. about 300 μm (see low magnification micrograph in Fig. 2.17a), shows only a moderate decrease to about 100 μm for a cooling rate of 100°C/min (Fig. 5.2b). The major decline in α colony size occurs then by definition between 100°C/min and 8000°C/min down to the width of individual martensite plates.

The presence of the continuous α layers at β grain boundaries can be seen clearly for all three cooling rates in the light micrographs of Fig. 5.2. Thus, the formation of the α layers cannot be avoided even for the very fast cooling rates (see also Fig. 2.15a for the Ti-6Al-4V alloy). It should be pointed out, that in slowly cooled material the α plates in the colonies have nearly the same width as the continuous α layers at β grain boundaries (Fig. 5.2a), for the Ti-6Al-4V alloy see Fig. 2.17a.

In colony type microstructures the remainder of the β matrix, the so-called β lamellae, can be seen clearly in the transmission electron micrographs of Figs. 5.3a and 5.3b in between the α plates. For martensitic microstructures an annealing treatment in the α+β phase field is necessary to change these structures to fine lamellar α+β microstructures, as will be discussed in the following.

In step IV (final annealing heat treatment) the temperature is more important than the time (see Table 5.1) because the temperature determines whether age-hardening of the α phase by Ti_3Al particles occurs or not (see Sect. 2.8.1). For example, in the Ti-6Al-4V alloy the Ti_3Al solvus temperature is about 550°C. Consequently, aging at 500°C will precipitate Ti_3Al particles whereas a final heat treatment at 600°C or above will be only a stress relieving treatment. In addition, depending on the details of the cooling process in step III, fine secondary α platelets can be precipitated in the β phase during the heat treatment in step IV. Although not being relevant for commercial application, it should be pointed out, that martensitic microstructures can be changed to fine lamellar α+β microstructures by annealing them in the temperature region of about 700-850°C. By this annealing treatment, β phase is formed as a continuous layer between martensitic plates. For the Ti-6Al-4V alloy an annealing treatment of 1h at 800°C was used in many laboratory experiments to test very fine lamellar α+β microstructures [48, 120, 143].

Another type of fully lamellar microstructures, namely a so-called β processed condition, will be described, although for α+β alloys this condition is not used extensively in commercial application. In case of β processing, which is a very common processing route for β alloys (see Sect. 7.1), the β recrystallization (step III in Fig. 5.1) is completely omitted (Fig. 5.4) and the material remains therefore in an unrecrystallized condition. Critical parameters to avoid recrystallization in step II of the processing route (Fig. 5.4) are strain rate as well as temperature and time in the β phase field and the cooling rate after the deformation process (Table 5.2). The resulting shape of the unrecrystallized β grains depends on deformation mode (rolling, forging, pressing, etc.) and on deformation degree, the latter determining also the width of the deformed β grains (Table 5.2). The main advantage of this unrecrystallized β processed condition is that the α colony size is limited at least in one direction by the β grain width and that the continuous α layers at β grain boundaries are "broken up", i.e. they adopted the zigzag shaped form of the deformed β grain boundaries. These two points are illustrated in the example of Fig. 5.5 showing a β processed microstructure of the Ti-6242 alloy. The cooling rate after the deformation process naturally affects the α lamellae width and also the α colony size in the faster cooling regime (Table 5.2). The α lamellae width is not significantly smaller in the β processed condition as compared to the lamellar (β annealed) structure, compare Figs. 5.5b and 5.3b. Apparently, the β processed material recovers to a large extent before the α plates are nucleated.

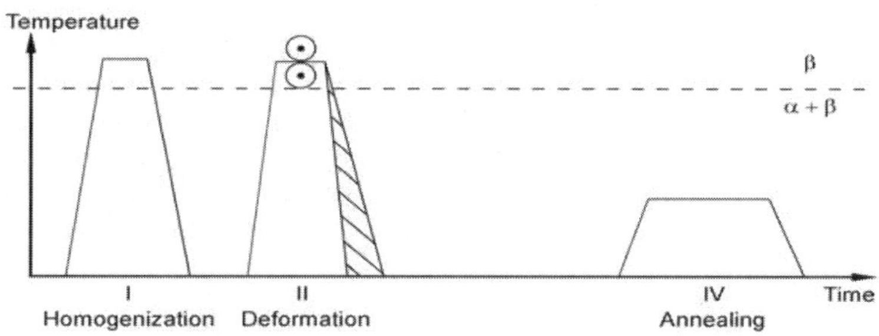

Fig. 5.4. Processing route for β processed microstructures of α+β titanium alloys (schematically)

Table 5.2. Important processing parameters and resulting microstructural features for β processed microstructures

Processing Step (see Fig. 5.4)	Important Parameters	Microstructural Features
II	Deformation Time Deformation Mode Deformation Degree Cooling Rate	Unrecrystallized Structure Shape of β Grains – Width of β Grains (→ α Colony Size) – Geometry of GB α Layer – Unrecrystallized Structure – Size of α Colonies – Width of α Lamellae
IV	Annealing Temperature	– Ti_3Al in α – Secondary α in β

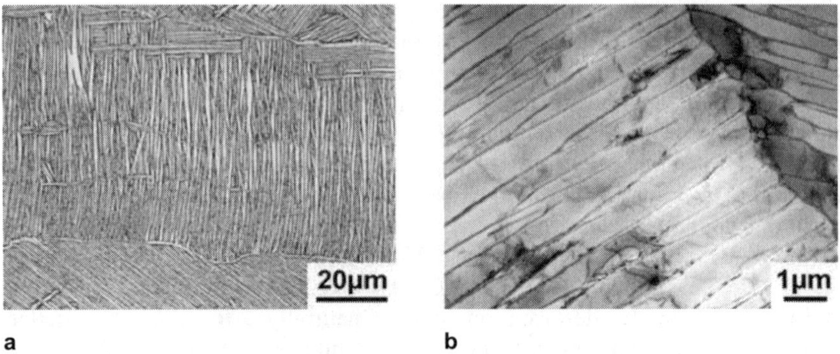

Fig. 5.5. Microstructure of the Ti-6242 alloy in the β processed condition, cooling rate about 100°C/min: (a) LM (b) TEM

5.1.2 Bi-Modal Microstructures

The processing route for obtaining the so-called bi-modal (duplex) microstructures is shown schematically in Fig. 5.6, where the process is again divided into four different steps: homogenization in the β phase field (I), deformation in the (α+β) phase field (II), recrystallization in the (α+β) phase field (III), and the final aging and/or stress relieving treatment (IV). The important parameters of this processing route and the resulting microstructural features of the bi-modal microstructure are summarized in Table 5.3.

A critical parameter in the processing route is the cooling rate from the homogenization temperature in the β phase field (step I) because the cooling rate determines the width of the α lamellae (see Figs. 5.2 and 5.3). These α lamellae are then deformed in step II and recrystallized in step III. The relation between the prior α lamellae width resulting from step I and the equiaxed primary α size is illustrated in Fig. 5.7 comparing two bi-modal structures. The only difference in the processing route was the cooling rate after the β homogenization treatment in step I during a compressor disk fabrication process [144].

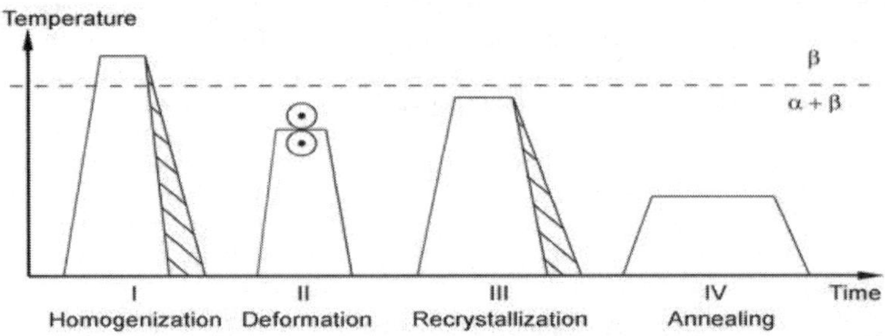

Fig. 5.6. Processing route for bi-modal microstructures of α+β titanium alloys (schematically)

Table 5.3. Important processing parameters and resulting microstructural features for bi-modal microstructures

Processing Step (see Fig. 5.6)	Important Parameters	Microstructural Features
I	Cooling Rate	Width of α Lamellae ($\rightarrow \alpha_p$ Size)
II	Deformation Temperature Deformation Degree Deformation Mode	Texture Type – Texture Intensity – Dislocation Density Texture Symmetry
III	Annealing Temperature Cooling Rate	– Vol% of α_p ($\rightarrow \beta$ Grain Size) – Alloy Element Partitioning Width of α Lamellae
IV	Annealing Temperature	– Ti_3Al in α – Secondary α in β

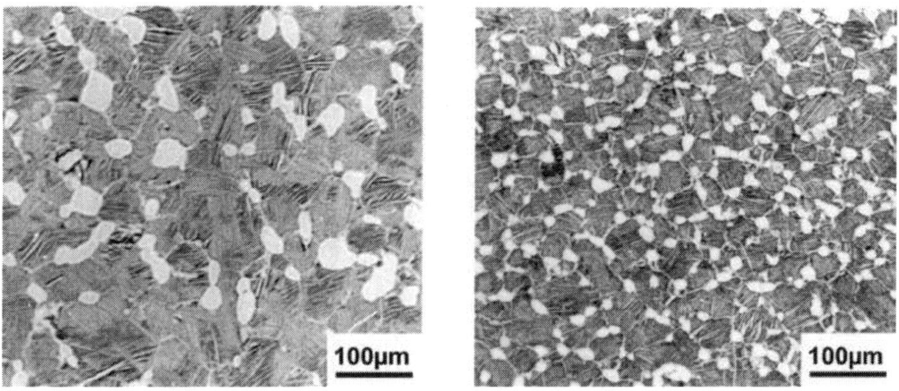

a b

Fig. 5.7. Bi-modal microstructures of the IMI 834 alloy cooled differently from the β phase field in step I of the processing route, LM: (a) Bi-modal 1, slow cooling rate (b) Bi-modal 2, fast cooling rate

In the next step II, the deformation process in the $(\alpha+\beta)$ phase field, the lamellar structure is deformed plastically (not broken up). The plastic deformation should be as high as possible but at least high enough to introduce enough stored energy (dislocations) to obtain complete recrystallization of the α and β phases during step III. During the deformation process in step II, crystallographic textures in the hexagonal α phase and in the bcc β phase can develop, which in turn can influence mechanical properties (Sect. 5.2.6). Fig. 5.8 shows schematically the different textures that can be formed during unidirectional rolling. The deformation temperature determines the texture type. At "low" deformation temperatures (a high volume fraction of α phase is present during deformation) an α-deformation texture, a so-called basal/transverse (B/T) type of texture, develops whereas at "high" deformation temperature in the (α+β) phase field (a high volume fraction of β phase is present during deformation) a β-deformation texture develops. During the subsequent cooling process (α phase formation in the textured β phase) only one out of the six possible $\{110\}$ planes of the β phase is selected preferentially by the α phase to fulfill the Burgers relationship, $(110)_\beta \parallel (0002)_\alpha$. This results in a so-called transverse (T) type of transformation texture. Frederick [145] proposed that this selection process in the β to α transformation is due to the superimposed strains from the deformation process. In between these two deformation temperature regions, a fairly weak texture of the α phase can be obtained (see Fig. 5.8) because the intensity of the α-deformation texture (B/T) decreases with increasing deformation temperature and the transformation texture (T) has a low intensity due to a low volume fraction of β phase present during deformation [146]. The deformation mode (unidirectional rolling, cross rolling, pancake forging, etc.) determines the texture symmetry (Table 5.3). The resulting textures of the hexagonal α phase will not change significantly during the subsequent recrystallization step III.

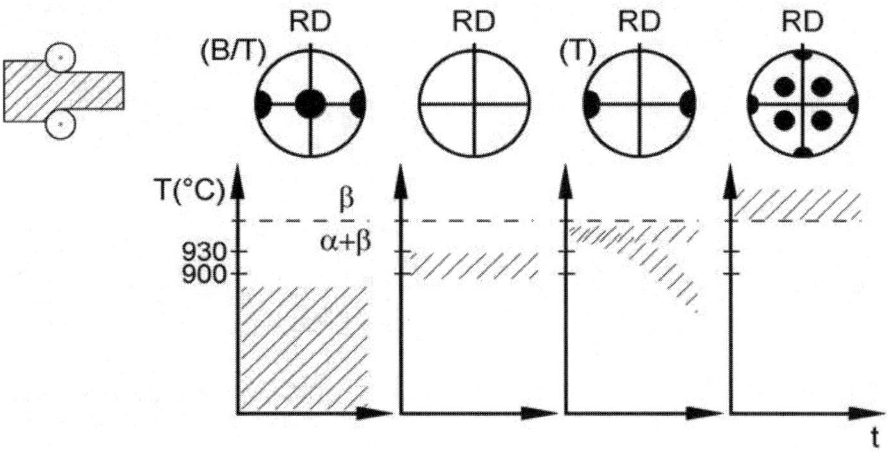

Fig. 5.8. Crystallographic textures (0002 pole figures) formed in α+β titanium alloys (schematically)

The most important parameter of the recrystallization annealing step III (Fig. 5.6) is the temperature, which determines the volume fraction of the recrystallized equiaxed primary α (α_p) located at the "triple-points" of the recrystallized β grains (Fig. 5.7). This α_p volume fraction together with the α_p size defines the most important microstructural feature of the bi-modal structure, namely the β grain size which is about equal to the distance between primary α (Fig. 5.7).

The annealing time during step III is not very critical (Table 5.3), as long as the time is sufficient for the generation of isolated, equiaxed α_p grains. Grain growth is very sluggish in these two-phase mixtures of α_p and β grains. The mechanism by which the lamellar "starting" structure (step I, see also Fig. 5.3), being deformed in step II, is changed to equiaxed α and β grains during the recrystallization process is illustrated in Fig. 5.9 [48]. It can be seen that the recrystallized β phase penetrates into the recrystallized α lamellae along α/α grain boundaries causing the separation into individual α_p grains.

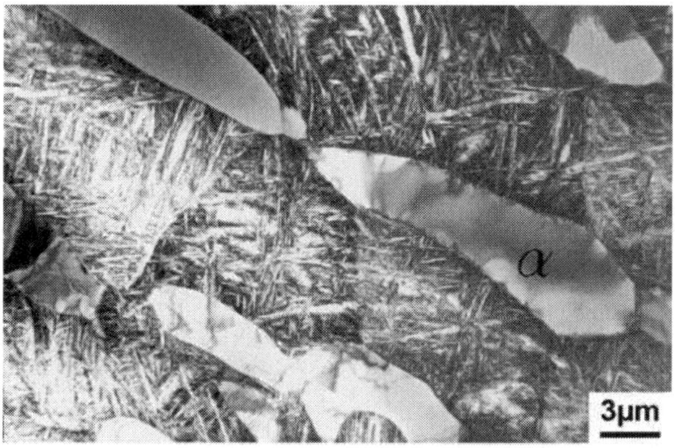

Fig. 5.9. Recrystallization mechanism to a bi-modal microstructure at 950°C, Ti-6Al-4V, TEM [48]

During separation into α_p and β an alloying element partitioning effect also takes place (see also Sect. 2.8.1). Those elements which are either strong α stabilizers (aluminum, oxygen) or strong β stabilizers (for example Mo, V) will partition into the two phases, respectively. This alloy element partitioning effect has the consequence that the α lamellae formed in the β grains upon cooling from the recrystallization annealing temperature, have a lower concentration of those elements (aluminum, oxygen) that promote age-hardening by the formation of coherent Ti_3Al particles in step IV as compared to α_p or to the fully lamellar microstructure.

In bi-modal microstructures, the cooling rate from the recrystallization annealing temperature in the (α+β) phase field (step III in Fig. 5.6) mainly influences the width of the individual α lamellae (Table 5.3), whereas the α colony size and the length of the continuous α layers at β grain boundaries are largely determined by the β grain size. Within the normal range of commercial cooling rates of about 30-600°C/min the α colony size in bi-modal microstructures is about equal to the β grain size. At slower cooling rates both the size and the volume fraction of α_p increase, as will be discussed in Sect. 5.1.3.

5.1.3
Fully Equiaxed Microstructures

There are two possibilities to obtain a fully equiaxed microstructure. In the first case (Fig. 5.10) the processing route is identical to the processing route for obtaining a bi-modal microstructure up to the recrystallization process in step III. If now the cooling rate from the recrystallization annealing temperature is sufficiently low, only the α_p grains will grow during the cooling process and no α lamellae are formed within the β grains, resulting in a fully equiaxed structure with the equilibrium volume fraction of β phase located at the "triple-points" of the α grains. An example for such a fully equiaxed microstructure is shown in Fig. 5.11 for the Ti-6242 alloy. In this case the α grain size will be fairly large and always larger than the α_p size of the corresponding bi-modal structure which would have been formed upon faster cooling.

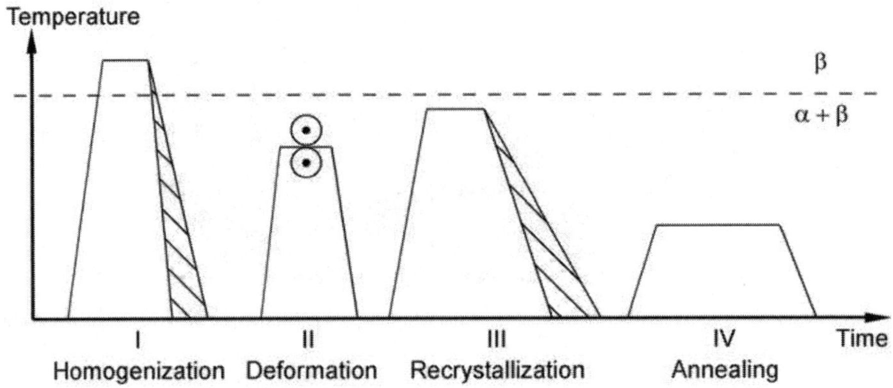

Fig. 5.10. Processing route for fully equiaxed microstructures of $\alpha+\beta$ titanium alloys slowly cooled from the bi-modal recrystallization annealing temperature (schematically)

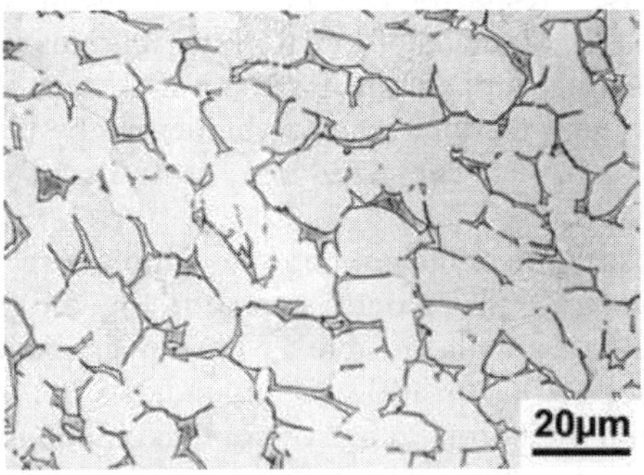

Fig. 5.11. Fully equiaxed microstructure of the Ti-6242 alloy slowly cooled from the bi-modal recrystallization annealing temperature, LM

The second possibility to obtain a fully equiaxed microstructure is to recrystallize in step III of the processing route (Fig. 5.12) at such a low temperature that the equilibrium volume fraction of α phase at that temperature is high enough to form the fully equiaxed microstructure directly from the deformed lamellar structure during the recrystallization process. The mechanism that changes the deformed lamellar "starting" structure to equiaxed grains is the same mechanism as described already for the bi-modal microstructure (see Fig. 5.9) but in the opposite direction: α phase penetrates along β/β grain boundaries into the recrystallized β lamellae (Fig. 5.13) causing the separation of the β grains in the final microstructure [48]. Using this second processing route with the low recrystallization annealing temperature (for example for Ti-6Al-4V between 800-850°C) smaller α grain sizes can be achieved as compared to the first method described above. For example, in case of the Ti-6Al-4V alloy using a recrystallization annealing temperature of 800°C it was possible to obtain on a laboratory scale fully equiaxed microstructures with α grain sizes of about 2 μm [146]. An example of such a microstructure is shown in Fig. 5.14, the small α grain size can only be seen clearly by TEM.

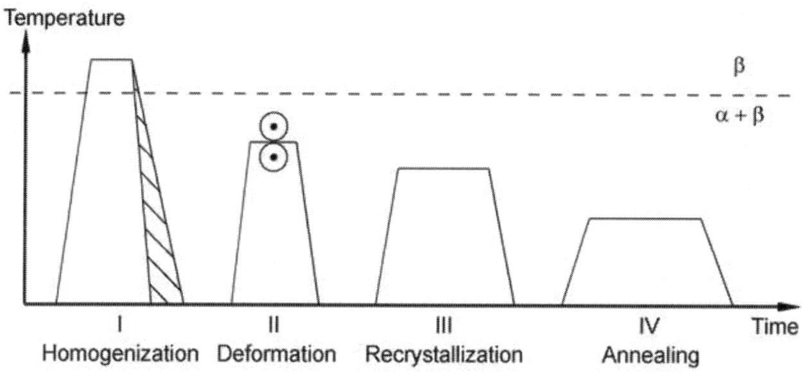

Fig. 5.12. Processing route for fully equiaxed microstructures of α+β titanium alloys recrystallized at low temperatures (schematically)

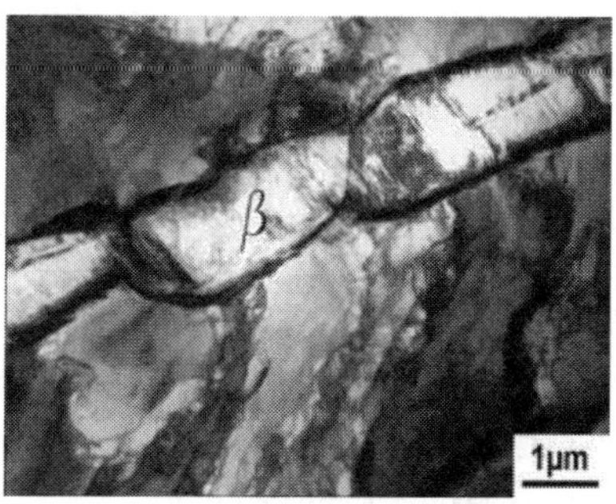

Fig. 5.13. Recrystallization mechanism to a fully equiaxed microstructure at 800°C, Ti-6Al-4V, TEM [48]

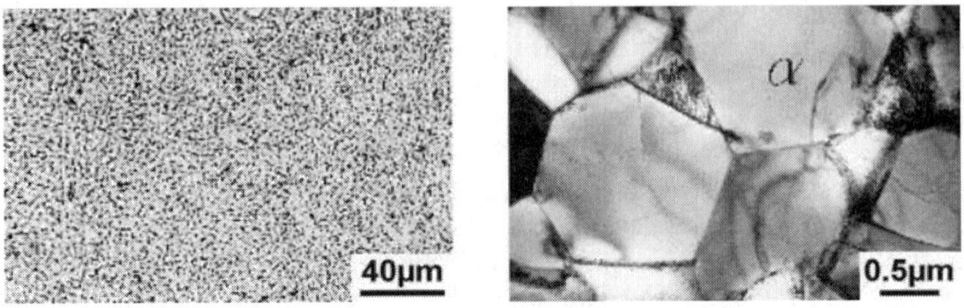

a **b**

Fig. 5.14. Fine grained, fully equiaxed microstructure of the Ti-6Al-4V alloy recrystallized at 800°C: (a) LM (b) TEM

The important processing parameters and the resulting microstructural features for both processing routes that result in fully equiaxed microstructures are summarized in Table 5.4. It should be emphasized that a fast cooling rate from the β phase field (step I) is a necessity to obtain fully equiaxed microstructures with small α grain sizes via the low recrystallization annealing temperature route (Fig. 5.12). During the final annealing treatment (step IV) secondary α platelets can be formed in the β phase even in fully equiaxed microstructures depending on the difference between the recrystallization annealing temperature and the final annealing temperature in the low recrystallization temperature route (Fig. 5.12) or on the details of the cooling procedure from the recrystallization annealing temperature in the slow cooling route (Fig. 5.10).

Table 5.4. Important processing parameters and resulting microstructural features for fully equiaxed microstructures

Processing Step (see Figs. 5.10 and 5.12)	Important Parameters	Microstructural Features
I	Cooling Rate	Width of α Lamellae (→ α Grain Size)
II	Deformation Temperature Deformation Degree Deformation Mode	Texture Type – Texture Intensity – Dislocation Density Texture Symmetry
III	Slow Cooling Rate (Fig. 5.10) Low Annealing Temperature (Fig. 5.12)	Fully Equiaxed Structure Fully Equiaxed Structure
IV	Annealing Temperature	– Ti_3Al in α – Secondary α in β

It should be pointed out that fully equiaxed microstructures can be changed to bi-modal microstructures by simply heating the material up to the temperature in the (α+β) phase field corresponding to the desired α_p volume fraction and subsequently cooling with a sufficiently high rate to form α lamellae within the β grains. The reverse microstructural change from a bi-modal structure to a fully equiaxed structure also is possible. In this case, the bi-modal material has to be heated in the (α+β) phase field at least up to a temperature at which all α lamellae in the β grains dissolve and then the material has to be cooled at a rate slow enough to allow growth of α_p. In a similar way, the volume fraction of α_p in bi-modal microstructures can be changed. However, it is important to realize that after each heat treatment performed to change the microstructure both the α grain size and the β grain size will be increased.

A very common, but less defined, processing route leads to a so-called mill-annealed condition. In this case the recrystallization step III is completely omitted (Fig. 5.15). Consequently, the details of the deformation procedure in step II (one or two heats, warm up time in the second heat, cooling rate after deformation, etc.) determine the details of the resulting microstructure, especially the degree of recrystallization. Since, in most cases, the details of the deformation procedure are beyond the control of the customer or end-user and the details will vary between producers or even between different lots, the mill-annealed condition is usually not a well defined microstructure. One example of a mill-annealed microstructure is shown in Fig. 5.16 for the Ti-6Al-4V alloy. In this case, the mill-annealed microstructure looks fairly well recrystallized containing only a few stringers with a lamellar, unrecrystallized appearance. Another example of a mill-annealed microstructure can be seen in Fig. 3.54. In this case the mill-annealed microstructure still has the appearance of a deformed lamellar microstructure, although the recrystallization process could be already in an advanced stage [48]. The final annealing step in the processing route of mill-annealed materials is typically done at temperatures higher than the Ti_3Al solvus temperature. In that case, this step is a pure stress relieving treatment and Ti_3Al precipitation in α is therefore not listed in Table 5.5 summarizing the processing parameters and the resulting microstructural features for the mill-annealed condition.

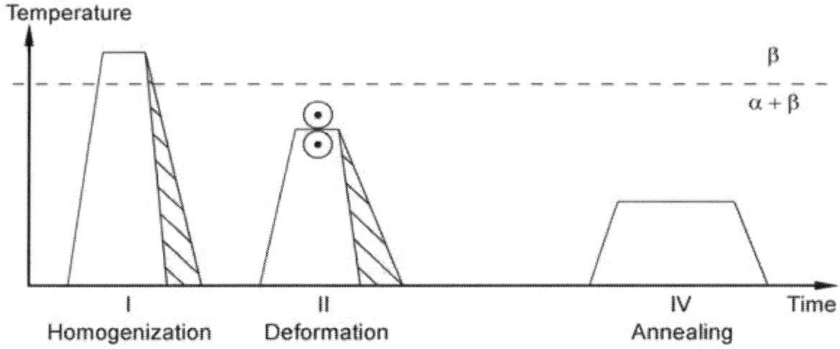

Fig. 5.15. Processing route for mill-annealed microstructures of α+β titanium alloys (schematically)

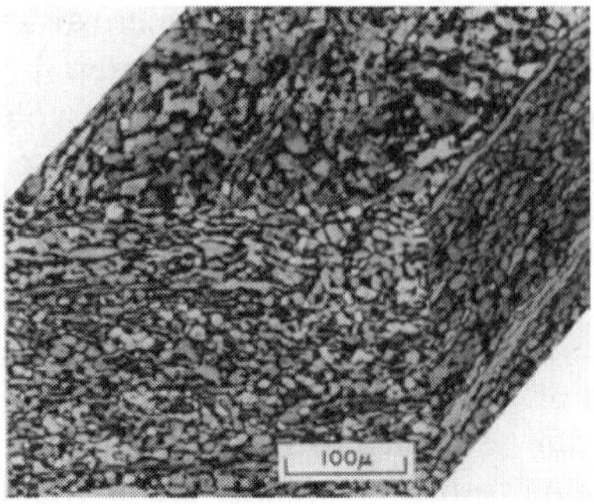

Fig. 5.16. Mill-annealed microstructure of Ti-6Al-4V alloy plate, LM

Table 5.5. Important processing parameters and resulting microstructural features for mill-annealed microstructures

Processing Step (see Fig. 5.15)	Important Parameters	Microstructural Features
I	Cooling Rate	Width of α Lamellae (→ α Grain Size)
II	Details of Deformation Procedure Deformation Degree Deformation Mode	Degree of Recrystallization (Vol% of Equiaxed Grains) Texture Intensity Texture Symmetry
IV	Annealing Temperature	Secondary α in β

5.2 Microstructure and Mechanical Properties

In this section the basic correlations between microstructural features and mechanical properties of α+β titanium alloys will be outlined. The mechanical properties emphasized in this section are tensile, fatigue, and fracture toughness properties. The creep properties will be discussed in the chapter on high temperature alloys (Sect. 6.2). Among the various microstructural features outlined in Sect. 5.1, only the most important ones will be discussed in detail. Others will be mentioned, but only briefly. A qualitative summary of the basic microstructure/property correlations is shown in Table 5.6 and this table will be used as guideline throughout this section. The qualitative trends (+,0,–) point out the direction in which a specific mechanical property will change when a microstructural feature is changed according to the footnote in Table 5.6. In case of tensile properties, yield stress $\sigma_{0.2}$ and ductility (reduction in area or in some cases tensile elongation) are used for the correlation chart in Table 5.6. The high cycle fatigue (HCF) strength at 10^7 cycles is taken as a measure for the resistance against fatigue crack nucleation. The resistance of a microstructure against fatigue crack propagation is abbreviated by ΔK_{th} in Table 5.6, meaning the low da/dN-ΔK region and not a "threshold" value. Small, self-initiated cracks are called microcracks in contrast to large through-thickness cracks (macrocracks) characteristically studied using fracture mechanics specimens such as compact tension (CT) specimens. For macrocracks two different columns are shown in Table 5.6 differentiating between crack propagation without (R = 0.7) and with (R = 0.1) crack closure. Fracture toughness K_{Ic} also does not involve crack closure and, therefore, is placed between the two fatigue crack growth columns. The creep strength which is evaluated for small plastic strains (0.2%), i.e. in the primary creep region, is included in Table 5.6 but as mentioned above will be discussed in the chapter on high temperature alloys (Sect. 6.2). The trends for tensile, fatigue, and fracture toughness properties shown in Table 5.6 are basically room temperature correlations but in principle they are also valid for high temperatures. One exception, the HCF strength of bi-modal microstructures, will be discussed in detail in Sect. 5.2.2.

Table 5.6. Qualitative correlation between important microstructural parameters and mechanical properties for α+β titanium alloys

	$\sigma_{0.2}$	ε_F	HCF	Micro-cracks ΔK_{th}	Macrocracks ΔK_{th} R = 0.7	Macrocracks K_{Ic}	Macrocracks ΔK_{th} R = 0.1	Creep Strength 0.2%
Small α Colonies, α Lamellae [a]	+	+	+	+	–	–	–	+/–
Bi-Modal Structure [b]	+	+	–	+	–	–	–	–
Small α Grain Size [c]	+	+	+	+	–	–	–	–
Aging (α_2), Oxygen	+	–	+	–	–	–	+	+
Secondary α in β	+	–	+	+	0	0	0	+
Texture: Stress ‖ c-Axis	+	0	+ Vac – Air	0 Vac – Air	0 Vac – Air	0	0 Vac – Air	+

[a] Compared to coarse lamellar structure.
[b] Compared to fully lamellar structure with same cooling rate.
[c] Compared to large α grain size of fully equiaxed structures.

The mechanical properties shown as examples in Chaps. 5 and 6 are basically all obtained on microstructures with a final aging treatment (24h 500°C for Ti-6Al-4V, 8h 595°C for Ti-6242, and 2h 700°C for IMI 834). The few exceptions are explicitly stated (Figs. 5.3.6 and 5.3.8 and Sect. 5.2.4 describing the effect of aging).

5.2.1 Fully Lamellar Microstructures

The most influential microstructural parameter on the mechanical properties of lamellar microstructures is the α colony size, which is controlled by the cooling rate from the β heat treatment temperature (see Fig. 5.1 and Table 5.1), because the α colony size determines the effective slip length in lamellar microstructures. Although the two phases, the α plates and the β matrix, have to deform independently, slip can be fairly easily transferred across the incoherent α/β interface, because as can be seen from Fig. 2.18 (Sect. 2.6.2) two slip systems in each phase are exactly parallel, $(110)\,[1\bar{1}1]_\beta \parallel (0002)\,[11\bar{2}0]_\alpha$ and $(\bar{1}12)\,[1\bar{1}1]_\beta \parallel (\bar{1}100)\,[11\bar{2}0]_\alpha$, and two others are off by only 10°, $(110)\,[1\bar{1}2]_\beta$ and $(0002)\,[1\bar{2}10]_\alpha$ as well as $(1\bar{1}2)\,[\bar{1}11]_\beta$ and $(10\bar{1}0)\,[1\bar{2}10]_\alpha$.

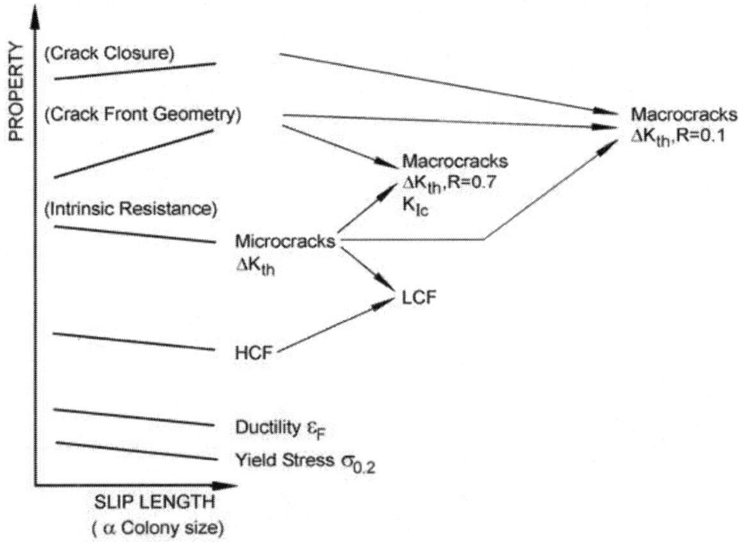

Fig. 5.17. Influence of slip length (α colony size) on mechanical properties (schematically)

The general effect of slip length (α colony size) on mechanical properties is shown schematically in Fig. 5.17. With increasing cooling rate the α colony size is decreased with a commensurate reduction in effective slip length and a corresponding increase in yield stress. This is illustrated in Fig. 5.18 for three α+β titanium alloys (Ti-6Al-4V, Ti-6242, IMI 834). It can be seen that, in the commercially feasible cooling rate regime (up to 1000°C/min), the effect on yield stress is only moderate (50-100 MPa), whereas a large increase in yield stress is observed when the colony structure is changed to a martensitic type of microstructure (slip length and "colony" size equal to the width of individual α plates). The yield stress ranking in this condition with the fastest cooling rate can be correlated with the fineness of the different martensitic structures, e.g. Ti-6242 has the finest martensitic structure, followed by the Ti-6Al-4V alloy, and the IMI 834 material exhibited fairly coarse martensitic plates [147].

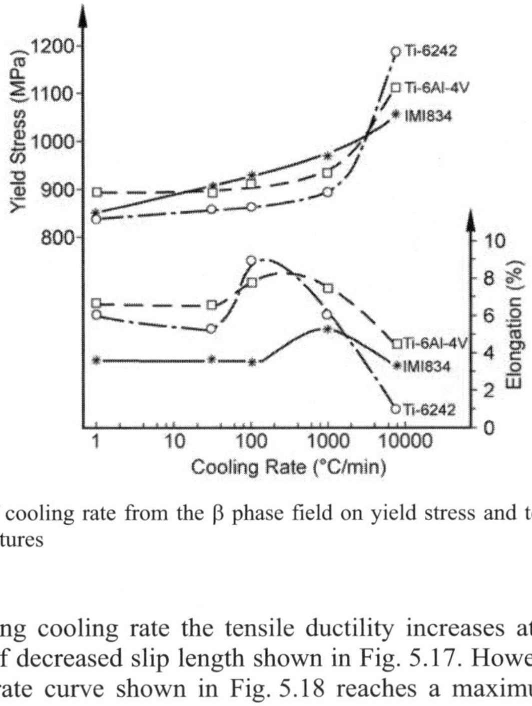

Fig. 5.18. Effect of cooling rate from the β phase field on yield stress and tensile elongation of lamellar microstructures

With increasing cooling rate the tensile ductility increases at first, consistent with the effect of decreased slip length shown in Fig. 5.17. However, the ductility versus cooling rate curve shown in Fig. 5.18 reaches a maximum and then declines. The variation in ductility which is shown in Fig. 5.18 in terms of tensile elongation is much larger for the corresponding RA values. The maximum in ductility corresponds to a change in fracture mode. At low cooling rates a ductile transcrystalline dimple type of fracture is observed, whereas at high cooling rates a ductile intercrystalline dimple type of fracture occurs along the continuous α layers at β grain boundaries. This change in fracture mode is shown in Fig. 5.19 [148]. The effect of the continuous α layers on ductility is related to the preferential plastic deformation in these areas and concomitant premature crack nucleation in these areas. The magnitude of the ductility decline depends primarily on the strength difference between these areas and the matrix and in addition on the grain boundary length (β grain size). The positive effect of reducing the β grain size is illustrated in Fig. 5.20 for the Ti-6Al-4V alloy. In this example, a large increase in tensile elongation can be achieved by reducing the β grain size from 600 μm to

100 μm by rapid heating [148]. To further illustrate this point, the tensile elongation for a bi-modal microstructure with a β grain size of 25 μm also is shown in Fig. 5.20. These data show that the bi-modal microstructure exhibits much higher ductility values in the fast cooling regime as compared to the two fully lamellar structures, although the appearance of the continuous α layers at β grain boundaries was similar for all three microstructures. As will be shown later (Sect. 7.2), the effect of continuous α layers on ductility and other mechanical properties is much more pronounced for high strength β alloys because of the higher strength difference between matrix and α layer regions for these alloys compared to α+β alloys. For this reason, the effect of continuous α layers on mechanical properties will be treated in detail in Sect. 7.2 and the α layers in α+β alloys are not listed as important microstructural parameter in Table 5.6.

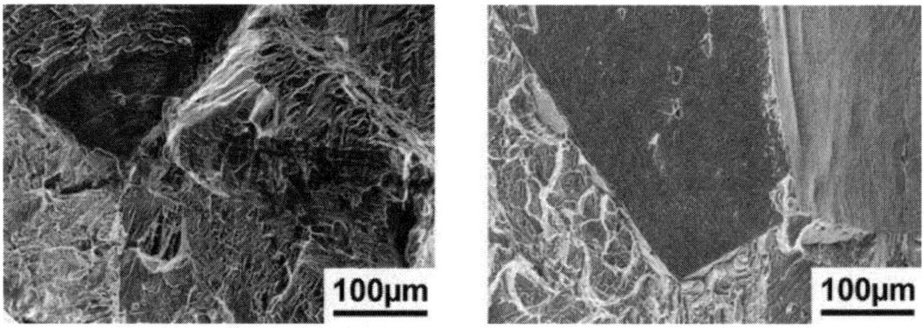

a b

Fig. 5.19. Tensile fracture surfaces of lamellar microstructures with different cooling rates from the β phase field (compare with Fig. 5.18), Ti-6242, SEM: (a) 100°C/min (b) 8000°C/min

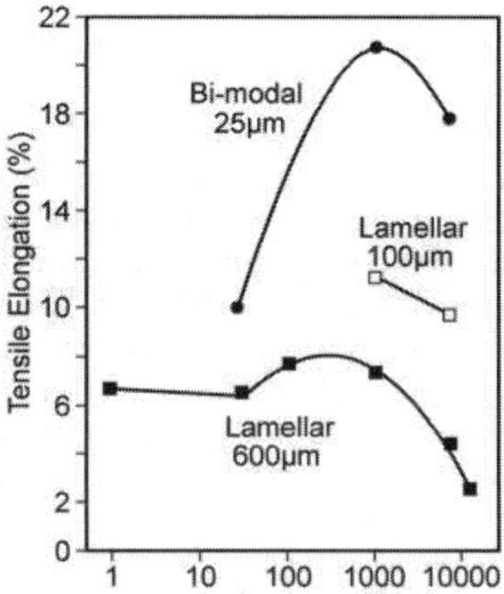

Fig. 5.20. Effect of cooling rate on tensile elongation of Ti-6Al-4V

The HCF strength (resistance to crack nucleation) depends primarily on the resistance to dislocation motion. Consequently, the HCF strength dependence on slip length and α colony size is qualitatively similar to that of the yield stress (see Fig. 5.17). As a function of cooling rate from the β phase field, the HCF strength shows the same tendency as the yield stress seen earlier in Fig. 5.18. Increased cooling rates in the slow to medium cooling rate regime results in a moderate HCF strength increase, but much larger increases are seen at faster cooling rates. This is illustrated for lamellar microstructures of the Ti-6Al-4V alloy in Fig. 5.21 where the HCF strength at 10^7 cycles is plotted as a function of cooling rate from the β phase field. The ratio of HCF strength (R = −1) to yield stress, which is typically about 0.5 for α+β alloys, can be as low as 0.45 for very coarse microstructures (for example for a cooling rate of 1°C/min) and as high as 0.60 for very fine microstructures. It should be pointed out that for fully lamellar microstructures the absolute values for both the HCF strength and the yield stress depend, in addition to the cooling rate effect, on the details of the final annealing/aging treatment (step IV in Fig. 5.1) as will be discussed in Sect. 5.2.4.

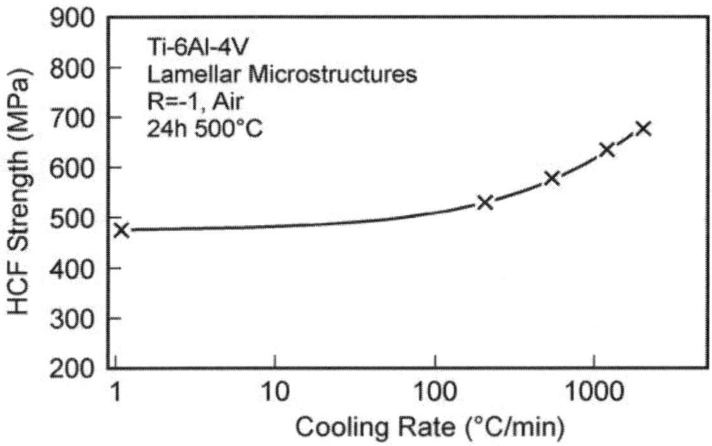

Fig. 5.21. Dependence of HCF strength (R = −1) of lamellar microstructures on cooling rate from the β phase field, Ti-6Al-4V

The fatigue cracks nucleate in the colony type of microstructures, formed at slow to medium cooling rates, either within pronounced slip bands, which extend over almost the whole colony width, or at the intersection of these slip bands with the adjacent α colony boundary. An example for this slip band crack nucleation mechanism is shown in Fig. 5.22a. In microstructures being composed more of individual α plates, characteristic of faster cooling rates, the fatigue cracks usually nucleate at the longest and widest α plates. An example for this crack nucleation mechanism is shown in Fig. 5.22b. It has been suggested [149] that the cracks nucleate due to preferred slip band activity within these coarse α plates. Fatigue cracks are only occasionally nucleated at the continuous α layers at β grain boundaries in these fast cooled microstructures of α+β alloys. This is a surprising result considering the observed fracture along the continuous α layers at β grain boundaries in tensile specimens (Fig. 5.19b) and the concomitant decrease in tensile ductility (Fig. 5.18).

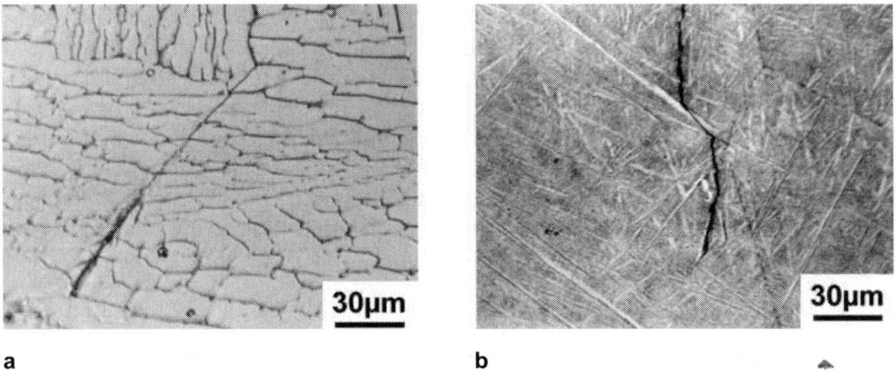

Fig. 5.22. Crack nucleation in lamellar microstructures with different cooling rates from the β phase field, Ti-6242, LM: (a) 1°C/min (b) 8000°C/min

The cooling rate from the β phase field also has a large influence on the propagation rate of the small, self-initiated surface cracks (microcracks). The data are shown in Fig. 5.23 for the Ti-6Al-4V alloy. Also indicated in this figure are the da/dN-ΔK curves for macrocracks which will be discussed later. In this figure, the two extreme microstructures are compared, a very coarse lamellar structure (1°C/min cooling rate) and a very fine lamellar structure (8000°C/min cooling rate). The scale for the surface crack length 2c of the microcracks is shown in the top part of Fig. 5.23 for additional information. It is evident from the curves in Fig. 5.23 that the microcracks (the two curves on the left) propagate much faster and grow at lower ΔK values than the macrocracks. Figure 5.23 also shows that the microcracks propagate faster in the coarse lamellar microstructure as compared to the fine lamellar microstructure. In the slowly cooled colony type of microstructure the microcracks propagate very fast within the intense slip bands (Fig. 5.22a). Only colony boundaries and β grain boundaries are strong obstacles because the microcracks have to change their direction when propagating across those boundaries. The density of colony boundaries increases with increasing cooling rate causing the propagation rate of microcracks to decrease. In case of very fine lamellar microstructures, the microcracks, which nucleated at the coarsest α plates as described above (Fig. 5.22b), tend to propagate initially along the plate interface but later depart from the interface and propagate through the matrix. In this fine lamellar microstructure, all individual martensite plates are strong obstacles and the resulting microcrack propagation rate is very slow, as can be seen from Fig. 5.23.

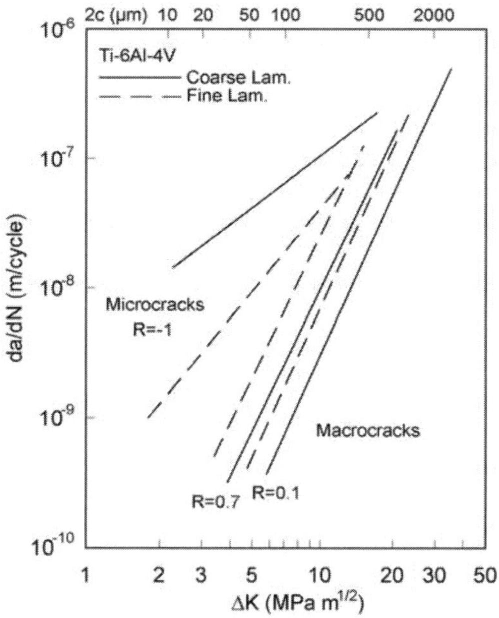

Fig. 5.23. Fatigue crack propagation of lamellar microstructures with different cooling rates from the β phase field; coarse lamellar (1°C/min) and fine lamellar (8000°C/min), Ti-6Al-4V

With increasing crack size the microcracks encounter more and more specific microstructural obstacles along the crack front. This leads to a lower average crack propagation rate as the cracks become larger. Crack nucleation and the first stages of microcrack propagation occur in a very selective way in the weakest areas of the microstructure, where no strong barriers to crack growth exist. As the cracks grow larger, it is statistically inevitable that the crack front will encounter stronger barriers and the crack propagation rate of microcracks slows down with increasing crack size due to this averaging process. The extent of crack propagation rate reduction with increasing crack front length due to this averaging process is different for coarse and fine lamellar microstructures. If in-plane obstacles were the only obstacles to crack growth, the crack propagation rate reduction would be greater for finer structures. Figure 5.23 shows that this is not the case, suggesting that additional factors must be considered. Two adjacent areas with different crystallographic orientations along the crack front, for example two α colonies in coarse structures or two individual α plates in fine structures, can cause the crack front to move locally out of the main propagation plane in directions determined by the different slip planes in the two adjacent areas. This is shown schematically in Fig. 5.24. Thus, the crack locally deviates from the average propagation plane over a distance which is proportional to the size of the microstructural feature. Thus, the crack front becomes bifurcated and the local crack tips in the two adjacent areas are separated by a distance Z perpendicular to the main propagation direction (Fig. 5.24). The distance Z is proportional either to the α colony size in coarse structures or to the size of individual α plates in fine structures. In order to move this bifurcated crack front forward, the distance Z has to be traversed by propagating the crack in a very unfavorable direction along very unfavorably oriented crystallographic planes. The strength of this crack propagation obstacle is directly proportional to the distance Z. Thus, the effect is large for coarse microstructures and small for fine microstructures. The resulting crack front profile (crack front geometry) determines to a large extent the position and order of the macrocrack propagation curves (average propagation behavior) in Fig. 5.23, as will be discussed later.

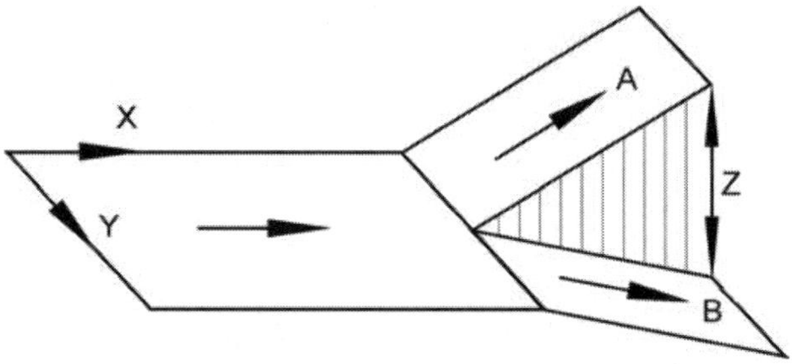

Fig. 5.24. Local crack front configuration with two adjacent areas A and B having different crystallographic orientations and therefore differently oriented major slip planes, i.e. crack planes (schematically)

The two factors described above lead to the often observed effect that the measured microcrack propagation curve depends on the stress level used in the experiment (LCF test) [150]. This is because microcracks do not have one unique da/dN-ΔK curve as a function of crack size. In principle, each microcrack size has its own characteristic da/dN-ΔK curve [151, 152]. With increasing crack size these curves are shifting in the direction of lower da/dN and higher ΔK values, i.e. the curves are shifting to the right in a da/dN-ΔK diagram. The magnitude of this shift is large for coarse lamellar microstructures and small for fine lamellar microstructures. This behavior is already incorporated in microcrack propagation curves such as those shown in Fig. 5.23. The crack size at which the average crack propagation behavior for a given microstructure is reached and from which on the crack propagation behavior of the microcracks will not change anymore with a further increase in crack size, i.e. the microcracks are now macrocracks, requires a large enough number of those microstructural barriers along the crack front which determine the crack propagation behavior, for example the α colony boundaries in case of the coarse lamellar microstructure. A number of twenty is usually assumed to be necessary because with this number it is often possible to describe the intersection point of the microcrack propagation curve with the corresponding macrocrack propagation curve at high R-ratio (Fig. 5.23).

For the overall discussion of the relationship between slip length and mechanical properties it is more convenient to plot ΔK_{th} in Fig. 5.17 instead of using the propagation rate to describe the microcrack propagation behavior. As a function of slip length the microcrack propagation behavior follows qualitatively the dependence of tensile ductility. This correlation is only valid if the crack propagation mechanism is basically the same in both the tensile and the fatigue test. The following might serve to illustrate this point. The fine lamellar microstructure has better microcrack propagation resistance than the coarse lamellar structure (Fig. 5.23) and both exhibit a transgranular propagation mechanism in fatigue. This does not correlate with the low tensile ductility of fine lamellar microstructures (Fig. 5.20, fast cooling rates). This is because fine lamellar structures with the low tensile ductility exhibit an intercrystalline tensile fracture mechanism as was shown in Fig. 5.19. If this intercrystalline fracture mechanism of fine lamellar structures is changed to a transcrystalline mechanism, for example by a reduction in β grain size, a high tensile ductility is observed (see Fig. 5.20) and the proposed correlation between tensile ductility and microcrack propagation resistance (Fig. 5.17) is valid.

The low cycle fatigue (LCF) strength of a material is a result of two contributing factors (see Fig. 5.17), its resistance to crack nucleation and its resistance to microcrack propagation. As outlined above, both factors increase with increasing cooling rate from the β phase field. As a result, the LCF strength of lamellar microstructures will increase with increasing cooling rate from the β phase field.

Understanding the measured propagation rates of large cracks (macrocracks), obtained from fracture mechanics type specimens, requires the knowledge of the basic microcrack propagation resistance and the additional factors related to the crack front geometry. At high R-ratio, i.e. without crack closure, the crack front geometry itself is the dominating additional factor. As pointed out earlier in the section on microcrack propagation, this factor also depends on slip length (α colony size) but with the opposite trend as compared to the basic microcrack propagation resistance, see Fig. 5.17. This geometrical term (often also called crack front roughness or crack front profile) increases with increasing slip length (α colony size), hindering crack propagation and increasing ΔK_{th}. Therefore, the crack propagation rate of macrocracks at high R-ratio can increase or decrease with increasing slip length depending on the slopes of the two contributing curves with negative and positive dependences on slip length as shown schematically in Fig. 5.17. For α+β titanium alloys it is usually observed that with increasing α colony size the propagation rate of macrocracks at high R-ratio is decreased (see Fig. 5.23), suggesting that the geometrical term is stronger than the "ductility" term describing the microcrack propagation behavior alone. This accounts for the opposite ranking of the propagation curves for microcracks and macrocracks (high R-ratio) as a function of cooling rate shown in Fig. 5.23.

For the propagation behavior of macrocracks at low R-ratios crack closure can be a third contributing factor. Crack closure results in higher ΔK_{th} values and increases with increasing roughness of the fracture surface and with increasing shear displacement at the crack tip (mode II contribution). Both of these increase with increasing slip length (α colony size). The resulting da/dN-ΔK curves for macrocracks at low R-ratios with the three contributions described in Fig. 5.17 show therefore even more than the curves at high R-ratios the tendency that with increasing α colony size (decreasing cooling rate) the propagation rate is lowered (Fig. 5.23). The comparison of the six curves in Fig. 5.23 illustrates the different dependence of microcrack and macrocrack propagation behavior on slip length as affected by cooling rate. It should be pointed out that, besides the effects mentioned so far, the normal mean stress effect in fatigue would have to be included in a more detailed and quantitative discussion of the fatigue crack propagation curves in Fig. 5.23. This factor is usually referred to as σ_{max} dependence. Such a discussion also would have to focus on the details of crack closure determination. It is felt that such a discussion exceeds the scope of this titanium book, especially since the qualitative correlations described in this section would not change.

The dependence of fracture toughness on α colony size (cooling rate) can be discussed qualitatively in a similar way as the dependence of ΔK_{th} for macrocracks without crack closure (Fig. 5.17). This analogy holds as long as the same basic fracture mechanism is occurring in both cases, i.e. for small and large plastic zone sizes ahead of the crack tip. As described already for fatigue crack propagation of macrocracks at high R-ratio (Fig. 5.23), the fracture toughness of α+β titanium alloys usually increases with increasing α colony size because the contribution from the rougher crack front profile dominates over the ductility term. Figure 5.25 shows as an example of this difference in crack path for a coarse lamellar microstructure (1°C/min cooling rate) and for a fine lamellar microstructure (8000°C/min cooling rate) in Ti-6Al-4V. The corresponding fracture toughness values are 75 MPa $m^{1/2}$ for the coarse structure and 50 MPa $m^{1/2}$ for the fine structure.

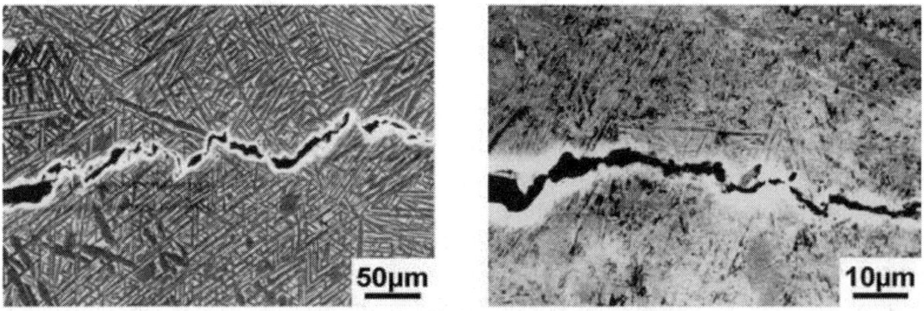

a b

Fig. 5.25. Crack path in center of fracture toughness specimens, Ti-6Al-4V, SEM: (a) Coarse lamellar (b) Fine lamellar

In the overview of the correlation between microstructure and mechanical properties (Table 5.6) all mechanical properties discussed in this section for lamellar microstructures exhibit a simple trend with respect to the cooling rate, see row "Small α Colonies and α Lamellae". These trends are either positive ($\sigma_{0.2}$, ε_F,

HCF, microcrack propagation resistance) or negative (macrocrack propagation resistance, fracture toughness) as a function of increasing cooling rate. As mentioned earlier, the last column in Table 5.6, the creep strength, will be discussed in the chapter on high temperature alloys (Sect. 6.2).

5.2.2 Bi-Modal Microstructures

The most influencial microstructural parameter on the mechanical properties of bi-modal microstructures is the relatively small β grain size. In correctly processed and well recrystallized structures this dimension is about equal to the distance between primary α (α_p) grains or particles. The β grain size is therefore determined by the volume fraction of α_p, i.e. the recrystallization annealing temperature (see Fig. 5.6 and Table 5.3), and by the α_p size, which is mainly influenced by the cooling rate from the β phase field prior to α+β processing as previously described (see Fig. 5.7 and Table 5.3). In commercially processed bi-modal microstructures, such as those shown in Fig. 5.7, the β grain size is in the range of about 30-70 μm. Excluded in the subsequent discussion of the mechanical properties of bi-modal microstructures are structures with such a high volume fraction of α_p (> 50 vol%) that the α_p grains start to interconnect. These structures will be included in the section on fully equiaxed structures (Sect. 5.2.3).

As pointed out in Sect. 5.1.2, for the whole range of commercial cooling rates (30-600°C/min) the α colony size of bi-modal microstructures is about equal to the β grain size, and therefore much smaller than in fully lamellar microstructures. The effect of slip length on mechanical properties of fully lamellar structures was discussed in Sect. 5.2.1 and is summarized in Table 5.6. Based on this, it can be postulated that if the slip length would be the only major parameter governing the mechanical properties of bi-modal microstructures, they should exhibit a higher yield stress, a higher ductility, a higher HCF strength, a slower fatigue crack propagation rate of microcracks, and a higher LCF strength than fully lamellar microstructures, when compared at constant cooling rate. Only the resistance to fatigue crack propagation of macrocracks and the fracture toughness should be better for fully lamellar microstructures and the creep resistance about equal for both microstructures. Looking in Table 5.6 at the row "Bi-Modal Structure", which indicates how the mechanical properties change comparing bi-modal structures with fully lamellar structures at a constant cooling rate, it is obvious that with the exception of HCF strength and creep strength the mechanical properties follow this general relation with slip length in a qualitative way.

The second important parameter for the mechanical properties of bi-modal microstructures is the alloy element partitioning effect which increases with increasing volume fraction of α_p as discussed in Sect. 5.1.2. This alloy element partitioning effect leads to a lower basic strength within the lamellar regions of the bi-modal microstructure as compared to a fully lamellar microstructure. The alloy element partitioning effect has only a very small effect on ductility and on the propagation behavior of microcracks and macrocracks including fracture toughness. The reason for this is that these fracture properties are not influenced by differences in yielding behavior, but are mainly determined by the first parameter, the α colony size.

The dependence of yield stress on α_p volume fraction is a combination of the contributions from α colony size and alloy element partitioning. The yield stress usually shows a maximum between 10-20 vol% α_p (Table 5.7). This indicates that, for small volume fractions of α_p, the α colony size effect dominates, whereas for large volume fractions of α_p, the alloy element partitioning effect dominates. At elevated temperature (600°C) the smaller decline in yield stress for high α_p volume fractions (Table 5.7) indicates that the alloy element partitioning effect is less pronounced at high test temperatures, probably because the strengthening effect of oxygen is reduced. Table 5.7 also shows clearly the higher ductility of the bi-modal microstructure as compared to the fully lamellar microstructure, see column for RA (reduction in area) in Table 5.7. This improved ductility is a result of the much smaller α colony size (slip length) of the bi-modal structure as compared to the lamellar structure.

Table 5.7. Tensile properties at room temperature and 600°C of the $\alpha+\beta$ alloy IMI 834

Microstructure	Test Temp.	$\sigma_{0.2}$ (MPa)	UTS (MPa)	σ_F (MPa)	El. (%)	RA (%)
Lamellar	RT	925	1 015	1 145	5	12
Bi-modal (20 vol% α_p)	RT	995	1 100	1 350	13	20
Bi-modal (30 vol% α_p)	RT	955	1 060	1 365	13	26
Lamellar	600°C	515	640	800	10	26
Bi-modal (10 vol% α_p)	600°C	570	695	885	10	30
Bi-modal (40 vol% α_p)	600°C	565	670	910	14	36

The HCF strength (crack nucleation resistance) is usually lowered with increasing α_p volume fraction (Fig. 5.26). The fatigue cracks are nucleated in the lamellar grains of the bi-modal structure (Fig. 5.27). These lamellar grains are softer than α_p as a consequence of the alloy element partitioning effect. This difference in strength can be clearly shown experimentally by microhardness measurements [153]. The continuous decline in HCF strength with increasing α_p volume fraction shows that, at low stress amplitudes with limited slip activity over short distances, the decline in basic strength of the lamellar regions (alloy element partitioning effect) is a stronger effect than the positive contribution due to the reduced α colony size. At the high test temperature of 600°C (Fig. 5.28) the HCF strength is equal or higher for bi-modal microstructures as compared to the fully lamellar microstructure, demonstrating again that the alloy element partitioning effect is less pronounced at high testing temperatures.

It should be emphasized that the above described influence of the alloy element partitioning effect on crack nucleation behavior is especially pronounced for aged microstructures. In unaged microstructures for which the final heat treatment is only a stress relieving treatment the effect would be much smaller. In addition, crack nucleation could then also occur in the primary α grains depending on α_p size. Furthermore, the alloy element partitioning effect also depends on alloy chemistry, e.g. the effect is large for IMI 834 and smaller for Ti-6Al-4V.

One practical way to eliminate the negative influence of the alloy element partitioning effect on the HCF strength at room temperature is to add an intermediate annealing treatment between the bi-modal recrystallization treatment and the final aging treatment. By such a treatment the α stabilizing elements, such as aluminum and oxygen, will diffuse from the primary α into the lamellar α regions resulting in an increase in strength of the lamellar α regions, which can be shown experimentally by microhardness measurements [153]. For example, an intermediate annealing treatment of 2 hours at 830°C improves the HCF strength of the bi-modal structure (Fig. 5.29), increasing it to the level of the fully lamellar structure or even slightly above (compare Fig. 5.29 with Fig. 5.26).

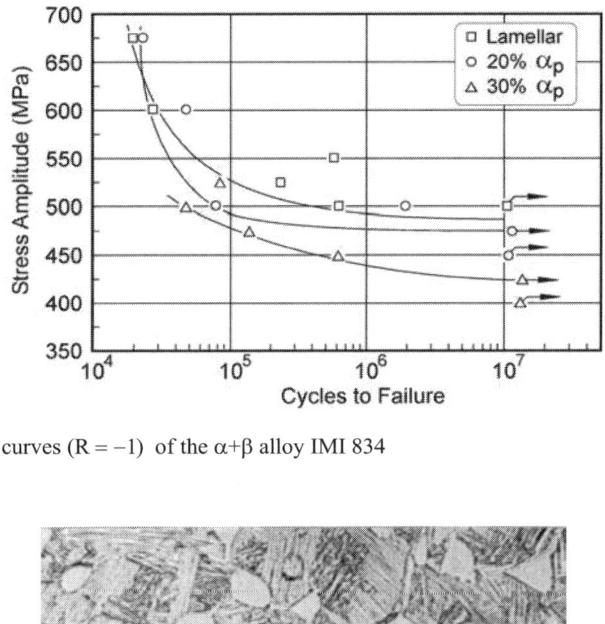

Fig. 5.26. HCF curves (R = −1) of the α+β alloy IMI 834

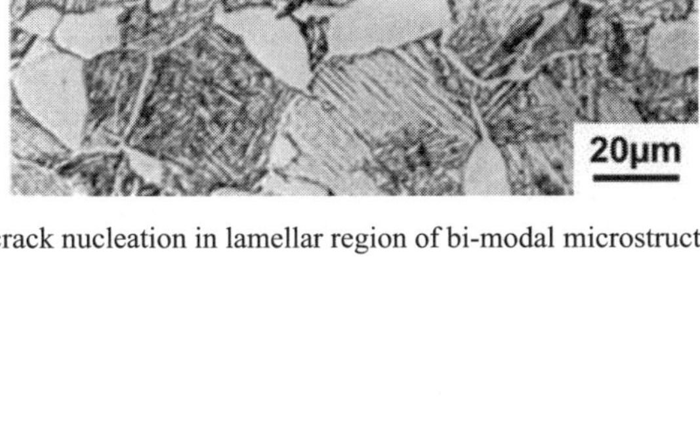

Fig. 5.27. Fatigue crack nucleation in lamellar region of bi-modal microstructure, IMI 834, LM

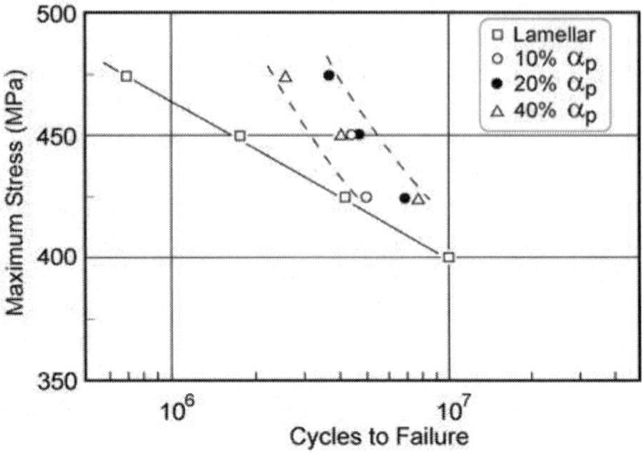

Fig. 5.28. HCF strength (R = 0.1) at 600°C of the α+β alloy IMI 834

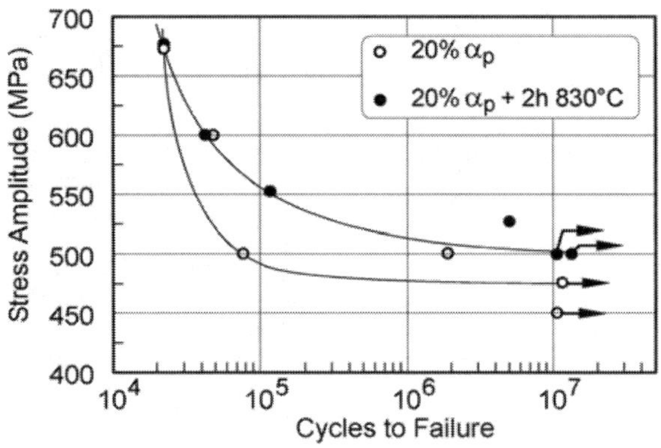

Fig. 5.29. Improvement in HCF strength of a bi-modal structure with 20 vol% α_p by an intermediate annealing treatment of 2 h at 830°C, IMI 834, see also Fig. 5.26

The two bi-modal microstructures shown previously in Fig. 5.7 and discussed in Sect. 5.1.2 provide a good example for the α colony size effect on HCF strength (Fig. 5.30) at a constant alloy element partitioning effect (α_p volume fraction is constant). For comparison, the results obtained on the corresponding fully lamellar microstructure (same cooling rate from the β phase field as the one used for the bi-modal microstructures from the α+β recrystallization temperature and identical final heat treatment of 2 h at 700°C) are also shown and discussed [144]. The comparison of the HCF strength in Fig. 5.30 of the two bi-modal microstructures demonstrates the effect of α colony size on crack nucleation. They are designated bi-modal 1 (large α colony size) and bi-modal 2 (small α colony size). The large alloy element partitioning effect can be seen again by comparing the HCF strength values of the coarse bi-modal 1 structure and the fully lamellar structure.

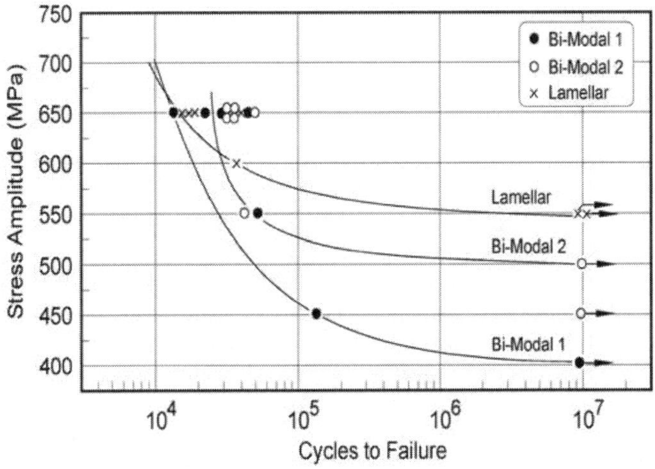

Fig. 5.30. HCF curves (R = −1) of the α+β alloy IMI 834, see Fig. 5.7 for microstructures bi-modal 1 and bi-modal 2

The LCF curves for these three different microstructures are shown in Fig. 5.31. It can be seen that the fully lamellar microstructure exhibits the lowest LCF strength. This behavior was already suggested by the partial crossover of the curves in Figs. 5.26 and 5.30 at high stress amplitudes. To separate the influence of crack nucleation and microcrack propagation on LCF life, the number of cycles until crack nucleation was investigated for the three microstructures and the results are shown in Table 5.8. It can be seen that the fatigue crack nucleation resistance at this relatively high stress amplitude of 650 MPa (see Fig. 5.30) was about the same for the lamellar and the coarser bi-modal 1 condition whereas the finer bi-modal 2 condition exhibited the highest crack nucleation resistance. Apparently, at high stress amplitudes and slip activity over long distances, the effect of α colony size (slip length) on crack nucleation resistance either compensates (bi-modal 1) or overcompensates (bi-modal 2) the alloy element partitioning effect on crack nucleation resistance.

The microcrack propagation curves for the coarser bi-modal 1 structure (Fig. 5.7a) and for the fully lamcllar structure are shown in Fig. 5.32. It can be seen that the microcrack propagation rate for the bi-modal structure was slower than for the lamellar structure demonstrating the positive effect of the smaller α colony size of the bi-modal structure on microcrack propagation resistance. No significant difference in microcrack propagation behavior existed for the two bi-modal structures shown in Fig. 5.7, which can also be deduced from the results shown in Table 5.8. Also shown in Fig. 5.32 are the macrocrack propagation curves exhibiting the anticipated opposite ranking in comparison to the micro-cracks because of the much rougher crack front profile of the lamellar structures as compared to the bi-modal structure. Also shown in Fig. 5.32 are the results for so-called short corner cracks measured on rectangular specimens including the corresponding scale for the crack depth "a" at the top of the diagram [144]. It can be seen that the propagation curves for the short corner cracks lie between the microcrack curves (self-initiated surface cracks on round, smooth specimens) and the macrocrack curves measured on CT specimens. The ranking and position of the corner crack curves for the two microstructures indicate that the range of crack front lengths covered by this data is slightly beyond the crossover of the two microcrack curves. In this case, the retarding effect of the developing rough crack front profile for the lamellar microstructure is about equal to or slightly larger than the intrinsically slower crack propagation behavior of the bi-modal microstructure.

As pointed out above, macrocracks have different crack front profiles in fully lamellar and bi-modal microstructures. Figure 5.33 illustrates this point by looking at the crack front in a plane perpendicular to the crack propagation direction. The large difference in crack front profile between the two microstructures is striking. The crack front profile for the coarse lamellar microstructure is very rough and the steps with the height Z, discussed earlier in Sect. 5.2.1 (see also Fig. 5.24), are clearly visible and lie in a very unfavorable orientation for crack propagation. In contrast, the crack front profile of the bi-modal microstructure is relatively smooth.

In agreement with this relatively smooth crack front profile of the bi-modal microstructure, the fracture toughness of bi-modal microstructures of the Ti-6Al-4V alloy is around 55 MPa $m^{1/2}$, being only slightly higher than the fracture toughness of the fine lamellar microstructure (50 MPa $m^{1/2}$) but much lower than the toughness of the coarse lamellar structure (75 MPa $m^{1/2}$), described already in Sect. 5.2.1.

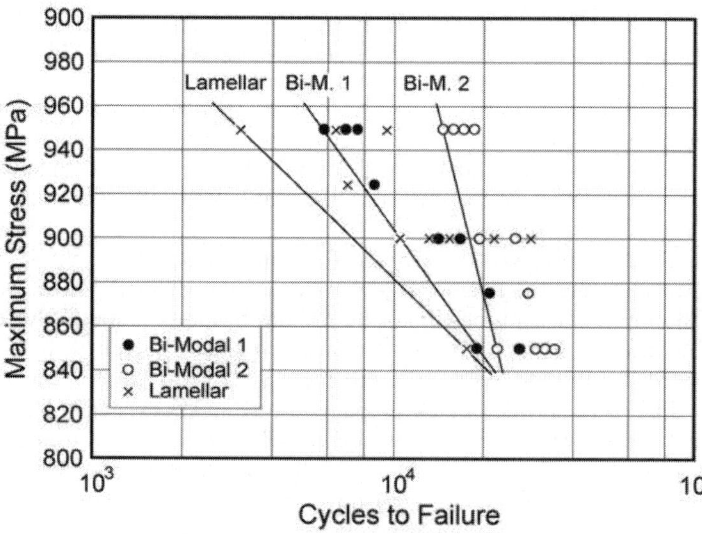

Fig. 5.31. LCF curves (R = 0.1) of the α+β alloy IMI 834, see Fig. 5.7 for microstructures bi-modal 1 and bi-modal 2

Table 5.8. Crack initiation (N_i), corresponding crack size (2c), and cycles to fracture (N_F) at 650 MPa stress amplitude (R = −1) of the α+β alloy IMI 834, see Fig. 5.7 for microstructures bi-modal 1 and bi-modal 2

Microstructure	σ_a (MPa)	N_i (Cycles)	2c (μm)	N_F (Cycles)
Lamellar	650	9 400	42	22 600
Bi-Modal 1	650	9 500	44	31 000
Bi-Modal 2	650	12 000	37	33 800

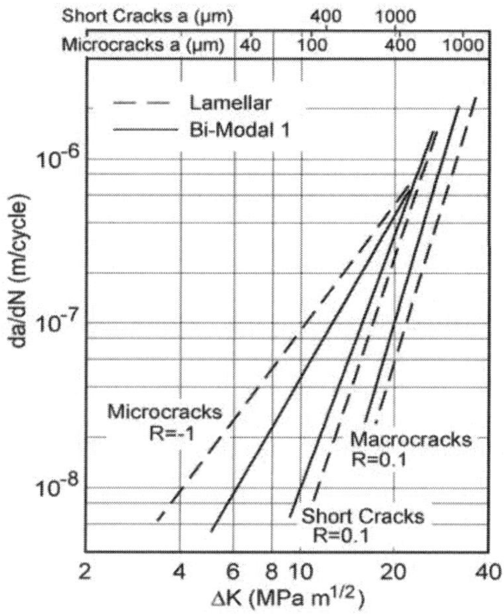

Fig. 5.32. Fatigue crack propagation of bi-modal 1 microstructure (Fig. 5.7a) and lamellar microstructure of the α+β alloy IMI 834, including short corner cracks

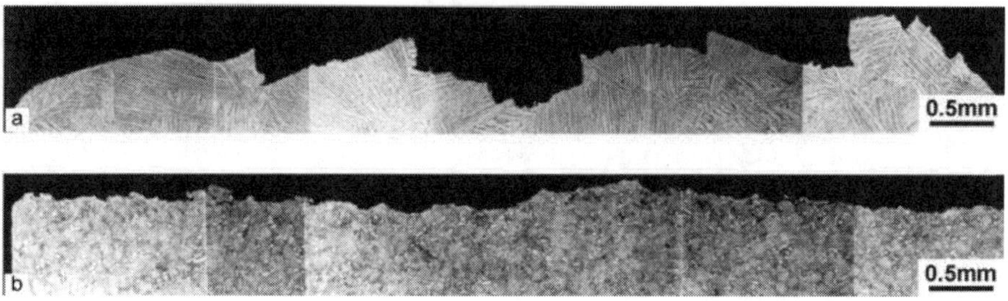

Fig. 5.33. Crack front profiles of macrocracks, Ti-6Al-4V, LM: (a) Coarse lamellar microstructure (b) Bi-modal microstructure

5.2.3 Fully Equiaxed Microstructures

The mechanical properties of fully equiaxed microstructures of α+β titanium alloys are primarily influenced by the α grain size. Since the α grain size determines the slip length, the qualitative correlation between slip length and mechanical properties, shown schematically in Fig. 5.17 for the example of α colony size of fully lamellar microstructures, can also be used for the α grain size of fully equiaxed microstructures. Due to this similarity, the trend symbols (+, –) in the summary table (Table 5.6) are the same for the rows "Small α Grain Size" in case of fully equiaxed microstructures and "Small α Colonies" in case of fully lamellar microstructures. Therefore, the correlation between mechanical properties and α grain size of fully equiaxed microstructures will not be discussed in detail, only a few examples will be given. It should be emphasized that fully equiaxed microstructures of α+β titanium alloys are similar to the microstructures of CP titanium materials and other α titanium alloys discussed in Chap. 4 (compare Figs. 5.11 and 5.14b with Fig. 4.6). Therefore, the general correlation between microstructure and mechanical properties is also similar.

The effect of α grain size on HCF strength of fully equiaxed microstructures is demonstrated in Fig. 5.34 for the Ti-6Al-4V alloy. It can be seen that high HCF strength values can be achieved in this range of small α grain sizes. The corresponding yield stress values are 1120 MPa (2 μm grain size), 1065 MPa (6 μm grain size), and 1030 MPa (12 μm grain size), respectively. The tensile ductility of these fully equiaxed microstructures are generally very high, i.e. equal or higher as compared to bi-modal microstructures. For example, the RA values for the 12 μm grain size is about 40% increasing to about 50% for the 2 μm grain size. It was already pointed out in Sect. 5.1.3 that the ultra-fine grain size of 2 μm can only be obtained on a laboratory scale and this microstructure is included only for comparison purposes.

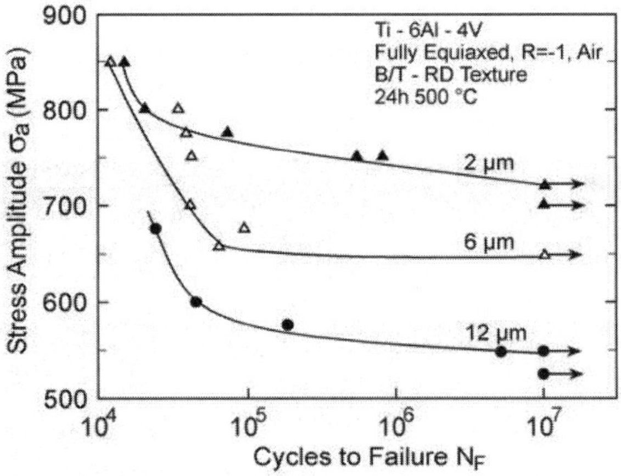

Fig. 5.34. Effect of α grain size on HCF strength of fully equiaxed microstructures, Ti-6Al-4V

When comparing the HCF strength values in Fig. 5.34 to values achievable for fully lamellar and bi-modal microstructures, the feasibility of obtaining these microstructures in commercial practice must be considered. When this is done, it can be concluded that for fully equiaxed microstructures the grain size region around 6 μm is very attractive for commercial application because the HCF strength (650 MPa) cannot be easily reached for fully lamellar and bi-modal microstructures. For example, in case of fully lamellar microstructures the required cooling rate from the β phase field would be in excess of 1000°C/min (see Fig. 5.21). This can only be obtained in thin section sizes. In case of bi-modal microstructures the temperature region for recrystallization is too high to achieve grain sizes below 10 μm in commercial practice.

An evaluation of the HCF strength of fully equiaxed microstructures in relation to bi-modal microstructures can best be done if the volume fraction of α_p is about 60 vol%. In this case, the α_p grains start to interconnect and are no longer separated by lamellar grains, as shown in Fig. 5.35. The expression "equiaxed" is used to describe this microstructure. A comparison of the HCF strength between a bi-modal microstructure and the "equiaxed" microstructure is shown in Fig. 5.36. In this case, the bi-modal microstructure was created from the "equiaxed" material by the method described in Sect. 5.1.3. The "equiaxed" material was heated to such a temperature in the (α+β) phase field that the volume fraction of α phase was reduced to about 40%. The subsequent cooling rate was identical to that used for the original "equiaxed" microstructure (100°C/min). It can be seen from Fig. 5.36 that the HCF strength of the "equiaxed" microstructure is lower than that of the bi-modal microstructure, i.e. the HCF strength of a bi-modal microstructure is reduced when the α_p grains start to interconnect. It should be pointed out that the crack nucleation site shifts from lamellar grains in case of a bi-modal microstructure (see Fig. 5.27) to interconnected α grains in case of the "equiaxed" microstructure. The relatively low HCF strength values in Fig. 5.36 are mainly a result of the final heat treatment (2h 700°C) which is only a stress relieving treatment and not an aging treatment. The importance of the final heat treatment for the mechanical properties will be discussed in the next section (Sect. 5.2.4). The yield stress values of the microstructures tested in Fig. 5.36 were 925 MPa for the bi-modal structures and 915 MPa for the "equiaxed" structure, respectively. The tensile ductility was the same for both microstructures (RA = 45%).

Comparing fully equiaxed and fully lamellar microstructures on the basis of thickness of α plates (lamellar structure) equal to the α grain size (equiaxed structure), then the HCF strength values of fully equiaxed microstructures are higher than the HCF strength values of fully lamellar microstructures. This is the expected result because in case of lamellar structures the α colony size is more appropriate for the slip length than the α plate thickness. It should be pointed out, however, that in case of fully equiaxed microstructures the slip length can also be larger than the α grain size, because the fully equiaxed structures exhibit in many cases strong crystallographic textures. The influence of crystallographic textures on mechanical properties for fully equiaxed and bi-modal microstructures will be discussed more in detail in a later section (Sect. 5.2.6).

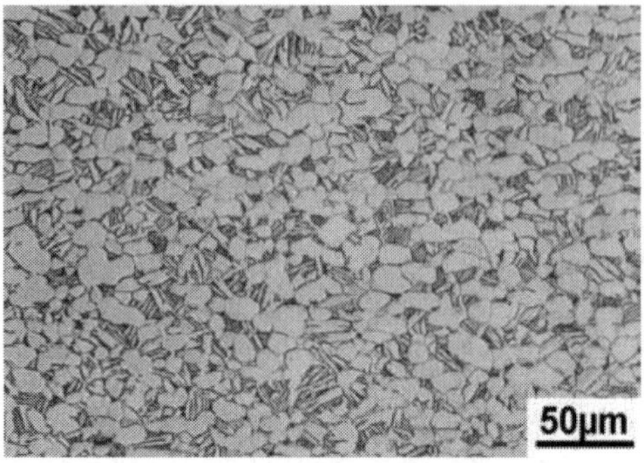

Fig. 5.35. A so-called "equiaxed" microstructure with such a high volume fraction of αp that α grains are interconnected, Ti-6Al-4V, LM

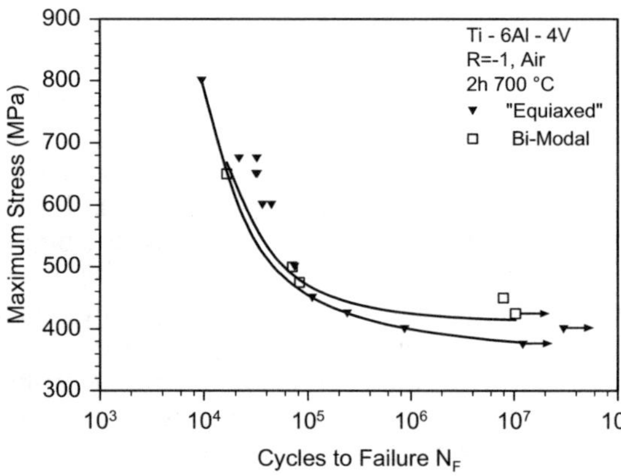

Fig. 5.36. HCF strength of a bi-modal microstructure and the "equiaxed" microstructure shown in Fig. 5.35, Ti-6Al-4V

An example for the fatigue crack propagation behavior of fully equiaxed microstructures is shown in Fig. 5.37 for the Ti-6Al-4V alloy with an α grain size of 6 µm. In this figure, the propagation rates of small, self-initiated surface cracks (microcracks) measured on round smooth specimens and large cracks (macrocracks) measured on fracture mechanics type specimens for two different R-ratios (R = 0.7 without crack closure, R = 0.1 with crack closure) are plotted in the same way as in Fig. 5.23 for fully lamellar microstructures. No systematic study of the effect of α grain size on microcrack propagation behavior has been done up to now for fully equiaxed microstructures. But taking the ranking observed for the Ti-8.6Al alloy [143], a model α titanium alloy with equiaxed α grain structure, and also the results on lamellar microstructures (Fig. 5.23), it can be concluded that the microcrack propagation rate of fully equiaxed microstructures will decrease with decreasing α grain size (see Table 5.6).

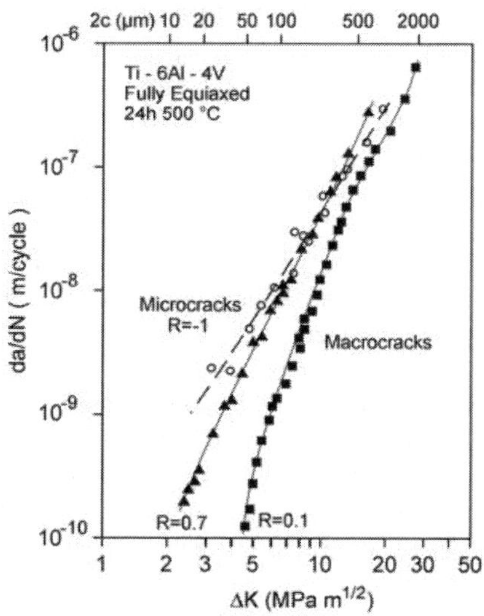

Fig. 5.37. Fatigue crack propagation of a fully equiaxed microstructure with 6 μm α grain size, Ti-6Al-4V

Since the crack front profile (crack front roughness) of macrocracks is fairly smooth for these fully equiaxed microstructures with relatively small α grain sizes, the difference between the microcrack curve and the two macrocrack curves in Fig. 5.37 is much smaller than for lamellar microstructures (see Fig. 5.23). The effect of α grain size on macrocrack propagation behavior of fully equiaxed microstructures has been studied on the Ti-6Al-4V alloy [154, 155] for grain sizes of 2 μm and 12 μm. The results show that the material with the larger α grain size exhibited slightly slower macrocrack propagation rates at both high and low R-ratios. These results indicate that, even for these relatively fine grained fully equiaxed microstructures, a sufficiently large grain size effect on crack front profile of macrocracks exists which explains the change in grain size ranking between microcracks and macrocracks (see Table 5.6).

To evaluate the differences in fatigue crack propagation behavior between fully equiaxed and bi-modal microstructures, the same two microstructures which were used for the HCF strength evaluation (Fig. 5.36) were tested. The results for microcracks and macrocracks are shown in Fig. 5.38. It can be seen that the microcracks in the bi-modal microstructure propagated at a slower rate as compared to the microcracks in the so-called "equiaxed" microstructure containing many interconnected α grains (Fig. 5.35). It was observed that the microcracks in the "equiaxed" microstructure showed the tendency to propagate through areas containing interconnected α grains whereas the microcracks in the bi-modal microstructure showed the tendency to propagate through the lamellar β grains. Although the macroscopic crystallographic texture measured by X-rays was similar for both microstructures, the slip length seems to be larger in areas with interconnected α grains ("equiaxed" structure) than in the lamellar β grains of the bi-modal structure. For the macrocracks, no measurable differences were found between the two microstructures (Fig. 5.38) indicating in a qualitative way that the crack front roughness was larger for the "equiaxed" structure in agreement with the larger slip length mentioned above.

The results obtained for fatigue crack nucleation resistance (HCF strength in Fig. 5.36) and for microcrack propagation resistance (Fig. 5.38) suggest that the LCF strength will be lower for fully equiaxed microstructures as compared to bi-modal microstructures. A similar evaluation of the effect of α grain size on LCF strength of fully equiaxed microstructures suggests that the LCF strength will increase with decreasing α grain size.

The effect of α grain size on fracture toughness of fully equiaxed microstructures was investigated by comparing microstructures with 2 μm and 12 μm grain size. The fracture toughness values were 45 and 65 MPa $m^{1/2}$, respectively. This result demonstrates the same tendency as found for lamellar microstructures, i.e. the crack front roughness term dominates over the inherent crack propagation resistance term, which was discussed already in detail at the end of Sect. 5.2.1.

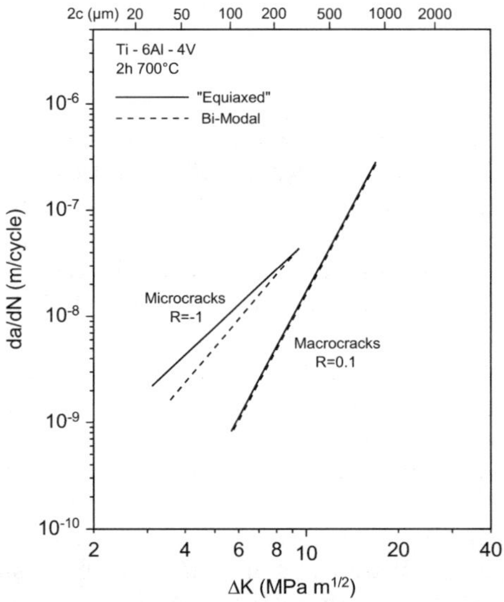

Fig. 5.38. Fatigue crack propagation of a bi-modal microstructure and the "equiaxed" microstructure shown in Fig. 5.35, Ti-6Al-4V

5.2.4 Effect of Aging and Oxygen Content

It was described in Chap. 2 (Sect. 2.8.1) that the α phase can be age-hardened by Ti_3Al (α_2) precipitates (see Fig. 2.20) if the alloy contains enough aluminum (about 6%). The elements Sn and especially oxygen also promote the α_2 phase formation. These coherent α_2 particles increase the yield stress but because they are sheared by the moving dislocations the presence of the α_2 particles results in the formation of intense, planar slip bands leading to easy crack nucleation (reduced tensile ductility) and also to fast crack propagation within these slip bands (fast microcrack propagation). The magnitude of these two effects depends on the slip length, i.e. the effects are very pronounced for coarse lamellar microstructures and less pronounced for bi-modal microstructures. For the Ti-6Al-4V alloy the solvus for α_2 precipitates lies between 550°C and 600°C depending on the exact aluminum and oxygen content. Therefore, this alloy forms α_2 particles if the common aging treatment of 24h 500°C is applied. If the final heat treatment is done at higher temperatures, for example in the region of 600-700°C, then the final heat treatment leads only to stress relieving and not to aging. It should be pointed out that the standard final heat treatment of typical high temperature $\alpha+\beta$ titanium alloys (Ti-6242, IMI 834) is below the α_2 solvus temperature (8h 595°C and 2h 700°C, respectively), i.e. these alloys always contain α_2 particles (see Sect. 2.8.1). Oxygen, besides promoting the formation of α_2, has an effect on mechanical properties which is qualitatively similar to the influence of aging. The effects of aging and oxygen on mechanical properties are therefore summarized by the same row in Table 5.6. The similarity is related to the observation that oxygen alone, i.e. in the absence of aluminum, can lead to the formation of intense slip bands (planar slip) as was shown for CP titanium in Chap. 4, see Fig. 4.9.

The influence of aging on the HCF strength is shown in Fig. 5.39 for a coarse lamellar microstructure of the Ti-6Al-4V alloy. This figure compares the aged condition (1h 800°C/WQ, 24h 500°C) with a stress relieved condition (1h 650°C). It can be seen that the HCF strength increased from about 350 MPa for the stress relieved condition to about 500 MPa for the aged condition. The corresponding yield stress values were 830 MPa and 930 MPa, respectively. The tensile ductility (RA values) dropped from 21% to 14%, respectively. A similar effect on tensile properties and HCF strength is observed for varying the oxygen content of the Ti-6Al-4V alloy. This is demonstrated in Fig. 5.40 by comparing the ELI grade (0.08% oxygen) to the regular grade (0.19% oxygen). This comparison was done on a fine lamellar microstructure. It can be seen that the HCF strength (10^7 cycles) is increased from about 480 MPa for the ELI material to about 580 MPa for the regular grade material. The corresponding yield stress values were 910 MPa and 990 MPa and the RA values 27% and 23%, respectively.

As mentioned earlier, the formation of intense slip bands should increase the microcrack propagation rate within these slip bands. Therefore, with increased aging and/or oxygen content, the microcrack propagation rate should increase. Since no results are available on $\alpha+\beta$ alloys, the effect of aging on microcrack propagation rate is demonstrated on the model α titanium alloy Ti-8.6Al (Fig. 5.41). It can be seen that the microcracks propagated at a faster rate in the aged condition (10h 500°C) as compared to the unaged condition. As pointed out previously in the section on fully equiaxed microstructures (Sect. 5.2.3), the tendency observed in Fig. 5.41 for the model α titanium alloy Ti-8.6Al should be applicable to $\alpha+\beta$ alloys. By varying the microstructure in $\alpha+\beta$ alloys, the increase in microcrack propagation rate with increased aging and/or oxygen content will be larger for coarse microstructures with large slip length as compared to fine microstructures, e.g. bi-modal microstructures.

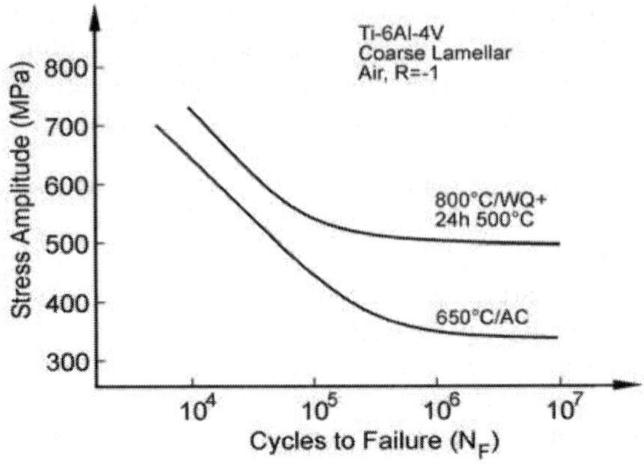

Fig. 5.39. Influence of final heat treatment on HCF strength, Ti-6Al-4V

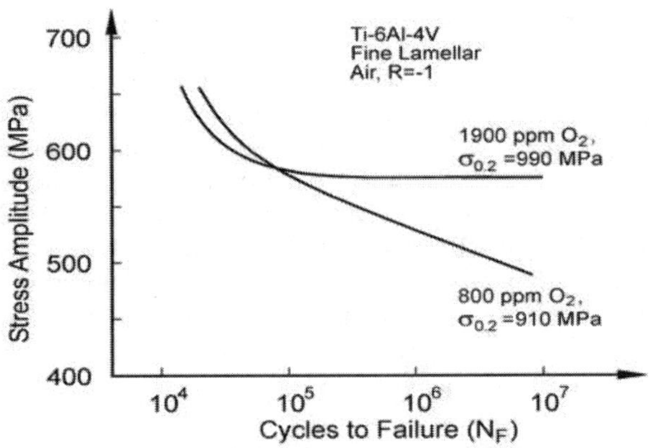

Fig. 5.40. Influence of oxygen content on HCF strength, Ti-6Al-4V

The LCF strength as a function of aging and oxygen content is difficult to predict because the resistance to crack nucleation (HCF strength) and the resistance to microcrack propagation are affected in opposite ways. Thus, it can come to a crossover in the S-N curves (see Fig. 5.40), i.e. the LCF strength is higher for the low oxygen material whereas the opposite is true for the HCF strength.

The influence of aging and oxygen content on the macrocrack propagation behavior is shown in Figs. 5.42 and 5.43, respectively. Again, the dependence is shown on the model α titanium alloy Ti-8.6 Al, because a complete set of curves at high and low R-ratios does not exist for $\alpha+\beta$ titanium alloys. The similarity of the influence of aging and oxygen content on macrocrack propagation behavior is obvious from the ranking of the curves in Figs. 5.42 and 5.43. At high R-ratio, i.e. without crack closure, the macrocrack propagation rate increases with increasing aging and increasing oxygen content. The same dependence existed for the propagation behavior of microcracks and this is in agreement with the observation that the retarding term for macrocrack propagation, the crack front roughness (see Sect. 5.2.1), was about the same for each set of curves shown in Figs. 5.42 and 5.43 [143]. All conditions exhibited crack propagation along slip bands and the slip length (α grain size) was kept constant for each set of curves resulting in about the same crack front roughness.

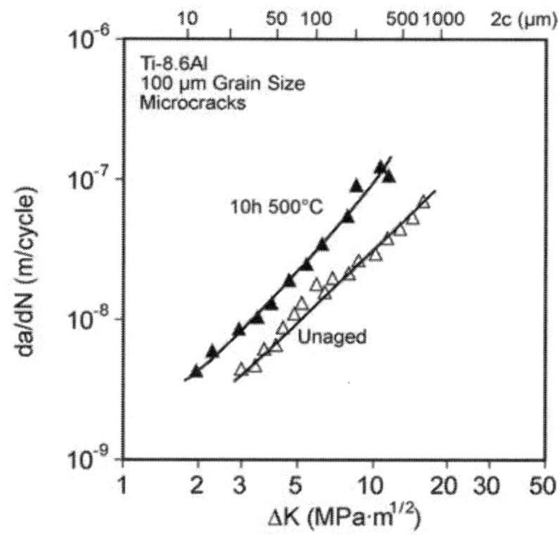

Fig. 5.41. Effect of aging on microcrack propagation behavior $(R = -1)$, Ti-8.6Al

For the macrocrack propagation behavior at low R-ratio, the ranking of the curves reversed as a function of aging and oxygen content (Figs. 5.42 and 5.43). The condition with the higher aging (260h 500°C) exhibited the slower crack propagation rate in the threshold region as compared to the condition with the lower aging (10h 500°C) and the macrocracks in the material with 0.1% oxygen propagated at a slower rate than in the material with 0.05% oxygen. The same ranking as a function of oxygen content was found on the Ti-6Al-4V alloy for macrocrack propagation at $R = 0.1$ [156]. The reversal in ranking between high R-ratio and low R-ratio can be explained by increasing crack closure values with increasing aging and increasing oxygen content. As pointed out in Sect. 5.2.1, crack closure will increase with increasing shear displacement at the crack tip which is consistent with higher aging and higher oxygen content.

The dependence of fracture toughness on aging and oxygen content should be qualitatively similar to the dependence of macrocracks at high R-ratio without crack closure, as outlined in Sect. 5.2.1. Therefore, the fracture toughness of the Ti-8.6Al material shown in Figs. 5.42 and 5.43 decreased with increasing aging and with increasing oxygen content [143]. In addition, there are numerous measurements on Ti-6Al-4V material comparing stress relieved versus aged conditions

[157] and materials with different oxygen contents [158]. Nearly all results show the tendency that the fracture toughness of α+β titanium alloys decreases with aging and with an increase in oxygen content. Besides the effect on fracture toughness just described, low temperature aging and oxygen have an even stronger negative effect on stress corrosion cracking susceptibility.

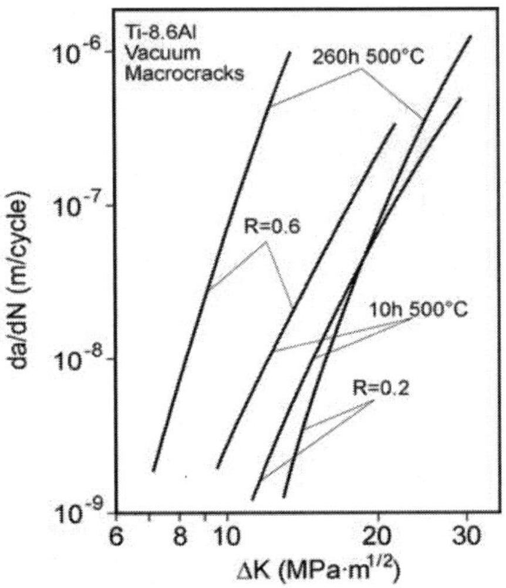

Fig. 5.42. Effect of aging on macrocrack propagation behavior, Ti-8.6Al

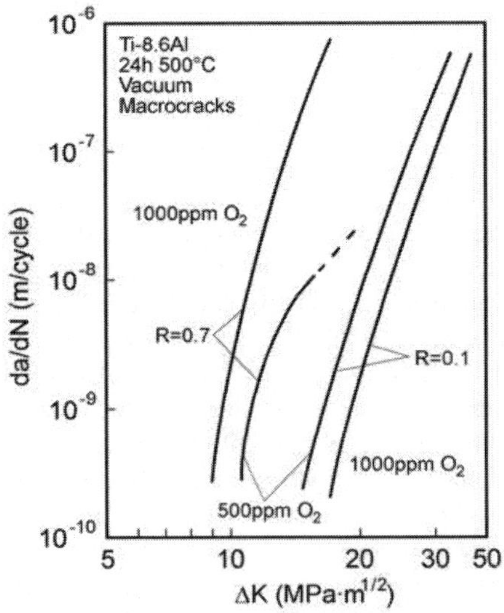

Fig. 5.43. Effect of oxygen content on macrocrack propagation behavior, Ti-8.6Al

5.2.5 Effect of Secondary Alpha in Beta Phase

As pointed out in Sect. 5.1.1, fine secondary α platelets can be precipitated in the β phase during the final heat treatment (step IV in the processing route, see for example in Fig. 5.1) depending on the details of the cooling process in the previous step of the processing route. The principal factor that governs the formation of secondary α platelets is the β phase composition which is directly connected to the volume fraction of β present in the microstructure before the final annealing treatment. As will be discussed, the hardening of the β phase by precipitated secondary α has a large, positive effect on mechanical properties. Therefore, it is essential to ensure that the α platelets form if the property benefits are to be realized. Since the cooling process is not selected to control the volume fraction of β present after cooling, it is difficult to predict whether or not precipitation of α platelets will occur during the final annealing treatment. Since the fine α platelets cannot be seen by light microscopy (TEM has to be done), in commercial practice it is simply not known in many cases whether the final microstructure contains secondary α platelets in the β phase or not.

A practical method to ensure that all β phase is hardened by fine secondary α platelets is to add an intermediate heat treatment before the final annealing treatment [159]. The temperature for this intermediate heat treatment must be high enough to create metastable β phase that can form α phase precipitates during the final aging treatment. An example of such microstructure is shown in Fig. 5.44 for the Ti-6242 alloy. The starting structure was a fully lamellar microstructure (see Fig. 5.3a for the starting microstructure without intermediate heat treatment) that has been changed in a controlled manner by an intermediate heat treatment in the (α+β) phase field. The resulting microstructure contains fine α platelets in all β "lamellae". The type of microstructure shown in Fig. 5.44 was called "bi-lamellar" to differentiate it from fully lamellar microstructure [160]. The greatest benefit of α precipitation in the β phase is for coarse fully lamellar microstructures. This is because the colony type structures have a long slip length and the change to a bi-lamellar structure reduces the slip length which has a large effect on mechanical properties. A good application for this special bi-lamellar microstructure are investment castings (see Sect. 5.3) which usually have a coarse lamellar microstructure and which can be heat treated but not thermo-mechanically processed to refine the microstructure. Details about the most important parameters for optimizing the bi-lamellar microstructure, such as the influence of starting microstructure, intermediate annealing temperature, and cooling rate can be found elsewhere [160]. It should be pointed out that α plates in the β "lamellae" can also be formed during cooling if the intermediate annealing temperature is high enough.

A summary of the mechanical properties of bi-lamellar microstructures is shown in Fig. 5.45 for the Ti-6Al-4V alloy as a function of cooling rate from the intermediate annealing temperature of 880°C. The bi-lamellar structure can be produced by cooling at rates as low as 17°C/min. Slower cooling rates result in a fully lamellar structure similar to the as-cast structure after hot isostatic pressing. The figure also includes the properties of the starting lamellar structure with a cooling rate of 1°C/min. The final heat treatment was the same for all structures

(24h 500°C). Compared to the fully lamellar starting structure, the yield stress and the HCF strength of bi-lamellar microstructures increased with increasing cooling rate because the hardened β "lamellae" in the bi-lamellar structure are effective barriers to slip transfer (reduction in effective slip length) and both the volume fraction and the strength of the hardened β "lamellae" are higher at fast cooling rates, i.e. no regrowth of the coarse α plates occurs at fast cooling rates. The tensile elongation is reduced from about 11% for the starting fully lamellar structure to about 4% for bi-lamellar structures with fast cooling rates. The mechanism for this ductility reduction is the same as discussed in Sect. 5.2.1. Due to the increase in strength of the matrix (interior of the β grains) the soft continuous α layers at β grain boundaries will deform preferentially leading to a ductile fracture along the α layers at β grain boundaries, similar to the fracture shown in Fig. 5.19. The propagation rate of microcracks is much slower in the bi-lamellar microstructures as compared to the fully lamellar structure (Fig. 5.45) because the hardened β "lamellae" are effective barriers for microcrack propagation. The LCF strength of bi-lamellar structures is also much higher as compared to fully lamellar structures because both the resistance to crack nucleation (HCF strength) and the resistance to microcrack propagation are increased. The resistance to macrocrack propagation is about the same for bi-lamellar structures as compared to the fully lamellar starting structure (Fig. 5.45). This is because the bi-lamellar structures exhibited a much smoother crack front profile than the fully lamellar structure and apparently this compensates roughly for the better inherent crack propagation resistance (microcrack propagation behavior) of the bi-lamellar structure. The same tendency as seen for the macrocrack propagation behavior in Fig. 5.45 is observed for the fracture toughness [161], see also the trend signs in the summary table (Table 5.6).

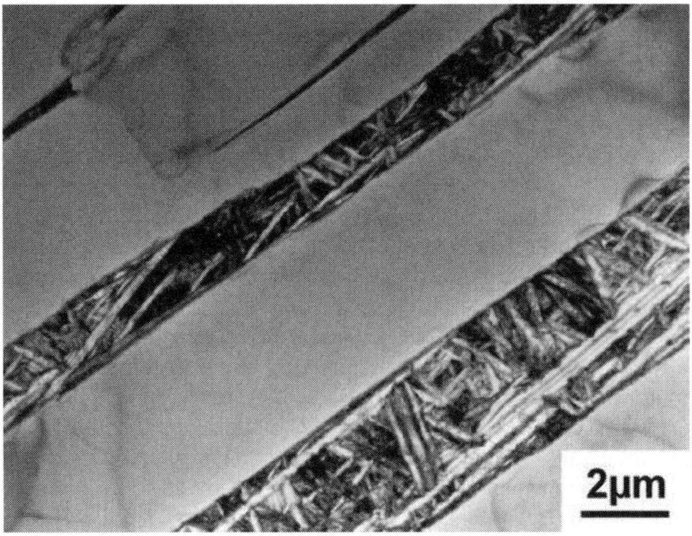

Fig. 5.44. Example of a so-called bi-lamellar microstructure showing secondary α platelets in β "lamellae", Ti-6242, TEM

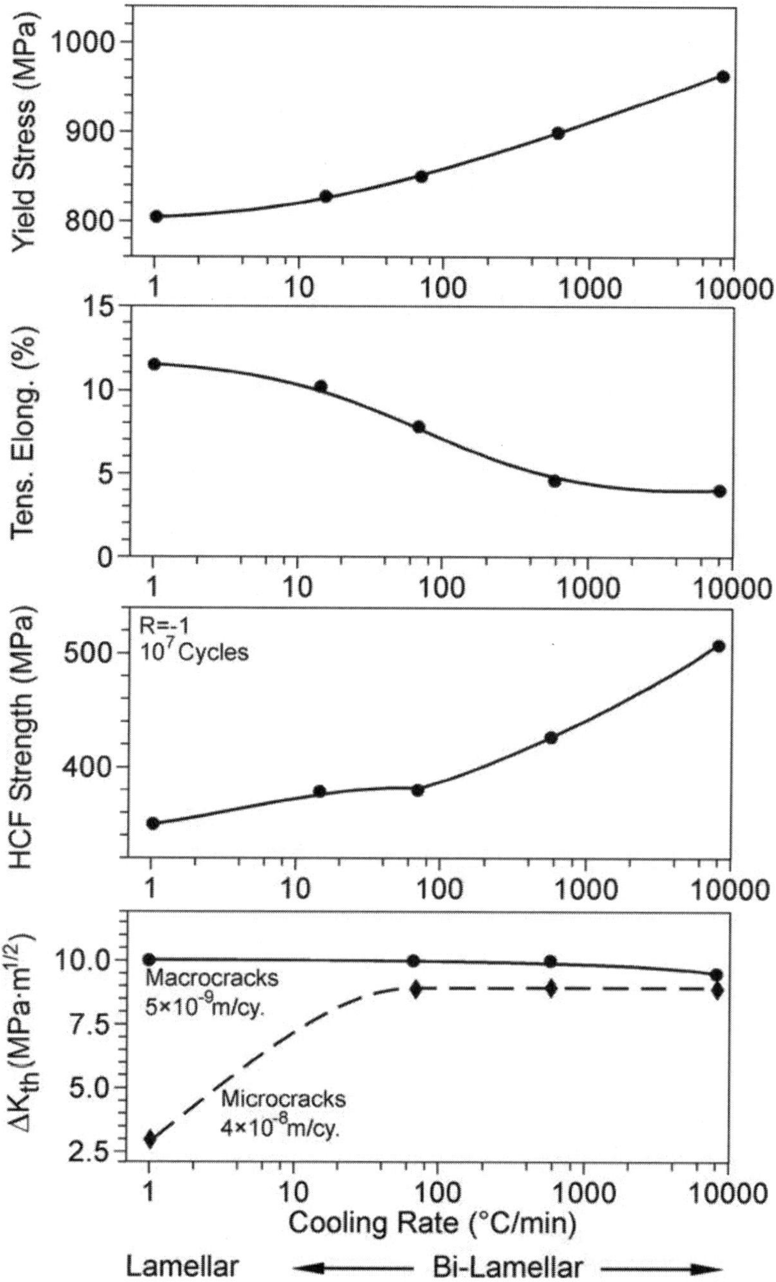

Fig. 5.45. Mechanical properties of so-called bi-lamellar microstructures as a function of cooling rate from the annealing temperature in the (α+β) phase field (880°C) including the starting lamellar structure at the cooling rate of 1°C/min, Ti-6Al-4V, final aging treatment: 24h 500°C

5.2.6 Effect of Crystallographic Texture

The effect of crystallographic texture on mechanical properties can be quite pronounced for both bi-modal and fully equiaxed microstructures. Both of the microstructures require a heavy deformation in the (α+β) phase field before recrystallization, which can lead to intense crystallographic textures. The development and the nature of the textures were described in Sect. 5.1.2 and the relation between deformation temperature and texture was shown in Fig. 5.8. The mechanical properties will be discussed for two common textures which are shown in Fig. 5.46. These are the transverse (T) texture (Fig. 5.46a) and the basal/transverse (B/T) texture (Fig. 5.46b). The difference in the processing route that leads to these two textures is the deformation temperature as discussed in Sect. 5.1.2 (see also Fig. 5.8). Deformation high in the (α+β) phase field results in the T texture, while deformation at lower temperature (for example for the Ti-6Al-4V alloy below about 900°C) leads to the B/T texture. The intensities of the textures in Fig. 5.46 (0002 pole figures) are quite high (8-12) because unidirectional rolled laboratory samples were used. For commercially produced material (sheet, plate, and forgings), the intensities would be much lower (about 4-6). In addition, the intensity of the T texture is usually lower as compared to the B/T texture because of the higher deformation temperature.

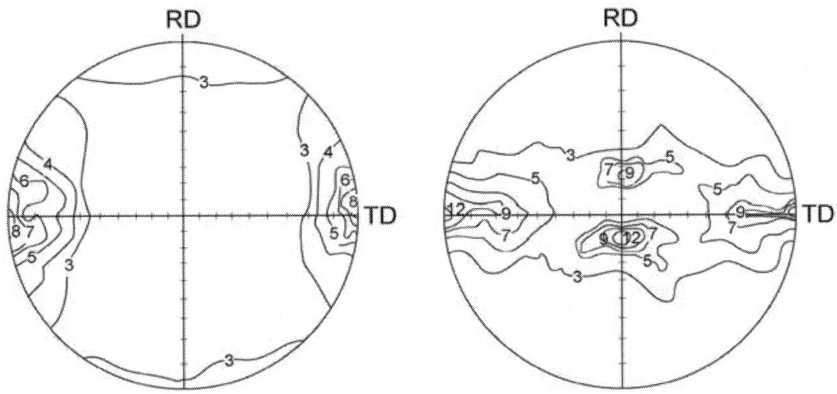

a b

Fig. 5.46. Common texture types of α phase (0002 pole figures) in Ti-6Al-4V with bi-modal or fully equiaxed microstructures: (a) Transverse texture (T) (b) Basal/Transverse texture (B/T)

The tensile properties for the two textures as a function of test direction are shown in Fig. 5.47 for a fine, fully equiaxed microstructure (α grain size about 2 μm) of the Ti-6Al-4V alloy. For an easier visualization of the angle between the basal planes and the test direction, the position of the hexagonal unit cell for the T texture is sketched in the lower part of Fig. 5.47. These sketches are also valid for the transverse component of the B/T texture, whereas for the basal component of the B/T texture the basal planes are in all cases nearly parallel to the test direction. For both texture types, the overall tendency in the variation of modulus of elasticity E as a function of test direction (Fig. 5.47a) is qualitatively the same as for α single crystals, see Fig. 2.3 [26]. The B/T texture has lower E values than the T texture because of the contribution from the basal (B) component having a low E value for all test directions. The more pronounced dependence on test direction for the B/T texture is due to the higher texture intensity as compared to the T texture. The E value for a random texture is about 119 GPa. The variation in yield stress as a function of test direction is shown in Fig. 5.47b. Both textures exhibit a minimum in yield stress for a test direction of 45° and the highest value in yield stress for testing in the transverse direction (TD). Again, the more pronounced dependence on test direction for the B/T texture is due to the higher texture intensity as compared to the T texture. The highest yield stress values for testing in TD direction can be explained by the fact that for the T texture as well as for the T component of the B/T texture the test direction is perpendicular to the basal plane. For such an orientation, dislocations with a Burgers vector $\vec{a}$ cannot be activated because this Burgers vector is parallel to the basal plane and the shear stress is zero in this case. Instead, dislocations with a Burgers vector $\vec{c}+\vec{a}$ have to be activated requiring a much higher stress to move. For both other test directions, RD and 45°, $\vec{a}$ slip can be readily activated. The low yield stress value for a test direction of 45° is due to the fact that one of the favorable slip planes, namely the basal plane, is oriented under 45° to the test direction, whereas in case of RD test direction all basal planes are parallel to the test direction and therefore cannot be activated for slip. The tensile ductility is not much affected by the loading direction and the texture type (Fig. 5.47b).

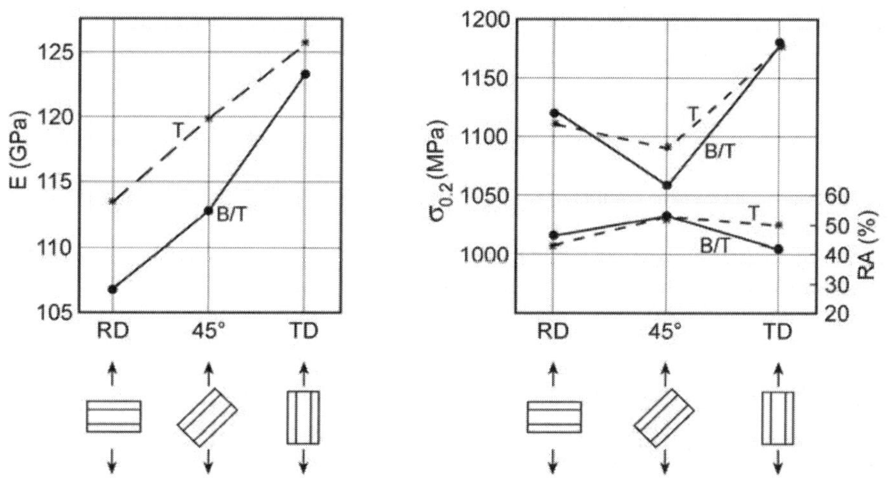

a b

Fig. 5.47. Influence of loading direction on tensile properties of fully equiaxed microstructures (α grain size: 2 μm) with B/T and T textures, Ti-6Al-4V: (a) Modulus of elasticity E (b) Yield stress $\sigma_{0.2}$ and ductility RA

The resistance to fatigue crack nucleation (HCF strength) is shown in Fig. 5.48 for the two texture types and for RD and TD loading directions. As pointed out in Sect. 2.9.2, crack nucleation and propagation along basal planes are strongly affected by the presence of hydrogen (relative humidity in laboratory air). Therefore, for an evaluation of the effect of crystallographic texture on HCF strength, S-N curves were measured in vacuum (Fig. 5.48a) and in laboratory air with about 40% relative humidity (Fig. 5.48b). The results in vacuum follow the tendency of the yield stress, i.e. tests in transverse direction (TD) exhibited a higher HCF strength than tests in rolling direction (RD). The material with T texture exhibited a lower HCF strength than the material with B/T texture because only one texture component is present in the T textured material resulting in large areas with similar grain orientations and, therefore, with a large effective slip length. In laboratory air (Fig. 5.48b) the fatigue strength in transverse loading direction (TD) is reduced drastically for both texture types whereas the fatigue strength in rolling direction (RD) is nearly unchanged (compare Fig. 5.48b to Fig. 5.48a). These results demonstrate the severe effect of hydrogen on crack nucleation along basal planes. The damaging effect is minimized for RD tests because all basal planes are aligned parallel to the loading direction, i.e. no shear stresses and no normal stresses are acting on the basal planes.

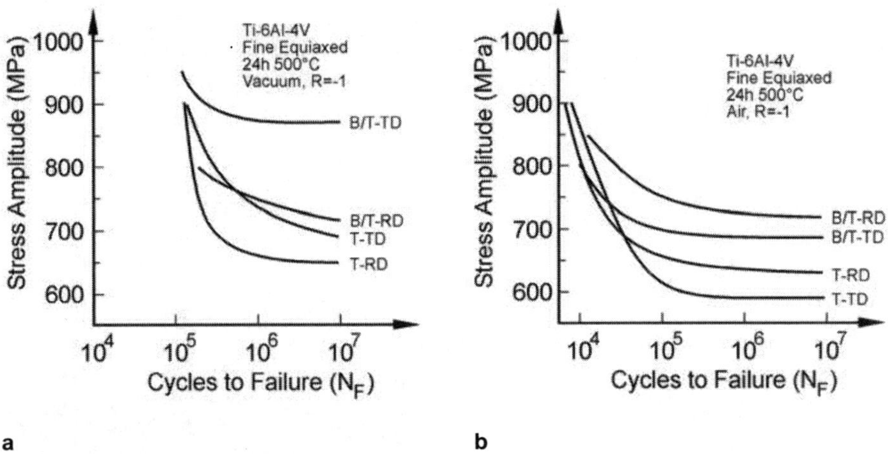

Fig. 5.48. Influence of texture and test direction on HCF strength, Ti-6Al-4V, fully equiaxed microstructures with α grain size of 2 μm: (a) Vacuum tests (b) Air tests

The effect of texture and loading direction on macrocrack propagation is shown in Fig. 5.49, again in vacuum (Fig. 5.49a) and in aggressive environment containing hydrogen, in this case 3.5% NaCl solution (Fig. 5.49b). It can be seen that there is not much of an effect of texture and test direction on macrocrack propagation in vacuum just as it was the case for tensile ductility (nucleation and propagation of cracks in the tensile specimen interior, i.e. also in vacuum). In the 3.5% NaCl solution (Fig. 5.49b), the macrocracks propagated much faster under

TD loading (basal planes perpendicular to the loading direction) as compared to RD loading (basal planes parallel to the loading direction). Although microcrack propagation behavior was not measured as a function of texture and loading direction, the effects should be qualitatively similar to those shown in Fig. 5.49, see also the trend sign in the summary table (Table 5.6). There is essentially no effect of texture and loading direction on fracture toughness (see Table 5.6) because unstable crack propagation is much too fast for any environmental/texture effect, but there is a texture effect on stress corrosion cracking susceptibility (K_{Iscc}) as pointed out already in an early overview article on stress corrosion and hydrogen embrittlement [162].

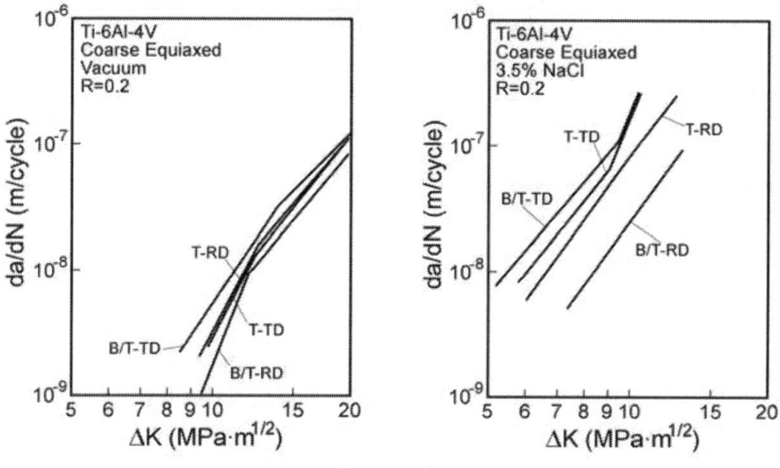

a b

Fig. 5.49. Influence of texture and test direction on macrocrack propagation behavior, Ti-6Al-4V, fully equiaxed microstructures with α grain size of 12 μm: (a) Vacuum tests (b) 3.5% NaCl tests

Other microstructural parameters that influence the effect of enviroment on mechanical properties include precipitation of coherent Ti_3Al particles, oxygen content, grain size (colony size), etc. All these microstructural parameters follow the general rule that with increasing tendency for planar slip and with increasing slip length the environmental effect on mechanical properties (fatigue crack nucleation and crack propagation) increases. For example, the environmental effects shown earlier for textured fully equiaxed microstructures are less pronounced for bi-modal microstructures with the same texture [120]. This behavior was explained by the fact that in bi-modal microstructures the strongly textured α_p grains are separated from each other by the lamellar grains whereas in fully equiaxed structures the textured α grains are interconnected [120].

Another effect which seems to be connected to crystallographic texture is the so-called "anomalous mean stress effect" on HCF strength [163, 164]. Lamellar microstructures and strongly textured bi-modal and fully equiaxed structures (for textures see Fig. 5.46) tested in RD direction exhibit a normal mean stress dependence of HCF strength whereas the textured materials tested in TD direction exhibit a much less pronounced increase in HCF strength with increasing mean stress. This "anomalous mean stress effect" was found for tests in air as well as for tests in vacuum, i.e. the effect is not depending on environment [163].

Noteworthy is that with increasing mean stress the fatigue crack nucleation site in the HCF regime shifted from the surface at ($R = -1$) to the interior of the specimens at $R \geq 0.1$ [163, 165]. Since this shift occurred for all microstructures (including lamellar microstructures) and test directions, it has no connection to the above described "anomalous mean stress effect". This shifting of the fatigue crack nucleation site can be explained by the preferred plastic deformation in the surface region resulting in case of tension-tension loading in residual compression stresses in the surface layer. These residual compressive stresses are balanced by residual tensile stresses deeper in the interior of the specimens favoring internal crack nucleation [163, 165].

The last two effects described ("anomalous mean stress effect" and interior crack nucleation) have been treated only briefly without illustrations because in both cases the basic mechanisms are not clarified completely, but both effects are worth mentioning here.

5.3 Properties and Application

This section on application of α+β titanium alloys principally deals with the application of the Ti-6Al-4V alloy because this is by far the most commonly used α+β titanium alloy, as can be seen in the alloy breakdown of the US market for 1998 in Fig. 1.3 [11]. The number of 20% deduced from this figure for all other α+β titanium alloys together as compared to 80% for Ti-6Al-4V would be even lower within the context of this book because this number includes the high temperature titanium alloys (e.g. Ti-6242, IMI 834), for which the application is separately addressed in this book in Sect. 6.3. In addition, published alloy breakdowns usually count alloys like Ti-17 and Ti-6246 as α+β alloys. These alloys are really β alloys and are treated as such in this book, see Chap. 7.

One major application area of α+β titanium alloys are aircraft structural parts. For such applications, α+β titanium alloys are selected over other competing metallic materials, such as high strength aluminum alloys, because of higher yield stress and fatigue strength (even on a density normalized basis), better corrosion resistance, higher modulus of elasticity, and higher temperature capability. Two heavy Ti-6Al-4V forgings which were discussed earlier in this book can be seen in Fig. 3.23 (rough forging of a landing gear beam for the Boeing 747 aircraft) and in Fig. 3.22 (bulkhead forging for a twin engine military aircraft with the machined part shown in Fig. 3.25). For such large structural parts, the most important mechanical property would be fatigue crack propagation of macrocracks resulting in a requirement for setting the service time between inspection. Because these large components are located well within the interior of the aircraft structure, considerable disassembly is required for inspection. Thus, good fatigue crack growth characteristics are highly desirable. In addition, a high fracture toughness is usually a requirement although the actual fracture toughness value has an insignificant influence on fatigue lifetime and essentially no influence on the inspection intervals. Thus, the high fracture toughness requirement can be viewed as largely redundant. The most economically processing route for large forgings typically consists of forging in the (α+β) phase field followed by mill-annealing. As outlined in Sect. 5.1.3, the resulting microstructure is not well defined but will consist of nominally equiaxed α with a varying extent of recrystallization. This mill-annealed structure exhibits much faster fatigue crack propagation rates of macrocracks than a fully lamellar microstructure (see Sect. 5.2.3). The fully lamellar structure can be produced by β annealing, which is a more expensive alternative. Although the β annealing treatment requires fixtures to support the part, the benefits of improved crack propagation resistance are in some cases sufficient to justify the added cost. This is sometimes done for safety critical aircraft structural components such as bulkheads, cockpit window frames, and attachment fittings for the fin and horizontal stabilizers.

Castings exhibiting in any case already the desired fully lamellar microstructure also have a size limitation resulting from the casting furnace equipment and the hot isostatic press facilities, see Sect. 3.5.1. Nevertheless, fairly large castings have been proposed for use as aircraft structural components. As an example, a cast Ti-6Al-4V wing attach fitting for the F-22 military aircraft is shown in Fig. 5.50. Due to the size of the wing attach fitting and the section sizes involved, the solidification rate will be quite slow resulting in a fairly coarse lamellar microstructure with excellent fatigue crack growth resistance for macrocracks and also with a high fracture toughness.

Another major application area of α+β titanium alloys are rotating and non-rotating parts in aero-engines. In the schematic of the GE-90 aero-engine (Fig. 1.6) some of the titanium parts are labelled. The major limitation of Ti-6Al-4V is the maximum application temperature of about 300°C which restricts the usage of this alloy in the rotating section of the engine to the fan stage, to the low pressure (LP) compressor section, and to the front stages in the high pressure (HP) compressor. Further back in the HP compressor the temperature is so high that the so-called high temperature titanium alloys (Ti-6242, IMI 834) have to be used (see Sect. 6.3).

Fan and compressor blades can experience large oscillatory stresses and, therefore, require a high fatigue crack nucleation resistance (HCF strength) in the airfoil section and good LCF strength in the attachment locations. Consequently, the α+β forged Ti-6Al-4V blades are usually recrystallized to a bi-modal microstructure that exhibits a higher HCF strength than equiaxed or mill-annealed microstructures (see Fig. 5.36). Typical large fan blades of Ti-6Al-4V are shown in Fig. 5.51. The length of the larger blade in Fig. 5.51 is about 1 m. If the fan blades have to be longer and wider because of engine design, for example in the recent GE-90 engine, then the solid Ti-6Al-4V blades are replaced by polymer composite blades avoiding too high stresses on the fan disk. Other engine manufacturers (RR and PW) use hollow Ti-6Al-4V blades for this kind of application.

Fig. 5.50. Cast Ti-6Al-4V wing attach fitting for the F-22 military aircraft (courtesy Neil Paton, Howmet Corporation)

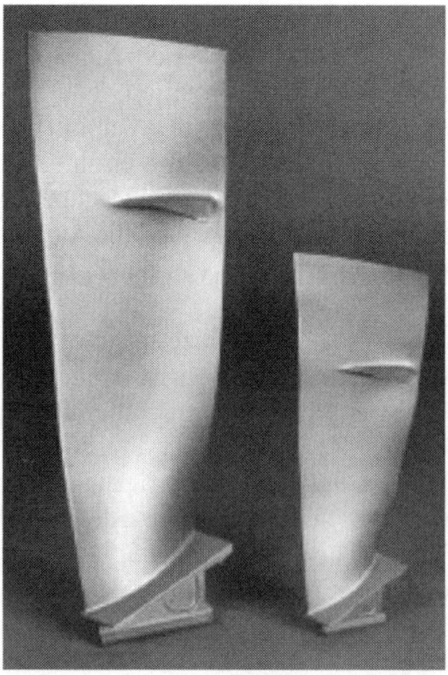

Fig. 5.51. Large fan blades (length of the larger blade about 1 m) of Ti-6Al-4V forged in the (α+β) phase field and recrystallized to a bi-modal microstructure (courtesy GE Aircraft Engines)

The Ti-6Al-4V alloy is also the most common disk material. For disks, LCF life is often limiting because each flight requires maximum power which counts as one cycle in the LCF life. Since the LCF strength is a combination of fatigue crack nucleation resistance and microcrack propagation resistance, both have to be taken into account. In the region of interest for the number of cycles to failure (about 2-$5x10^4$ cycles), both contributions are about 50%. From all possible microstructures, the bi-modal microstructure is selected for disk material because of the much better LCF strength and microcrack propagation resistance as compared to the fully lamellar microstructure (see Fig. 5.31 and 5.32) as well as better and more reproducible properties than fully equiaxed and mill-annealed structures (see discussion in Sect. 5.3.3 and Figs. 5.36 and 5.38). It should be pointed out, that for fatigue life calculations of disks based on fatigue crack propagation data, it is absolutely necessary to consider microcrack propagation data because macrocrack propagation data would result in a serious overestimation of cycles to failure [166].

The new technology of manufacturing integrated rotor components by direct attachment of the blades to the disk using solid state linear friction welding (see Sect. 3.6.2) is also done with Ti-6Al-4V material. The integrally bladed low pressure compressor rotor of Ti-6Al-4V, such as that shown in Fig. 3.59, is used in the new EJ 200 engine for the Eurofighter [167]. In general, the weld between the blade and the disk, both having bi-modal microstructures, consists of a very fine fully lamellar microstructure difficult to resolve by light microscopy [97]. Due to the high plastic strains involved, combined with the high temperature in the β phase field and the fast cooling rate, the resulting size of the recrystallized β grains is very small (about 5 µm). Such a microstructure (very fine fully lamellar with very small β grain size), which cannot be achieved by conventional processing methods, exhibits an extremely small slip length and therefore (see Fig. 5.17) superior fatigue crack nucleation and microcrack propagation resistance (HCF and LCF strength) as compared to the conventional bi-modal structures of the blades and the disk. Actually, the microstructure of the friction weld behaves like a bi-modal microstructure with small β grain size but without any alloying element partitioning effect, see discussion in Sect. 5.2.2 and Table 5.6. Therefore, although not directly relevant for the integrally bladed Ti-6Al-4V rotor, it should be pointed out that this friction welded microstructure will have a fairly low fracture toughness and a fairly low fatigue crack propagation resistance of macrocracks resulting from the very smooth crack front geometry due to the small β grain size (see Fig. 5.17).

The HP compressor spool for the GE CF6 class engine, shown in Fig. 5.52, is another part for which friction welding is used. In this case (axially symmetrical part), the individual stages are joined using rotational inertia welding (see Sect. 3.6.2). The front stages are made out of Ti-6Al-4V but the temperature is too high for Ti-6Al-4V for the last two stages and the high temperature alloy Ti-6242 has to be used to achieve the necessary creep resistance (see Chap. 6). Bi-modal type microstructures are used for both the Ti-6Al-4V and the Ti-6242 alloys because the LCF strength is the most important mechanical property for this kind of application and the bi-modal microstructure is the best choice for this application as outlined previously for disk materials. In some cases, the Ti-6242 portions are β forged because creep strength is more critical than LCF strength. This decision is very application specific. For rotational inertia welding, the microstructure in the weld is similar to the one described above for linear friction welding. Therefore, the above discussion about the mechanical properties of linear friction welds also applies to inertia welding. It should be mentioned, that connecting different titanium alloys by inertia welding, as done for the Ti-6Al-4V and Ti-6242 stages of the HP compressor spool shown in Fig. 5.52, is accomplished without difficulty and causes no problems.

Fig. 5.52. Compressor spool for GE CF6 class engine using inertia welding to connect the individual stages; front stages: Ti-6Al-4V, rear stages: Ti-6242 (courtesy GE Aircraft Engines)

Non-rotating parts in aero-engines made out of Ti-6Al-4V material can be casings, ducts, frames, stators, manifolds, etc., depending on the type of engine (see Fig. 1.6). The selection criteria over competing materials (mainly aluminum alloys) are higher stiffness, higher ratio of yield stress to density, higher temperature capability, and better resistance to burning. Fatigue properties are not so much of concern, although for some non-rotating parts vibrations can occur and have to be taken into account in that case.

As an example, a Ti-6Al-4V compressor casing (outer-duct) can be seen in Fig. 3.71. The casing is made from rolled plate and fabricated by conventional sheet forming methods (Sect. 3.5.4). The plate has a fine grained mill-annealed microstructure for good forming properties (high ductility). Another example, a Ti-6Al-4V manifold can be seen in Fig. 3.48. The manifold is used to distribute cooling air for the variable clearance control system of the LP turbine case of a turbofan aircraft engine. The manifold is made by superplastic forming and diffusion bonding (SPF/DB) (see Sect. 3.5.5) and has therefore a fine grained fully equiaxed microstructure because it is slowly cooled after the forming and bonding process. Non-rotating parts made from wrought material are often replaced by investment castings (Sect. 3.5.1), especially if they are of complex shape. Ti-6Al-4V investment cast fan frames (see Fig. 3.29) are a good example. In former times they were fabricated from multi-pieces wrought parts. Fan frames can be subjected to serious vibrations, therefore, besides stiffness and strength, resistance to fatigue crack nucleation has to be taken into account. Since cast fan frames have a fully lamellar microstructure, which has a relatively low resistance to fatigue crack nucleation and to microcrack propagation, cast fan frames might serve as a good example to increase these properties through the creation of a so-called bi-lamellar microstructure by an additional heat treatment (see Sect. 5.2.5). An even better example for the usefulness of bi-lamellar microstructures might be cast non-rotating parts in helicopter engines. As an example, a Ti-6Al-4V investment casting of a transmission adapter case for a large Bell Helicopter is shown in Fig. 5.53. Besides many similarities of aero-engines for airplanes and helicopters, all parts of a helicopter are subjected to serious vibration stresses due to the rotation of the main rotor. Consequently, fatigue strength is even more important.

The above mentioned application examples of Ti-6Al-4V in the aerospace area are by no means complete. They were selected to illustrate within the scope of this book the relationship between processing, microstructure, properties, and application. More examples can be found elsewhere [168, 169].

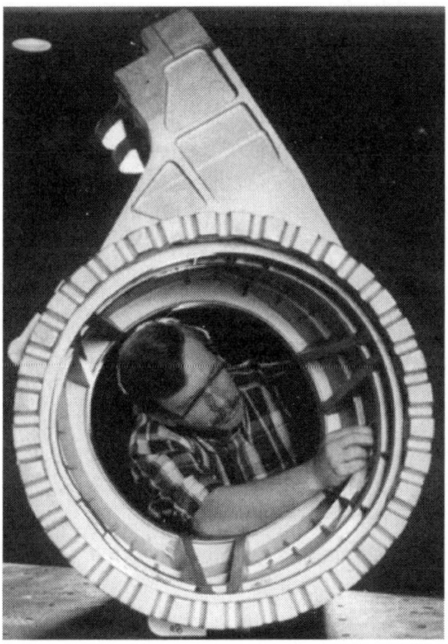

Fig. 5.53. Ti-6Al-4V casting of a Bell Helicopter transmission adapter case (courtesy Neil Paton, Howmet Corporation)

In the area of power generation, besides the extensive use of CP titanium as heat exchanger tubes and tube sheets (see Sect. 4.3), there is also a limited application of Ti-6Al-4V as steam turbine blading material [170]. The use of Ti-6Al-4V blades is limited to the long last stage and the second-from-last (L-1) stage of low pressure steam turbines replacing the common 12Cr steel blades because of lighter weight in case of the last stage and because of better pitting corrosion resistance in case of the L-1 stage. An example of a large steam turbine rotor is shown in Fig. 5.54. Large individual titanium blades can be seen in Fig. 5.55. The length of the Ti-6Al-4V blades can be up to about 1.4 m. The most relevant mechanical property for design is the HCF strength and the microstructure used in these long steam turbine blades is the bi-modal microstructure. The manufacturing process [171] of these long steam turbine blades with bi-modal microstructure resulted in better and more reproducible HCF strength values as compared to conventional mill-annealed blades. The bi-modal blades also exhibited a much smaller scatter-band in modulus of elasticity (below 3% as compared to 10% for mill-annealed blades) which is an important parameter for managing the vibration characteristics of the blades ("tuning of the rotor").

In the area of offshore oil and gas production, Ti-6Al-4V turbular riser systems offer a big advantage over competing steel material because of the low modulus of elasticity (50% of steel) combined with better seawater fatigue resistance. In particular, the low modulus of elasticity in combination with taper stress joint technology allows larger deflection of risers in deep-sea floating production systems (see the schematic illustration in Fig. 1.7a). An example for the scale of such taper stress joints is shown in Fig. 5.56 [19]. The ELI type version of Ti-6Al-4V is commonly used as material, although the addition of 0.1 Ru would further improve the basic corrosion resistance [19], as outlined already for CP titanium in Sect. 4.2. Usually, turbular sections of Ti-6Al-4V would be in the mill-annealed condition, but RMI [19] developed for this application a manufacturing process for β processing resulting in fully lamellar microstructures with higher fracture toughness and better fatigue crack propagation resistance for macrocracks as compared to the mill-annealed condition (see Sect. 5.2).

Ti-6Al-4V material has found some limited application in armor components, for example in the M2 Bradley Infantry Fighting Vehicle and in the Abrams M1 main battle tank [172]. The main advantage over conventional armor steel is the lower weight of the components reducing the total weight of the heavy battle tanks. For these applications, the ELI type version of Ti-6Al-4V is used in the mill-annealed condition. No significant differences in the ballistic response were found for plates with fully lamellar microstructure [173]. Clearly, cost is a major concern for armor. There is an ongoing effort to produce lower cost material that has comparable ballistic properties but at 35-50% lower cost. Elemental blended powder rolling and slab casting are two methods that show promise for meeting the performance requirements, but meeting the cost goals remains a greater challenge.

The application of Ti-6Al-4V in the biomedical field, in the area of sporting goods, and in automobiles will be covered in Chap. 10.

Fig. 5.54. Example of a large steam turbine rotor (courtesy R. I. Jaffee, EPRI, Palo Alto, USA)

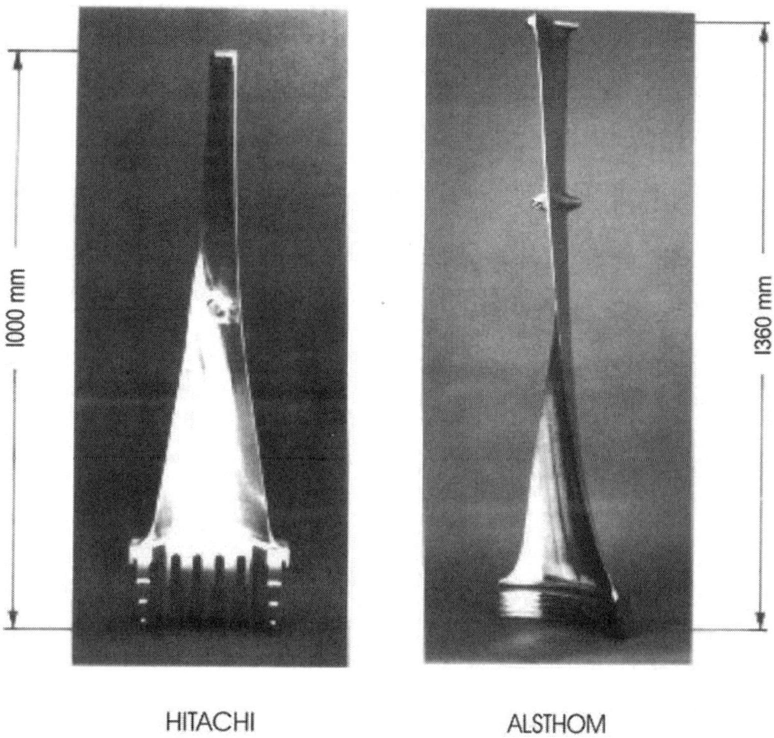

Fig. 5.55. Large Ti-6Al-4V steam turbine blades (courtesy R. I. Jaffee, EPRI, Palo Alto, USA)

Fig. 5.56. Example of long Ti-6Al-4V (ELI grade) taper stress joints used for deep-sea riser systems (courtesy RMI)

6 High Temperature Alloys

For long term applications at high temperatures, the Ti-6Al-4V alloy is limited to about 400°C. For higher temperatures, titanium alloys (such as Ti-6242 and IMI 834) have been formulated according to the following general principles.

The diffusion rates in the β phase are about two orders of magnitude faster than in the α phase. Therefore, the volume fraction of β phase is reduced in these high temperature alloys as compared to Ti-6Al-4V, for example at 800°C the Ti-6Al-4V alloy contains about 15 vol% β phase whereas the Ti-6242 alloy contains about 10 vol% and the IMI 834 alloy only about 5 vol%. This decrease in volume fraction of β phase in Ti-6242 and IMI 834 is achieved by reducing the total content of β stabilizing elements and by alloying in addition to the 6% Al the elements Sn and Zr which act as α stabilizing elements (see Table 2.6 and Sect. 2.5). Furthermore, the β stabilizing element vanadium is replaced by Mo and Nb which are slower diffusing elements [58]. In addition, the content of Fe, which is a very strong β stabilizer and also leads to very fast diffusion rates in α titanium (see Sect. 2.9.1), is reduced especially in IMI 834 to the very low level of 0.05%.

Due to the reduction in volume fraction of β phase, the thickness of the "β lamellae" in a colony structure is reduced in many areas to zero, i.e. only low angle boundaries are left separating parallel α lamellae. In this case, easy slip is possible over long distances with the negative consequences outlined in Sect. 5.2, especially for fatigue strength (HCF and LCF). To create new obstacles to dislocation motion at the α/α lamellae boundaries, silicon is added to high temperature alloys (about 0.1-0.5%). Silicon forms with titanium the intermetallic compound Ti_5Si_3 or in the presence of zirconium the compound $(Ti,Zr)_5Si_3$. This intermetallic compound has a complicated crystallographic structure. Therefore, the silicides are incoherent with respect to β and α and precipitate at the α/β lamellae boundaries and at grain boundaries. An example of the microstructure in a lamellar area is shown in Fig. 6.1 for the IMI 834 alloy. It can be seen that the former "β lamellae" have dissolved and $(Ti,Zr)_5Si_3$ particles have precipitated at the α/α plate boundaries which otherwise consist only of arrays of dislocations (low angle boundaries).

Since coherent precipitates are effective barriers for dislocation glide and climb, the volume fraction of Ti_3Al (α_2) particles is increased in high temperature titanium alloys, mainly due to the addition of Sn which promotes the formation of α_2, see Sect. 2.8.1. For example, the solvus temperature for α_2 in Ti-6Al-4V is around 550-600°C, in Ti-6242 around 650°C, and in IMI 834 around 750°C. In contrast to Ti-6Al-4V, the standard final heat treatment in Ti-6242 and IMI 834 is always an aging treatment in the (α+α_2) phase region, i.e. 8h 695°C for Ti-6242 and 2h 700°C for IMI 834. An example of the α_2 particles in IMI 834 was shown in Fig. 2.20 (Sect. 2.8.1).

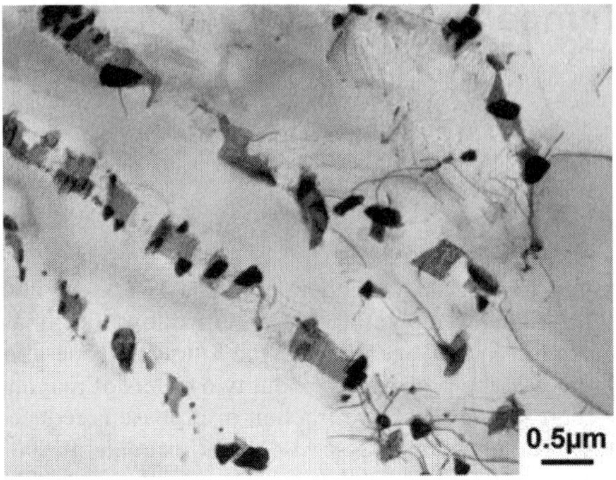

Fig. 6.1. Precipitation of $(Ti,Zr)_5Si_3$ particles at α/α plate boundaries in lamellar regions of bi-modal microstructures of IMI 834, aged 2 h at 700°C, TEM

6.1 Processing and Microstructure

The processing routes to generate different microstructures, for example fully lamellar or bi-modal microstructures, in high temperature titanium alloys (Ti-6242, IMI 834) are the same processing routes as those outlined in Chap. 5 (Sect. 5.1) for $\alpha+\beta$ titanium alloys. The only additional feature important for the processing route of high temperature titanium alloys is the solvus temperature of the silicides in relation to the other temperatures in the processing route. This is shown qualitatively in Fig. 6.2 on the example of the processing route for bi-modal microstructures. This processing route was selected as example, because the bi-modal microstructure is the most commonly used microstructure in high temperature titanium alloys for applications in the compressor section of aero-engines. From Fig. 6.2 it can be seen that both the deformation temperature (step II) and the recrystallization annealing temperature (step III) are above the solvus temperature for the silicides. Therefore, all the silicon is in solid solution at these temperatures (silicon partitions nearly exclusively into the β phase). The silicon is then precipitated as silicides at the lamellae boundaries in the lamellar regions of the bi-modal microstructure (Fig. 6.1) during the final annealing treatment. Also shown in Fig. 6.2 is the solvus temperature for the Ti_3Al (α_2) precipitates demonstrating that the final annealing treatment is performed in the ($\alpha+\alpha_2$) phase region and is, therefore, an aging treatment for the α phase. It should be mentioned that due to the alloying element partitioning effect (see Sect. 5.1.2) the volume fraction of α_2 particles is much higher within the α_p grains than within the α plates in the lamellar areas of the bi-modal microstructures [174].

Although the recently developed high temperature alloy Ti-1100 (Ti-6Al-2.7Sn-4Zr-0.4Mo-0.45Si) is no longer on the market, this alloy might serve as processing example for the case that the solvus temperature for $(Ti,Zr)_5Si_3$ is higher (1040°C) than the β transus (1015°C). In this case, the deformation in the (α+β) phase field has to be done in the presence of relatively coarse $(Ti,Zr)_5Si_3$ particles. In addition, the bi-modal microstructures would contain relatively coarse silicides because the recrystallization annealing treatment in the (α+β) phase field to create the bi-modal microstructure would be below the solvus temperature for the silicides. So, for Ti-1100 the alloy producer recommended to use for (α+β) processed material a subsequent recrystallization treatment in the β phase field above the solvus temperature of the silicides (β annealing) or to use β processing above the silicides solvus temperature leading in both cases to fully lamellar microstructures without coarse silicides.

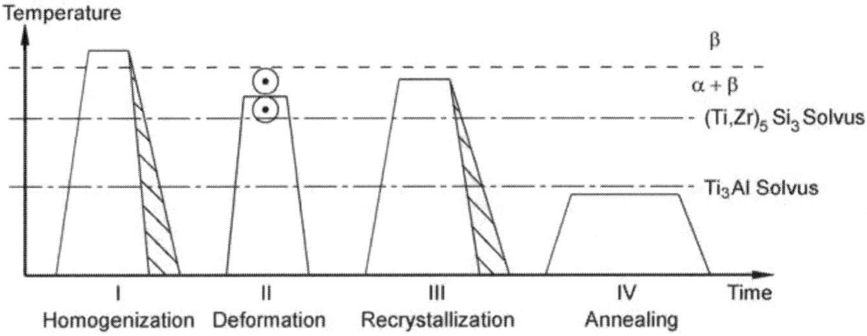

Fig. 6.2. Processing route for bi-modal microstructures of high temperature titanium alloys including the solvus temperatures for (Ti,Zr)5Si3 and Ti3Al (schematically)

6.2 Microstructure and Mechanical Properties

The dependence of the mechanical properties at room temperature on cooling rate either from the β phase field in case of fully lamellar microstructures or from the recrystallization annealing temperature in the (α+β) phase field in case of bi-modal microstructures was discussed in detail on the example of fully lamellar microstructures in Sect. 5.2.1 and summarized in Table 5.6 by the row "Small α Colonies and α Lamellae". This cooling rate also has a dominating influence on the mechanical properties at elevated temperatures. This is shown in Fig. 6.3 for the yield stress of the Ti-6242 alloy using the bi-modal microstructure (α_p about 20 vol%) as example [174]. It can be seen that the dependence of yield stress on cooling rate is similar for both the high temperature tests (450°C, 510°C) and the room temperature tests. The dashed lines in Fig. 6.3 point out that for the slow cooling rate of 1°C/min the bi-modal microstructure is changed to a fully equiaxed microstructure as described in Sect. 5.1.3. Comparing Fig. 6.3 with Fig. 5.18, it can be seen that qualitatively the dependence of yield stress on cooling rate is similar for bi-modal and fully lamellar microstructures in accordance with the overall discussion on slip length (Fig. 5.17).

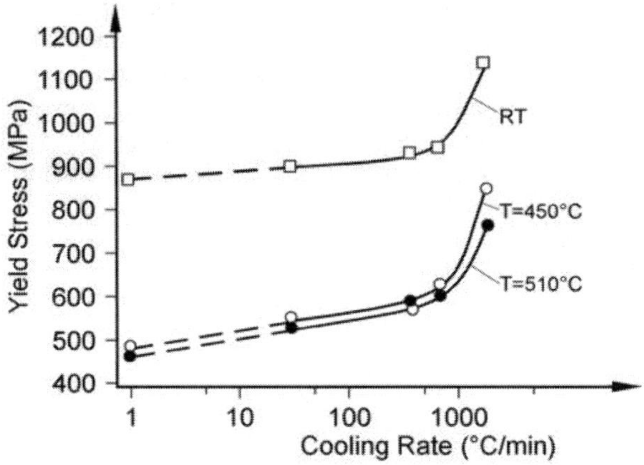

Fig. 6.3. Yield stress at different temperatures as a function of cooling rate from the recrystallization annealing temperature for bi-modal microstructures (fully equiaxed for a cooling rate of 1°C/min), Ti-6242, final heat treatment: 8h at 595°C

The ductility values of the tensile specimens in Fig. 6.3 were high at room temperature (RA about 40%) and increased to RA values of about 50% at high temperatures. The drastic decline in room temperature ductility at fast cooling rates that was observed for fully lamellar structures (see Fig. 5.18) did not occur for the bi-modal microstructures of the Ti-6242 alloy [174] in agreement with the results shown in Fig. 5.20 for the Ti-6Al-4V alloy.

The dependence of high temperature LCF strength at 450°C on cooling rate is shown in Fig. 6.4 for the same material conditions used in Fig. 6.3 (Ti-6242, bi-modal microstructure). In Fig. 6.4 the LCF strength after 10^4 cycles is plotted and it can be seen that the LCF strength shows qualitatively a similar dependence on cooling rate as the yield stress in Fig. 6.3. It should be mentioned again that the LCF strength involves both crack nucleation and microcrack propagation. In Sect. 5.2.1 it was shown for fully lamellar microstructures and room temperature testing that both the resistance to crack nucleation (HCF strength), see Fig. 5.21, and the resistance to microcrack propagation, see Fig. 5.23, were improved with increasing cooling rates. This shows that the discussion in Sect. 5.2.1 on the influence of cooling rate (fineness of the fully lamellar structure) on LCF strength at room temperature also applies for bi-modal microstructures and also for high temperature LCF testing at 450°C (Fig. 6.4), as long as there are no long holding times at peak stress involved in the LCF testing (creep/fatigue interactions).

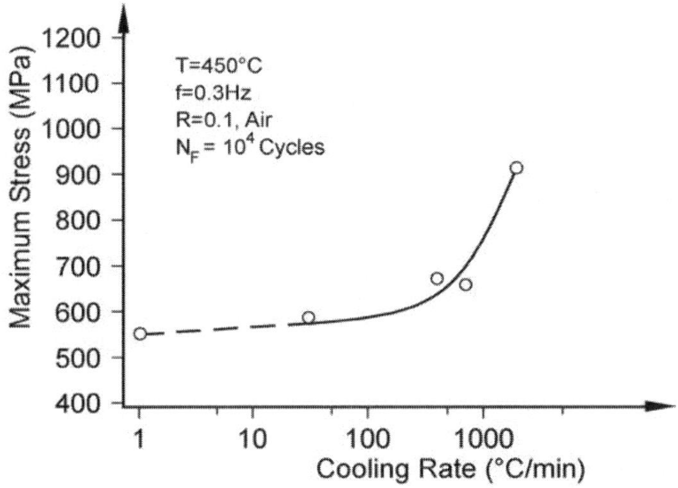

Fig. 6.4. LCF strength at 104 cycles as a function of cooling rate for bi-modal microstructures (same material conditions as in Fig. 6.3), Ti-6242, test temperature: 450°C

Fatigue crack propagation of microcracks and macrocracks at elevated temperatures are usually thought of being no concern because of the high ductility at high temperatures increasing the inherent resistance of the material to crack propagation. But there is a strong environmental effect at elevated temperatures even in laboratory air. An example is shown in Fig. 6.5 comparing for IMI 834 the fatigue crack propagation curves of macrocracks at room temperature and 500°C in vacuum and air [175]. It can be seen that the fatigue crack propagation rates in vacuum are only slightly higher at 500°C as compared to room temperature. In contrast, in laboratory air the fatigue crack propagation rates at 500°C in the medium to low ΔK regime are much faster than at room temperature. It was shown that this effect is caused by the presence of water vapor in the environment [176]. Typical for the curves at elevated temperatures in such environments is the appearance of a pronounced threshold below which no measurable crack propagation occurs (see Fig. 6.5). A possible explanation for the pronounced threshold is that the protective surface oxide layer at the crack tip is not ruptured by slip steps below a critical ΔK value [177] eliminating the transport of hydrogen into the material by dislocations. This hypothesis is supported by a result, which was obtained on an orthorhombic type of titanium alloy, that fatigue crack propagation in vacuum occurred below the threshold value in air, i.e. the two curves were intersecting each other [177]. In this case, the residual water vapor in the vacuum caused fatigue crack propagation at very low ΔK values because no protective surface oxide layer at the crack tip was present in the vacuum tests.

The resistance to unstable fracture, i.e. fracture toughness, generally is increased with increasing temperature because the material is more ductile and no environmental effect is present due to the fast crack propagation rate involved.

For high temperature applications, the creep resistance is a very important property in addition to fatigue resistance, especially in the primary creep regime. This is because usually only small plastic creep strains are allowed for many structural components, for example for compressor disks in aero-engines. For all α+β titanium alloys, the plastic creep strain as a function of cooling rate exhibits the characteristic dependence shown in Fig. 6.6 for lamellar structures of the Ti-6Al-4V alloy and in Fig. 6.7 for bi-modal structures of the Ti-6242 alloy. This is true whether the material is cooled from the β phase field (fully lamellar microstructures) or from the recrystallization annealing temperature in the (α+β) phase field (bi-modal microstructure). It is generally observed that the plastic creep strain exhibits a minimum at intermediate cooling rate, approximately between 100-500°C/min. The fully lamellar microstructures (Fig. 6.6) and the bi-modal microstructures (Fig. 6.7) follow the same general trend in creep resistance as a function of cooling rate. This suggests that the possible reasons for the maximum in creep resistance at intermediate cooling rates should pertain to both fully lamellar microstructures and bi-modal microstructures. In the overview of the correlation between microstructure and mechanical properties (Table 5.6) this maximum in creep resistance at intermediate cooling rates is expressed by +/– in the row "Small α Colonies and α Lamellae".

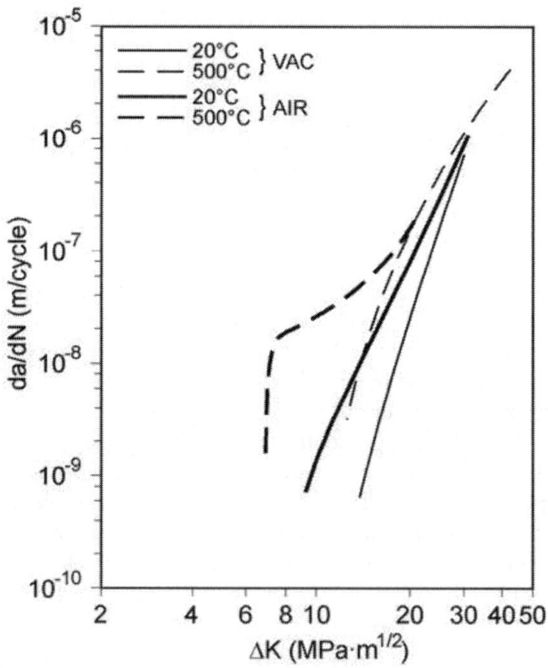

Fig. 6.5. Influence of environment and test temperature on fatigue crack propagation behavior of macrocracks (R = 0.1), IMI 834, bi-modal microstructure

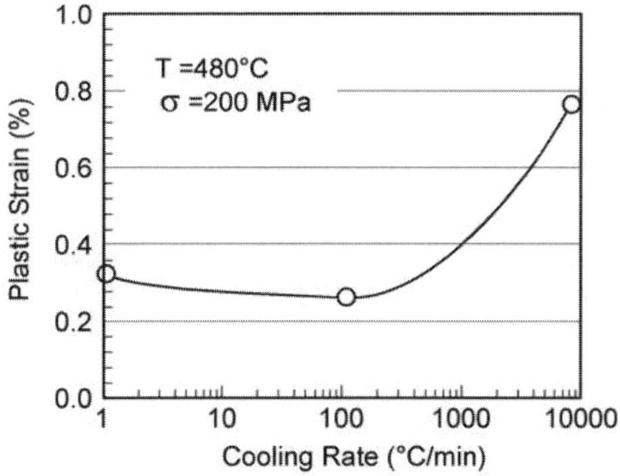

Fig. 6.6. Influence of cooling rate from the β phase field on creep strain after 100 hours for lamellar microstructures of Ti-6Al-4V

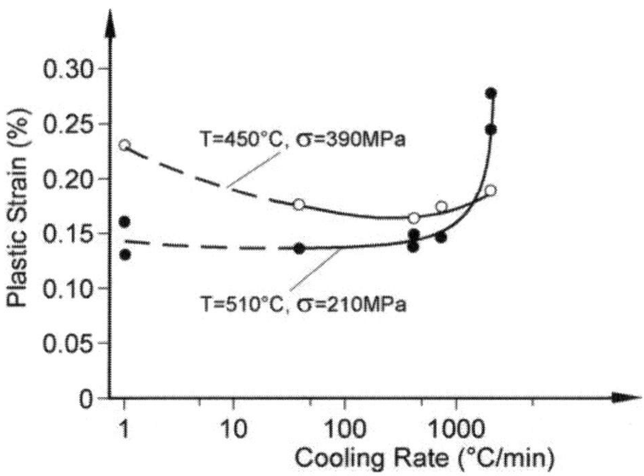

Fig. 6.7. Influence of cooling rate on creep strain after 100 hours for bi-modal microstructures (same material conditions as in Figs. 6.3 and 6.4), Ti-6242

For a discussion of the curves in Figs. 6.6 and 6.7 it is important to realize that the equivalence of slip length and α colony size used up to now in Sects. 5.2 and 6.2, see Fig. 5.17, does not apply for creep deformation. In case of creep deformation the dislocations move, for example by diffusion assisted climb, only over short distances, e.g. across a single α plate, and are annihilated in many cases at the next incoherent boundary, i.e. the α plate boundary.

The decrease in creep resistance with slower cooling rates (Figs. 6.6 and 6.7) is usually explained by the wider α lamellae (plates) leading to larger distances between obstacles for dislocation motion and to lower strain hardening. Another possible explanation is that, because the yield stress is decreasing with decreasing cooling rate, the ratio of applied stress in the creep test to the yield stress at test temperature becomes larger for slower cooled microstructures. The results shown in Fig. 6.7 for the two different stress levels and temperatures are consistent with this explanation because the above mentioned ratio is between 0.7-0.8 for the curve at 450°C (large effect), whereas the ratio is only 0.45 for the curve at 510°C (small effect). The reason for the sharp decrease in creep resistance in the fast cooling rate regime is still unclear. One possibility is that the annihilation process of dislocations at lamellae boundaries dominates over the strain hardening effect because of the large increase in boundary density in microstructures corresponding to faster cooling rates. Since the rate of annihilation of dislocations at lamellae boundaries would be strongly affected by the test temperature because diffusion is involved and less affected by the applied stress, the results in Fig. 6.7 would support this hypothesis, because a sharp decrease in creep resistance in the fast cooling rate regime is observed for 510°C test temperature, whereas only a very small effect is present at a test temperature of 450°C.

A direct comparison of the creep resistance as a function of cooling rate between fully lamellar and bi-modal microstructures is shown in Fig. 6.8 for the IMI 834 alloy. It can be seen that fully lamellar microstructures exhibit a better creep resistance than bi-modal microstructures. This decrease in creep resistance of the bi-modal microstructures with increasing volume fraction of α_p over the whole range of cooling rates shown in Fig. 6.8 can be explained by the alloy partitioning effect in the bi-modal microstructures. The basic strength of the lamellar part of the bi-modal microstructure will become continuously softer with increasing volume fraction of α_p and, in comparison, the reduction in β grain size (α colony size) has no or little influence on the creep resistance as outlined above. At very low volume fractions of α_p (see curve for 4 vol% α_p in Fig. 6.8) the alloy element partitioning effect is not yet in equilibrium, because of the large distances between α_p grains, but the β grain size (α colony size) is already decreased sufficiently for good fatigue properties. This combination of properties cannot be realized in any practical way because the necessary temperature accuracy required during the bi-modal heat treatment to obtain such a low volume fraction of α_p (and the required homogeneity of the material) cannot be achieved in commercial practice.

As outlined in Sect. 5.2.2, the only practical way to partially eliminate the alloy element partitioning effect for bi-modal microstructures seems to be to add an intermediate annealing treatment between the bi-modal recrystallization annealing treatment and the final aging treatment. By such a treatment, for example 2 h at 830°C, it was possible to eliminate the negative influence of the alloying element partitioning effect on HCF strength (see Fig. 5.29). An intermediate annealing treatment of 2 h at 830°C also improves the creep resistance of the bi-modal microstructure with 15 vol% α_p in the slow to medium cooling rate range (Fig. 6.9). This is because the α stabilizing elements such as aluminum and oxygen will diffuse from the primary α into the lamellar α regions during the intermediate annealing treatment. This redistribution of aluminum and oxygen results in a higher volume fraction of Ti_3Al particles in these lamellar α regions upon the final aging treatment of 2 h at 700°C. Consequently, the local yield stress within the lamellar α regions of the bi-modal microstructure is increased resulting in an improved creep resistance. In comparison, the creep resistance of the fully lamellar condition did not change by the intermediate annealing treatment (Fig. 6.9).

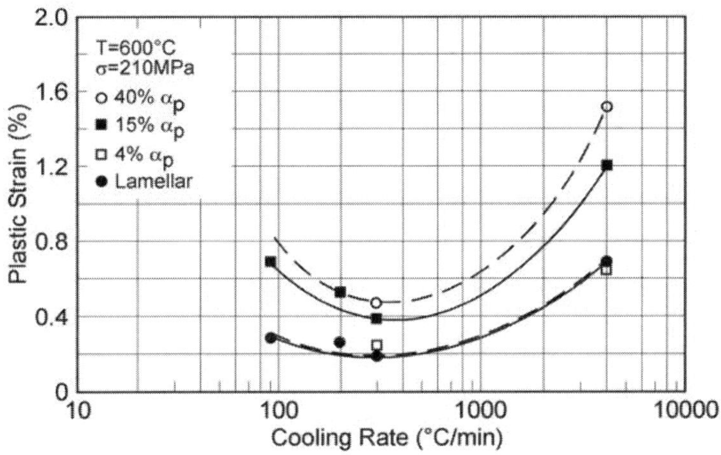

Fig. 6.8. Creep strain after 100 hours as a function of cooling rate for bi-modal and lamellar microstructures of IMI 834

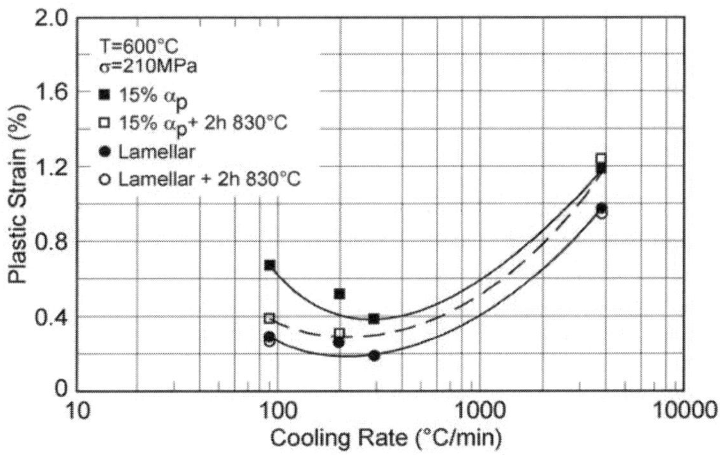

Fig. 6.9. Improvement in creep resistance of a bi-modal structure with 15 vol% α_p by an intermediate annealing treatment of 2 h at 830°C, IMI 834

The influence of α grain size in fully equiaxed microstructures on creep strength of the Ti-6Al-4V alloy is shown in Fig. 6.10. It can be seen that the creep strength is decreasing with decreasing grain size (see also Table 5.6). A drastic decrease in creep resistance is observed for small α grain sizes of about 2-3 μm. The whole curve in Fig. 6.10 is in qualitative agreement with the curves in the medium to fast cooling rate regime of Figs. 6.6 to 6.8. The decrease in creep strength which was observed in the medium to slow cooling rate regime in Figs. 6.6 to 6.8 is not present at large α grain sizes in Fig. 6.10. It should be pointed out that for the fully equiaxed structures in Fig. 6.10 the ratio of applied stress in the creep tests to the yield stress at 480°C was below 0.5.

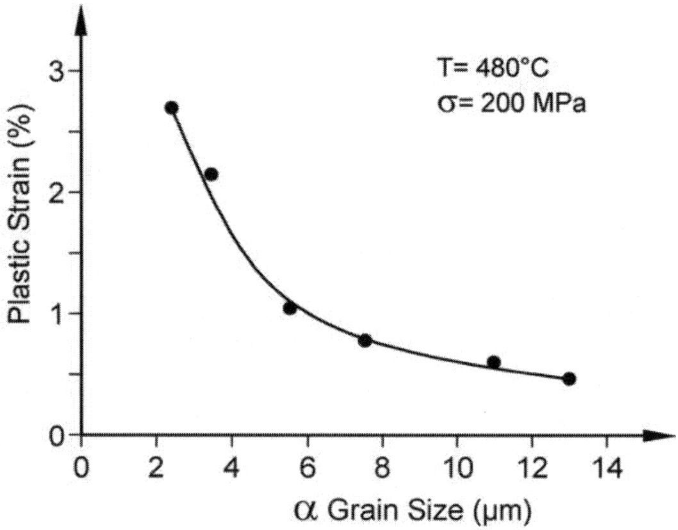

Fig. 6.10. Creep strain after 100 hours as a function of α grain size of fully equiaxed microstructures, Ti-6Al-4V, final aging treatment: 24h at 500°C

The influence of aging on the creep strength is shown in Fig. 6.11 for the IMI 834 alloy with a bi-modal microstructure [178]. It can be seen that with increasing aging from 2h 625°C to 2h 700°C and to 24h 700°C the creep strength is increasing (see Table 5.6). For reference, the yield stress values at room temperature [178] also are included in Fig. 6.11.

The effect of secondary α in the β phase, i.e. the so-called bi-lamellar microstructure, on creep resistance is demonstrated in Fig. 6.12 for the Ti-6Al-4V alloy. The creation of this bi-lamellar microstructure, see as example Fig. 5.44, was discussed in Sect. 5.2.5. The plot in Fig. 6.12 is identical to those shown in Fig. 5.45 for the other mechanical properties, i.e. the creep strain is plotted as a function of cooling rate from the intermediate annealing temperature of 880°C and the starting coarse lamellar microstructure is shown at a cooling rate of 1°C/min. It can be seen from Fig. 6.12 that the bi-lamellar structure exhibited a better creep resistance than the coarse lamellar starting structure, i.e. the hardening of the β phase by secondary α reduced effectively the contribution of the creep strain within the β phase to the total creep strain of the microstructure (see Table 5.6)

The effect of crystallographic texture and test direction on creep strength is directly related to the effect on yield stress (see Table 5.6). This was shown by testing Ti-6Al-4V material with a pronounced T type texture (see Fig. 5.46a) in RD and TD directions [179]. The creep resistance was higher in TD direction in agreement with the higher yield stress in this direction at creep testing temperature as well as at room temperature [179], see Fig. 5.47b.

All correlations between microstructural parameters and mechanical properties discussed up to now in this section do not depend on Si content of the material because they are valid for Ti-6242 and IMI 834, both containing silicon, as well as for Ti-6Al-4V, containing no silicon. The silicon in Ti-6242 and IMI 834, which is present, as described earlier in this chapter, as $(Ti,Zr)_5Si_3$ particles mainly at α "lamellae" boundaries and also at grain boundaries, raises the absolute creep resistance of these materials in comparison to Ti-6Al-4V. However, the silicides, besides improving the creep resistance, also can have a negative effect on mechanical properties, especially on room temperature ductility, if the volume fraction of silicides at α "lamellae" boundaries is too high. This was noted on IMI 834 material after prolonged creep testing (500 h) at 600°C [180]. The room temperature tensile elongation after complete removal of the oxidized layer was reduced from about 10% to about 7% for a bi-modal microstructure. This was attributed to an increase in volume fraction of silicides at the α "lamellae" boundaries [180]. For a fully lamellar microstructure with a much lower starting ductility of the virgin material, this drop in ductility would lead to unacceptable low ductility values.

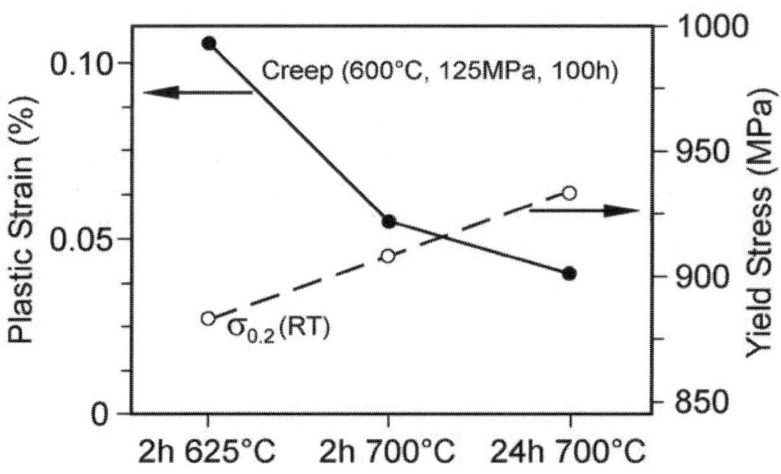

Fig. 6.11. Effect of aging on creep strain after 100 hours for IMI 834 with bi-modal microstructure; also shown are the yield stress values at room temperature

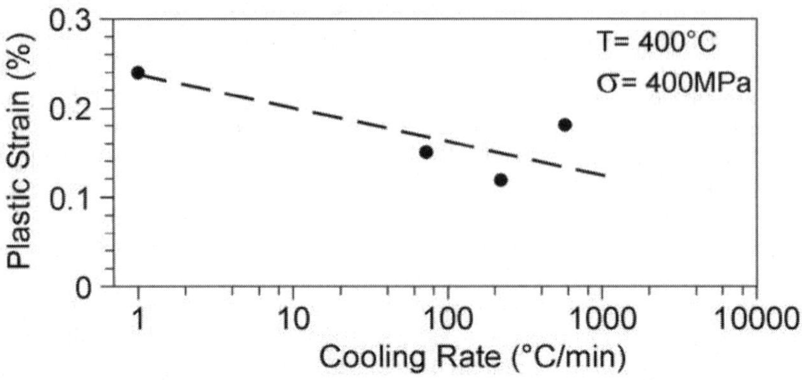

Fig. 6.12. Effect of secondary α in β phase (bi-lamellar microstructure) on creep strain plotting the cooling rate from the intermediate annealing temperature of 880°C (see Fig. 5.45), 1°C/min: lamellar, others: bi-lamellar, Ti-6Al-4V

6.3 Properties and Application

The main application of high temperature titanium alloys, such as Ti-6242 and IMI 834, are blades and disks in the compressor section of aero-engines. These alloys are applied in the high pressure (HP) compressor stages for which the temperature exceeds 350°C where Ti-6Al-4V cannot be used because of creep considerations.

A good example is the HP compressor spool, shown in Fig. 5.52 and discussed in Sect. 5.3, with five Ti-6Al-4V front stages followed by two Ti-6242 rear stages. Since the maximum temperature capability is about 500°C for Ti-6242 (550°C for IMI 834), the last stages in HP compressors of aero-engines are made out of Ni-Super-Alloys, i.e. in case of the spool example the nickel part is bolted to the titanium part shown in Fig. 5.52. As mentioned already in Sect. 5.3, a bi-modal microstructure is used for the Ti-6242 stages (as well as for the Ti-6Al-4V stages) of the compressor spool, because the LCF strength is the most important mechanical property for this type of application (besides adequate creep strength) and the bi-modal microstructure exhibits better LCF strength than a fully lamellar microstructure (see Fig. 5.31). Recently, Ti-6242 in the fully lamellar condition has been introduced in some applications where creep has become limiting. These fully lamellar structures are produced by β forging creating an elongated unrecrystallized β grain structure (see Sect. 5.1.1 and Fig. 5.5a). This elongated grain shape minimizes the negative effect of the continuous α layers at β grain boundaries on mechanical properties (see detailed discussion in Chap. 7).

Another example for the application of Ti-6242 material with bi-modal microstructure is the impeller shown in Fig. 6.13. Impellers are used as last compressor stage in small, low flow aero-engines and in auxiliary power units (APUs). The impeller shown in Fig. 6.13 has a diameter of 350 mm and is installed in small engines of regional jets, which have several axial stages in between the fan and the impeller.

The newer IMI 834 alloy which can be applied up to about 550°C is used for blades and blisks in the EJ 200 aero-engine [167] as well as in the Rolls-Royce TRENT 800 engine. The component shown in Fig. 6.14 is an IMI 834 blisk used in the HP compressor section of the EJ 200 aero-engine. The blisk is manufactured starting with an oversized disk forging and machining out the airfoils at the rim. For the blisk shown in Fig. 6.14 with a diameter of 480 mm the blades are too small for using economically the new technology of solid state linear friction welding to attach the blades to the disk (see Sect. 3.6.2). The IMI 834 alloy is also used as impeller material in the PW 300 engine of PW Canada.

For all these applications of the IMI 834 alloy a bi-modal microstructure with a relatively low volume fraction of α_p (about 15-20 vol%) is used. This microstructure exhibits a very good balance of LCF, HCF, and creep properties. This is because the low volume fraction of α_p reduces the negative alloy element partitioning effect on the latter two properties, see Sect. 5.2.2 (Fig. 5.26) and Sect. 6.2 (Fig. 6.8). Bi-modal microstructures of the Ti-6242 and Ti-6Al-4V alloys, which are usually used with α_p volume fractions of about 35-40%, would also benefit from reducing the volume fraction of α_p, but are difficult to heat treat on industrial scale to lower volume fractions of α_p due to the relatively small annealing temperature "window". In case of IMI 834, this annealing temperature "window" was widened by the addition of 0.06C [181, 182].

A more detailed list of compressor parts made out of high temperature titanium alloys including the "older" alloys IMI 685 and IMI 829 can be found elsewhere [167, 168].

Fig. 6.13. Impeller used in a small engine for regional jets, Ti-6242, bi-modal microstructure (courtesy J. A. Hall, Honeywell Engines and Systems)

Fig. 6.14. High pressure compressor blisk used in the EJ 200 aero-engine, IMI 834, bi-modal microstructure (courtesy D. Helm, MTU)

7 Beta Alloys

In contrast to α+β alloys, β alloys do not transform martensitically upon quenching to room temperature, resulting in a metastable β phase. The α phase can be precipitated from the metastable β phase as very fine, undeformable particles (platelets) with a high volume fraction, see Sect. 2.8.2. Therefore, the main characteristic of β alloys is that they can be hardened to much higher yield stress levels than α+β alloys. Another advantage of β alloys is that they can be processed at lower temperatures than α+β alloys, and some heavily stabilized β alloys are even cold deformable. Further, the corrosion resistance of β alloys is equal to or better than that of α+β alloys. The β alloys are especially good in environments where hydrogen pickup is possible because the β phase has a higher hydrogen tolerance than the α phase. It is not surprising that the usage of β alloys has been slowly but steadily increasing over the recent years.

Within the class of β alloys, a distinction can be made between so-called "high strength" β alloys which are located with their chemical composition close to the β to α+β alloy boundary (see Fig. 2.19) and contain therefore a high volume fraction of α phase and so-called "heavily stabilized" β alloys which are located more to the right in the pseudo-binary phase diagram of Fig. 2.19. These latter alloys contain a much lower volume fraction of α phase and the maximum achievable strength is therefore lower as compared to the high strength group of β alloys. From the β alloys listed in Table 2.6, typical examples for the first group of alloys are Ti-6246, Ti-17, β-CEZ, and Ti-10-2-3, whereas the alloys Beta 21S, Ti-15-3, and Beta-C belong to the second group of alloys. Since the main emphasis in this book is placed on the group of high strength β alloys and most examples shown are from this group, the term "β alloys" is used for this group of alloys. If the group of heavily stabilized β alloys is discussed, this will be explicitly stated.

The structure of Chap. 7 on β alloys follows the structure of Chap. 5 on α+β alloys. The various processing routes and the resulting microstructures will be outlined in Sect. 7.1. The influence of these different microstructures on mechanical properties will be discussed in Sect. 7.2, and at the end of the chapter some application examples will be given (Sect. 7.3).

7.1 Processing and Microstructure

The so-called β annealed microstructure resulting from a pure thermal treatment will be discussed at the beginning (Sect. 7.1.1). This β annealed structure nicely illustrates the key microstructural feature of all β alloys, namely the continuous α layers at β grain boundaries. The main goal of the three subsequently described processing routes for high strength β alloys is to create microstructures which restrict the formation of the continuous α layers, namely β processing (Sect. 7.1.2) and through-transus processing (Sect. 7.1.3), or limit the influence of these α layers on mechanical properties, namely a bi-modal microstructure (Sect. 7.1.4).

7.1.1 Beta Annealed Microstructures

The basic processing route for β annealed microstructures is shown in a schematical way in Fig. 7.1. It can be seen that the β annealed microstructure is obtained in a simple way, i.e. recrystallization in the β phase field (step III) and aging in the (α+β) phase field (step IV) to precipitate the α phase in form of fine α platelets. Such a microstructure is shown in Fig. 7.2. The main characteristic of all β alloys is that the α phase nucleates preferentially at β grain boundaries and forms a continuous α layer, as has been already discussed in Chap. 2 (Sect. 2.8.2). Adjacent to this continuous α layer is a so-called PFZ (precipitate free zone) which does not contain any α platelets (see Fig. 7.2b) and which is therefore soft with respect to the age-hardened matrix. For the mechanical properties (Sect. 7.2), both the strength difference between the soft zone and the matrix (i.e. yield stress) and the slip length in the soft zone (i.e. β grain size) are important. Both of these parameters are influenced by alloy composition, i.e. the effect of the continuous α layers is large for high strength β alloys and smaller for heavily stabilized β alloys. This is the reason why high strength β alloys are not used in the β annealed condition for any application. This condition will still be discussed in Sects. 7.1 and 7.2 because it nicely demonstrates the effect of the continuous α layers on mechanical properties and serves as a useful comparison. For heavily stabilized β alloys the β annealed condition is the most commonly used microstructure and this group of alloys will be described first in this section.

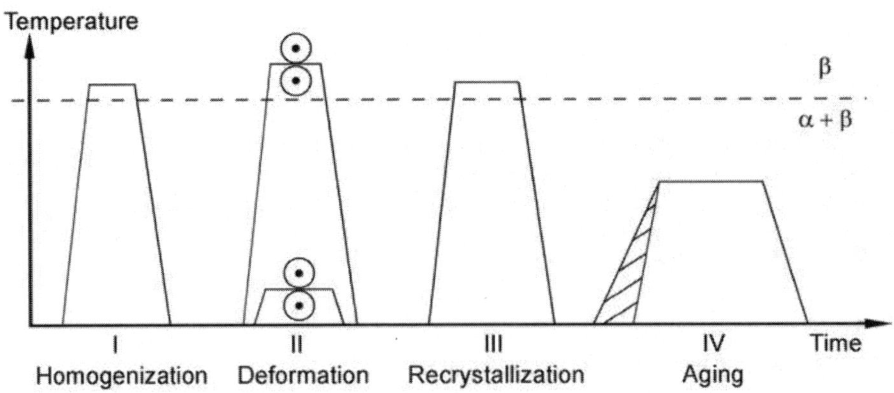

Fig. 7.1. Processing route for β annealed microstructure of heavily stabilized β titanium alloys (schematically)

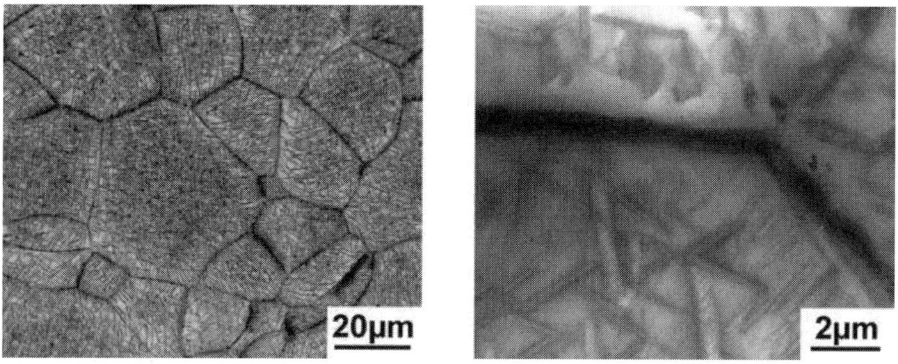

Fig. 7.2. Example of β annealed plus aged microstructure of heavily stabilized β alloys, Beta 21S: (a) LM (b) TEM

For heavily stabilized β alloys, the main characteristic is the relatively low β transus temperature of 800°C or even lower (see Table 2.6). This means, that the deformation process (step II in Fig. 7.1), which is usually done slightly above the β transus, can be performed at relatively low temperatures. Some heavily stabilized β alloys can even be cold deformed. Hot deformation in the (α+β) phase field is usually avoided because of the presence of α phase increasing the flow stress. Recrystallization (step III in Fig. 7.1) is done slightly above the β transus and the time can be fairly short (below one hour) because no α phase has to be dissolved. For example, during commercial processing of Ti-15V-3Cr-3Al-3Sn annealing times as short as five minutes are used. Because grain growth is exponentially dependent on absolute temperature once the β transus is exceeded, the β grain size of heavily stabilized β alloys with recrystallization temperatures around 800°C is much smaller than in fully lamellar structures of α+β alloys which are recrystallized above 1000°C. For example, in the Beta 21S alloy grain sizes of 40-50 µm are common (see Fig. 7.2a). The aging treatment (step IV in Fig. 7.1) is usually done in the temperature region between 500°C and 600°C, i.e. above the solvus temperature of the metastable coherent particles ω and β' (see Chap. 2, Sect. 2.8.2). But as pointed out already in Sect. 2.8.2, these metastable particles serve as precursors for the nucleation of the α platelets, so their presence during the heating period to the aging temperature is beneficial for obtaining a uniform distribution of fine α platelets. Therefore, the heating rate to the aging temperature can be an important parameter for heavily stabilized β alloys [183]. Table 7.1 summarizes the important processing parameters and resulting microstructural features of heavily stabilized β alloys. The role of the metastable precursors also explains the observed difference between direct aging and the normal aging practice with cooling to room temperature before aging [183]. An important parameter of the aging treatment (step IV in Fig. 7.1) is the choice of the aging temperature because this determines the volume fraction of α platelets which influences the yield stress level (see Sect. 7.2). Since the growth of the α platelets is sluggish, the aging temperature is much more important in step IV than the aging time (Table 7.1). The formation of the continuous α layers at β grain boundaries during the aging treatment cannot be avoided (see Table 7.1).

Table 7.1. Important processing parameters and resulting microstructural features for β annealed microstructures of heavily stabilized β alloys

Processing Step (see Fig. 7.1)	Important Parameters	Microstructural Features
III	Recrystallization Temperature	β Grain Size
IV	Heating Rate Aging Temperature	Distribution of α Platelets – Size and Vol% of α Platelets – GB α Layer – Distribution of α Platelets

As pointed out already in Chap. 2 (Sect. 2.8.2), it is sometimes difficult for heavily stabilized β alloys (especially for relatively high aging temperatures) to obtain a homogeneous distribution of α platelets by the normal one-stage aging treatment shown in Fig. 7.1. This is because the necessary formation of the precursors (ω or β') can be too sluggish to occur during the heating period to the aging temperature. It was shown already in Fig. 2.28 on the example of Beta C that in this case a pre-aging at lower temperatures can create a more homogeneous distribution of α platelets [57]. The processing route including this two-stage aging process is shown schematically in Fig. 7.3. During the pre-aging (step IVa in Fig. 7.3) a homogeneous distribution of very fine α platelets is formed. These fine α platelets then only grow or coarsen to the desired level during the normal aging treatment (step IVb in Fig. 7.3). In this case, the heating rate to the aging temperature in step IVb is unimportant, see Table 7.2. Another example for the influence of the pre-aging treatment on the resulting microstructure is shown in Fig. 7.4 for the heavily stabilized β alloy Beta 21S [184]. It can be seen that the aging of 8h 690°C results in an inhomogeneous distribution of fairly coarse α plates (Fig. 7.4a). The additional 8h 650°C aging treatment indicated in the figure caption is recommended by the alloy producer to stabilize the microstructure of this alloy for high temperature application. The pre-aging treatment of 8h 500°C creates a homogeneous distribution of α platelets even for a final aging treatment of 24h 725°C (Fig. 7.4b). The pronounced effect of pre-aging on microstructure can be seen even more clearly when comparing the resulting microstructure to the one-step aging treatment of 24h 725°C (Fig. 7.4c). Nonuniform recrystallization also leads to an inhomogeneous distribution of α platelets after aging. This is because of the higher density of nucleation sites in the unrecrystallized regions.

For high strength β alloys the processing route for β annealed microstructures is different from that of heavily stabilized β alloys because of the higher β transus temperature and the higher volume fraction of the α phase. The processing route is shown schematically in Fig. 7.5 which illustrates the various processing steps. Table 7.3 summarizes the important parameters of the processing steps and the resulting microstructural features.

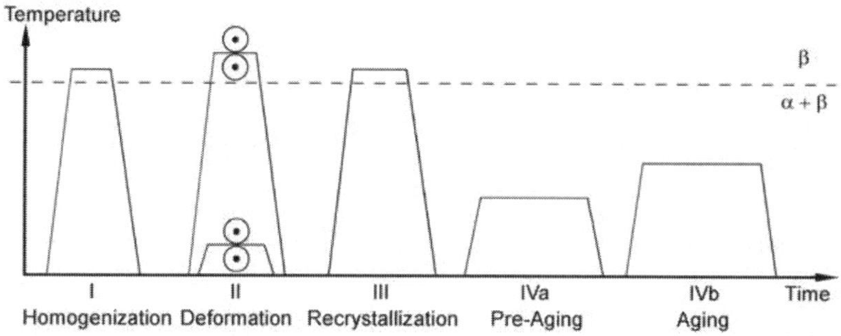

Fig. 7.3. Processing route for β annealed microstructures of heavily stabilized β titanium alloys with pre-aging step (schematically)

Table 7.2. Important processing parameters and resulting microstructural features for β annealed microstructures of heavily stabilized β alloys with pre-aging step

Processing Step (see Fig. 7.3)	Important Parameters	Microstructural Features
III	Recrystallization Temperature	β Grain Size
IVa	Pre-Aging Temperature	Homogeneous Distribution of α Platelets
IVb	Aging Temperature	– Size and Vol% of α Platelets – GB α Layer

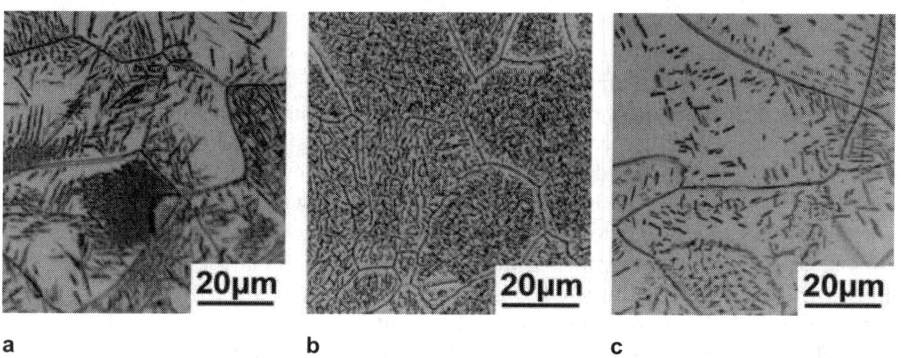

Fig. 7.4. Effect of pre-aging (8h 500°C) on microstructure of heavily stabilized β alloys, Beta 21S, LM: (a) 8h 690°C + 8h 650°C (b) 8h 500°C + 24h 725°C (c) 24h 725°C

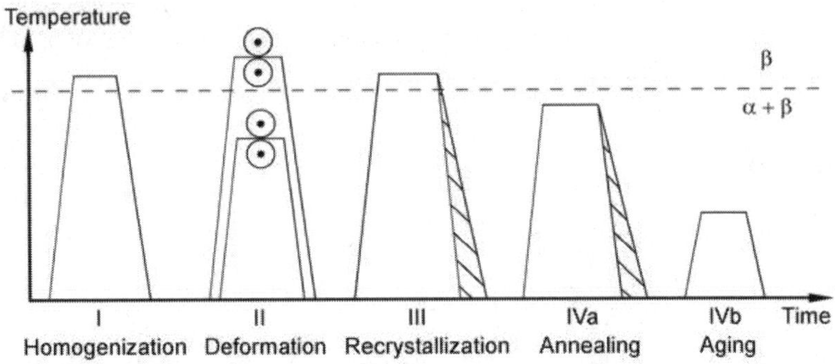

Fig. 7.5. Processing route for β annealed microstructures of β titanium alloys (schematically)

Table 7.3. Important processing parameters and resulting microstructural features for β annealed microstructures

Processing Step (see Fig. 7.5)	Important Parameters	Microstructural Features
III	Cooling Rate	– GB α Layer – "Side Plates" at GB
IVa	Annealing Temperature and Cooling Rate	Size and Vol% of α Plates (→ Vol% of α Platelets in Step IVb)
IVb	Aging Temperature	Size and Vol% of α Platelets

After homogenization (step I), the deformation (step II) can be done either in the β phase field or in the (α+β) phase field. The latter has the advantage of creating smaller β grain sizes in the recrystallization step III. The grain size of β annealed microstructures is somewhat smaller than the β grain size in fully lamellar microstructures of α+β alloys. A typical grain size of β alloys is about 400 µm as compared to about 600 µm for α+β alloys for equivalent processing histories and for a recrystallization temperature of 30-50°C above the β transus. The smaller grain size of β alloys is a result of the lower β transus. The most important parameter in step III is the cooling rate from the recrystallization temperature because it controls the width and extent of the continuous α layers at β grain boundaries (Table 7.3). It should be emphasized that even for fast cooling rates in commercial practice (e.g. 600°C/min) the formation of the continuous α layers cannot be avoided. An example of the continuous α layers at β grain boundaries is shown in Fig. 7.6a for the β-CEZ alloy cooled at a rate of 100°C/min. In these high strength β alloys, so-called "side plates" at β grain boundaries can be formed upon slow enough cooling (Table 7.3). The critical cooling rate for the formation of "side plates" is alloy chemistry dependent, but is about in the range of 30-

50°C/min. “Side plates” are parallel α plates that grow out of the grain boundary α layer. There are two possibilities for the formation mechanism. One is that they form by sympathetic nucleation and growth at the incoherent α/β interface of the continuous α layers. In this case, the “side plates” have a different variant of the Burgers orientation relation than the grain boundary α layer. The other is that they grow from the grain boundary α layer as a result of an instability in the α/β boundary. In this case, the “side plates” have the same crystallographic orientation as the grain boundary α layer. The factors that determine which of the two possibilities occurs are the nature of the α/β interface and the degree of undercooling. Some of these “side plates” can be seen in the upper right corner of the micrograph in Fig. 7.6a for β-CEZ, but they are much more pronounced in Ti-6246 using a cooling rate of 50°C/min, see for example Fig. 7.8a.

The final heat treatment in the (α+β) phase field is usually done for high strength β alloys in two steps, see Fig. 7.5. The first step IVa is performed high in the (α+β) phase field as an intermediate annealing step to precipitate a desired volume fraction of coarse α plates (Fig. 7.6a). It should be pointed out, that with decreasing cooling rate from the annealing temperature, the size and volume fraction of the coarse α plates will increase (Table 7.3). In the subsequent aging step IVb at temperatures of about 500-600°C the fine α platelets are formed (Fig. 7.6b) which determine the yield stress level of the material. Since the total volume fraction of α phase (coarse α plates plus fine α platelets) is fixed by the alloy chemistry, the volume fraction of coarse α plates has a direct influence on the volume fraction and size of the fine α platelets and therefore on the resulting yield stress level of the material (Table 7.3).

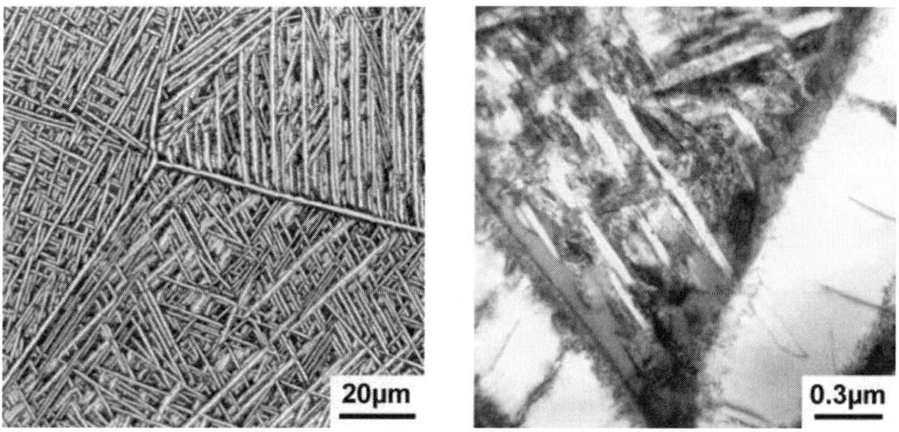

a b

Fig. 7.6. Example of β annealed microstructure, β-CEZ: (a) LM (b) TEM

7.1.2 Beta Processed Microstructures

The processing route for creating β processed microstructures is shown schematically in Fig. 7.7 and the important parameters of the processing steps are listed together with the resulting microstructural features in Table 7.4. From Fig. 7.7, it can be seen that the recrystallization step III is completely omitted in the processing route with the intention to create an unrecrystallized structure with highly deformed β grain boundaries. The α layers which form on the β grain boundaries during cooling from the β deformation temperature will then take on the local shape of the deformed grain boundaries. An example is shown in Fig. 7.8a for the Ti-6246 alloy. In this example it can be seen that the α layers are still fairly continuous but have a pronounced wavy shape with only short straight segments. In other cases, the α layers are broken up more into individual segments on most β grain boundaries with only a few long segments remaining on some boundaries. Irrespective of processing history, it is nearly impossible to completely avoid the α layers in β processed materials of high strength β alloys. This is because the β grain boundaries are such strong heterogenuous nucleation sites for α formation.

Since the final deformation process must be a continuous operation without reheating, a good control of the processing temperature and time is necessary. Critical parameters are the total time of the process, determined by deformation rate and any holding time both between deformation steps and after the deformation process, and the deformation degree (Table 7.4).

Table 7.4. Important processing parameters and resulting microstructural features for β processed microstructures

Processing Step (see Fig. 7.7)	Important Parameters	Microstructural Features
II	Deformation Time Deformation Mode Deformation Degree Cooling Rate	Unrecrystallized Structure Shape of β Grains – Width of β Grains (→ Max. Size of α Plates) – Geometry of GB α Layer – Unrecrystallized Structure – GB α Layer – "Side Plates" at GB
IVa	Annealing Temperature and Cooling Rate	Size and Vol% of α Plates (→ Vol% of α Platelets in Step IVb)
IVb	Aging Temperature	Size and Vol% of α Platelets

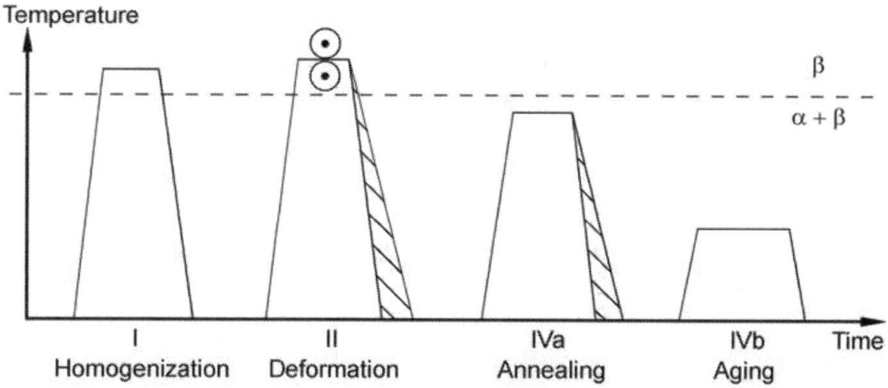

Fig. 7.7. Processing route for β processed microstructures of β titanium alloys (schematically)

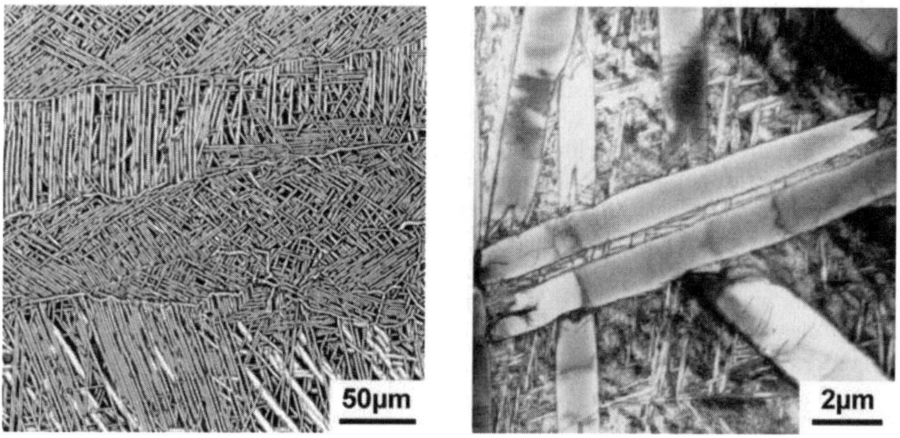

a b

Fig. 7.8. Example of β processed microstructure, Ti-6246: (a) LM (b) TEM

The deformation mode determines the shape of the unrecrystallized β grains, for example ellipsoidal shaped in case of unidirectional rolling or pancake shaped in case of cross rolling and axisymmetric upsetting. Consequently, it is obvious that those mechanical properties which are influenced by the α layers will be anisotropic, as will be shown later in Sect. 7.2.

The cooling rate from the deformation temperature including the transfer time to the quenching media is critical in determining the extent of α layers and "side plates" at β grain boundaries (Table 7.4). As pointed out already earlier, the formation of "side plates" is a critical parameter in Ti-6246, see Fig. 7.8a. The final two steps for β processed materials, the intermediate annealing step IVa and the aging step IVb, are in principle identical to those of β annealed materials (compare

Figs. 7.5 and 7.7 and Tables 7.3 and 7.4). An example of the small α platelets in between the coarse α plates for β processed material of the Ti-6246 alloy is shown in Fig. 7.8b. It also should be noted that in β processed materials with highly elongated β grains, the width of the β grains limits the maximum size of the coarse α plates.

7.1.3 Through-Transus Processed Microstructures

Extensive development work on the through-transus processing route was done for the β-CEZ alloy by CEZUS in France [185]. The intention of this processing route is to change the continuous α layers at β grain boundaries to individual globular α particles. From the through-transus processing route (Fig. 7.9) it can be seen that with the exception of the details of the deformation step II, the route is identical to that for β processed materials (Fig. 7.7), i.e. an unrecrystallized β grain structure is anticipated by omitting the recrystallization step III. The control of deformation temperature and time in step II is for the through-transus processing route even more critical than for the β processing route [186]. Stringent control is needed during through-transus processing because the material should only remain in the (α+β) phase region for a short time interval near the end of the deformation process. During this time α phase is precipitated at the β grain boundaries. The α phase is then also deformed and recrystallizes to globular particles at the boundaries of the unrecrystallized, elongated β grains. An example for such a microstructure after the through-transus deformation step II is shown for the Ti-6246 alloy in Fig. 7.10a. It can be seen that α particles are present only at the β grain boundaries and not in the interior of the elongated β grains. This kind of microstructure was called "necklace" microstructure by CEZUS in case of their β-CEZ alloy [187]. The higher magnification micrograph in Fig. 7.10b for a fully heat treated material shows clearly the globular α particles at the β grain boundaries and the α plate structure in the grain interior.

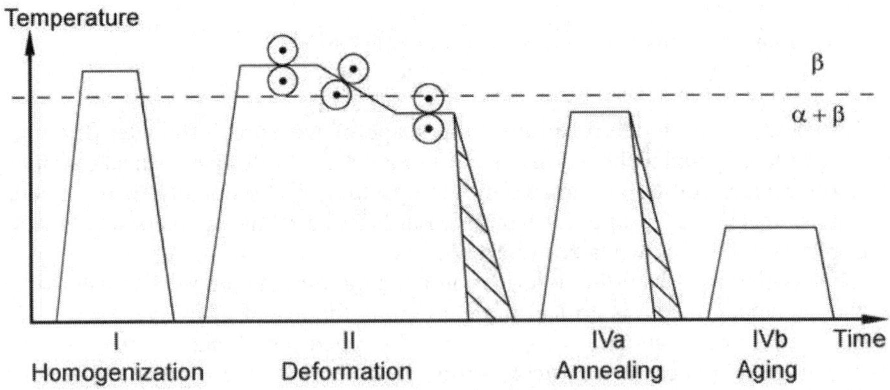

Fig. 7.9. Processing route for through-transus processed microstructures of β titanium alloys (schematically)

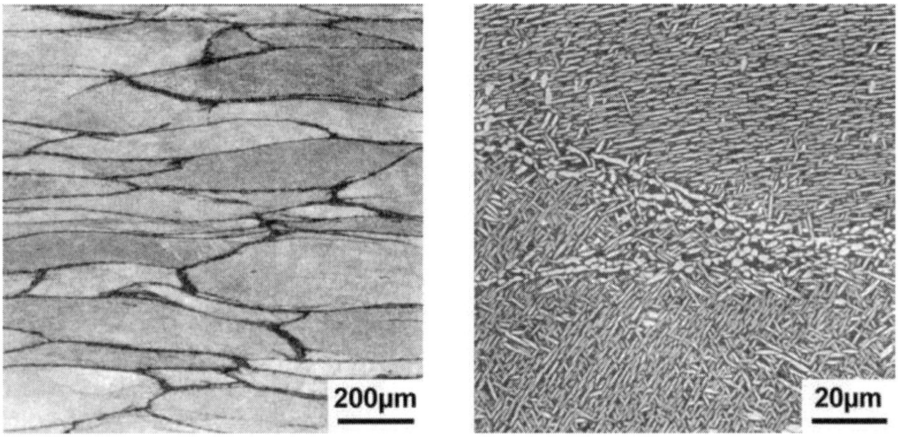

a b

Fig. 7.10. Example of through-transus processed microstructure ("necklace" microstructure), Ti-6246, LM: (a) As deformed (b) Heat treated

The most critical processing parameter for obtaining a necklace type microstructure is the deformation time in the (α+β) phase region in step II (Table 7.5). This point is illustrated schematically in a time-temperature-transformation diagram for step II (Fig. 7.11). The time should be long enough to cross the boundary line for α precipitation at the deformed β grain boundaries, but the deformation process should be finished before crossing the boundary line for α precipitation in the β matrix. If intragranular α precipitation occurs during deformation of the β matrix, then the α phase also is deformed and recrystallizes. This results in globular α particles instead of the desired plate shape, see Fig. 7.12a [186]. Another issue associated with through-transus processing is illustrated in Fig. 7.12b. The time of the deformation process must be long enough to allow α precipitation to occur at β grain boundaries. It is important to restrict the time in the β phase field to avoid recrystallization of the deformed β grains, as can be seen in Fig. 7.12b [186]. Recrystallization of the β grains is less common in β processed materials because in this case the deformation time does not have to conform to an interval (Fig. 7.7). Therefore, the time can be as short as feasible, but cooling rate is critical.

The only advantage in the through-transus processing route as compared to β processing is that the cooling rate after the deformation process is less critical because the α phase is already precipitated at the β grain boundaries in the necklace microstructure, see Tables 7.5 and 7.4. Again, the final heat treatments for the necklace microstructure, the intermediate annealing treatment in step IVa and the aging treatment in step IVb, are the same as for β processed and β annealed microstructures.

Table 7.5. Important processing parameters and resulting microstructural features for through-transus processed microstructures

Processing Step (see Fig. 7.9)	Important Parameters	Microstructural Features
II	Deformation in β Phase Region:	
	– Deformation Time	Unrecrystallized Structure
	– Deformation Mode	Shape of β Grains
	– Deformation Degree	Width of β Grains (→ Max. Size of α Plates)
	Deformation Time in (α+β) Phase Region	"Necklace" Microstructure
IVa	Annealing Temperature and Cooling Rate	Size and Vol% of α Plates (→ Vol% of α Platelets in Step IVb)
IVb	Aging Temperature	Size and Vol% of α Platelets

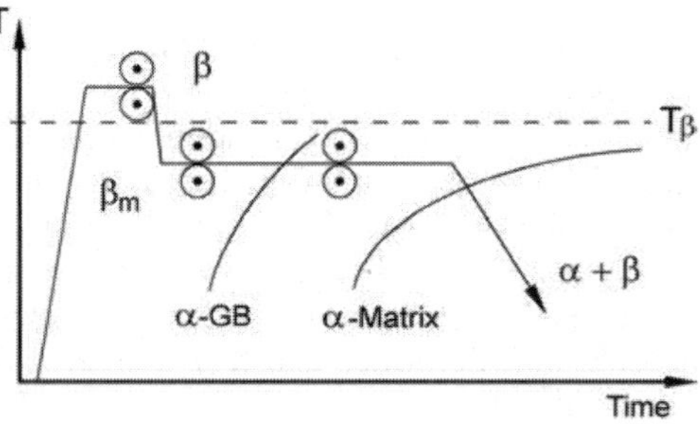

Fig. 7.11. Time/temperature control of through-transus deformation (schematically)

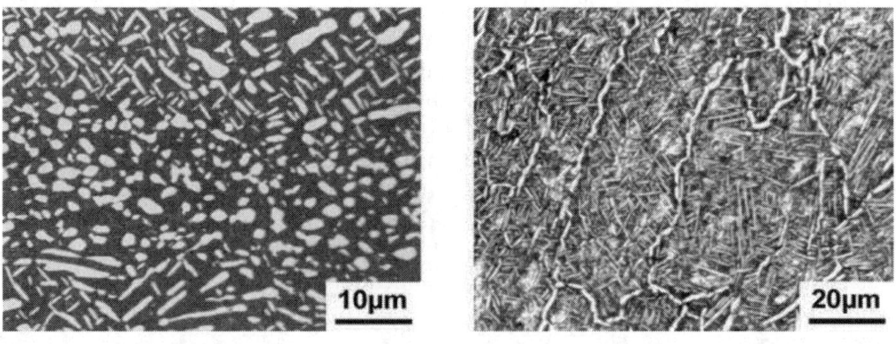

a **b**

Fig. 7.12. Problems associated with through-transus processing, β-CEZ, LM: (a) Equiaxed α phase inside β grains because of too long deformation time in (α+β) phase region (b) Start of β recrystallization because of too long deformation time in β phase region

7.1.4 Bi-Modal Microstructures

The benefit of a bi-modal microstructure is to compensate for the formation of the continuous α layers at β grain boundaries by creating a small enough β grain size that the α layers only have a negligible effect on mechanical properties. The processing route for obtaining a bi-modal microstructure is shown schematically in Fig. 7.13 and the important processing parameters together with the resulting microstructural features are summarized in Table 7.6. An example for the bi-modal microstructure in the fully heat treated condition is shown in Fig. 7.14a for the β-CEZ alloy.

Table 7.6. Important processing parameters and resulting microstructural features for bi-modal microstructures

Processing Step (see Fig. 7.13)	Important Parameters	Microstructural Features
I	Cooling Rate	GB α Layer
II	Deformation Degree	Dislocation Density
III	Annealing Temperature	Vol% of α_p (→ β Grain Size and Vol% of α Plates in Step IVa)
IVa	Annealing Temperature and Cooling Rate	Size and Vol% of α Plates (→ Vol% of α Platelets in Step IVb)
IVb	Aging Temperature	Size and Vol% of α Platelets

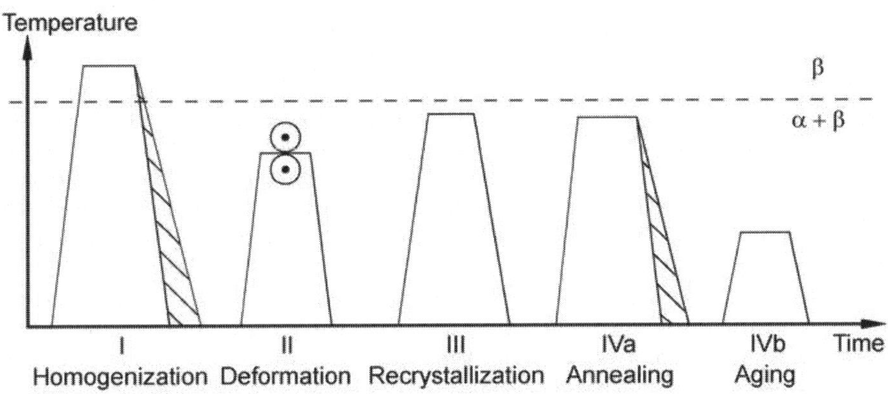

Fig. 7.13. Processing route for bi-modal microstructures of β titanium alloys (schematically)

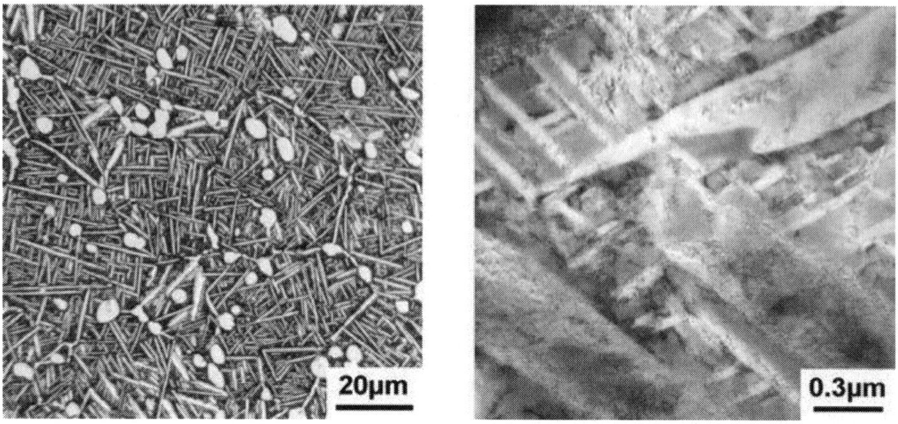

a b
Fig. 7.14. Example of bi-modal microstructure, β-CEZ: (a) LM (b) TEM

One problem with the bi-modal structure in high strength β alloys can arise from a very early step in the processing route, e.g. from a too slow cooling rate after the β homogenization treatment (see Fig. 7.13, step I). As pointed out in Sect. 7.1.1, the thickness of the continuous α layers at β grain boundaries increases with decreasing cooling rate from the β phase field and thick α layers are more difficult to eliminate during subsequent processing. Changing thick grain boundary α layers to isolated equiaxed α particles by deformation and recrystallization in the (α+β) phase field (steps II and III in the processing route, Fig. 7.13) is much more difficult than for thin α layers because of the diffusion distances involved and the relatively low processing temperatures of all β alloys. Figure 7.15 shows an α layer that remained at a prior β grain boundary even though the rest of the material has been converted to a uniform bi-modal structure [186]. If the importance of the cooling rate in step I is overlooked during processing, unnecessary and costly repetitions of steps II and III are required to eliminate any remaining extensive α layers, such as the one shown in Fig. 7.15.

The recrystallization annealing temperature (step III in Fig. 7.13) determines the volume fraction of primary α and therefore also has an influence on the size of the recrystallized, equiaxed β grains which is equal to the distance between the equiaxed primary α. The cooling rate from the recrystallization annealing temperature is not a critical parameter because continuous α layers at the boundaries of the small β grains (see Fig. 7.14a) are less detrimental to properties and can be tolerated.

The intermediate annealing treatment in step IVa is done to precipitate coarse α plates. When selecting the intermediate annealing temperature for bi-modal microstructures it should be realized that in this case the volume fraction of coarse α plates is determined by the temperature difference between the recrystallization annealing temperature (formation of primary α) and the intermediate annealing temperature.

The final aging treatment for bi-modal microstructures to form the small α platelets (Fig. 7.14b) is identical to the other already discussed microstructures. Clearly, the combined volume fraction of globular primary α and coarse α plates determines the volume fraction of fine α platelets that can be formed during aging. In turn, this volume fraction of α platelets determines together with the size of the α platelets the strength of the alloy in its final condition.

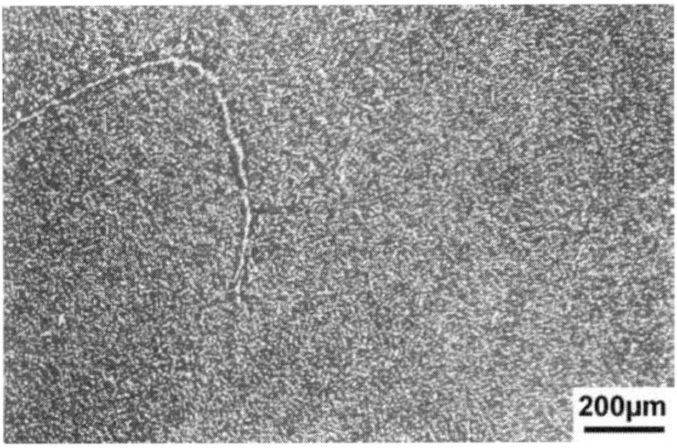

Fig. 7.15. Example of remaining α layer at prior β grain boundaries in fully heat treated bi-modal microstructure because of too slow cooling rate in step I of processing route (see Fig. 7.13 and Table 7.6), β-CEZ, LM

7.2 Microstructure and Mechanical Properties

In this section the mechanical properties of the different microstructures which were outlined in Sect. 7.1 will be discussed. A qualitative summary of the microstructure/property correlation is shown in Table 7.7, which is structured similarly to Table 5.6 used in Sect. 5.2 for α+β alloys. For an explanation of the symbols used in Table 7.7 to describe the mechanical properties, the reader is referred to the introductory remarks in Sect. 5.2 explaining these symbols for Table 5.6. Again, qualitative symbols (+, 0, –) are used for indicating qualitatively the direction in which the mechanical properties are changing when the microstructure is changed. The first row in Table 7.7 describes the basic effect of the continuous α layers in β annealed structures on mechanical properties. The second row compares the mechanical properties of bi-modal structures to those of β annealed structures both with continuous α layers. In the next row, which actually consists of two subrows "a" and "b", the effects of the unrecrystallized structures on mechanical properties are shown. No distinction was made between β processed and through-transus processed ("necklace") structures, because the properties of these two structures exhibit the same general trends. In subrow "a" the properties of the unrecrystallized structures are compared to those of β annealed structures and in subrow "b" to those of bi-modal structures. It should be noted, that for this comparison in Table 7.7 the mechanical properties in longitudinal direction (L direction) are taken for the unrecrystallized structures, i.e. in the mechanical tests the stress axis was parallel to the longitudinal processing direction and therefore parallel to the long dimension of the elongated β grains. This was done because the mechanical properties of the unrecrystallized structures are anisotropic and the longitudinal test direction, which usually results in the best mechanical properties, is the most relevant stress direction for most applications. Nevertheless, this anisotropy in mechanical properties of unrecrystallized β processed microstructures will be treated in detail in this Sect. 7.2 to point out that caution should be exercised in using this microstructure in components where more complicated stress states are present. This comparison in mechanical properties between the microstructures resulting from the different processing routes of high strength β alloys (β annealed, bi-modal, and β processed) is outlined in Sect. 7.2.1. These data are presented on the basis of a constant final aging treatment, i.e. at a constant yield stress level.

Table 7.7. Qualitative correlation between important microstructural parameters and mechanical properties for β titanium alloys (both high strength and heavily stabilized as noted)

		$\sigma_{0.2}$	ε_F	HCF	Micro-cracks ΔK_{th}	Macrocracks ΔK_{th} R = 0.7	Macrocracks K_{Ic}	Macrocracks ΔK_{th} R = 0.1	Creep Strength 0.2%
GB α Layers in β Annealed Structure		0	–	–	–	0	+	0	0
Bi-Modal Structure [a]		0	+	+	+	–	–	–	0
"Necklace" or β Processed Structure in L Direction	a	0	+	+	+	–	+	–	0
	b	0	–	–	–	+	+	+	0
Decreasing Age-Hardening		–	+	–	+	+	+	+	–
Small β Grain Size in β Annealed Structure [c]		0	+	+	+	–	–	–	0

[a] Compared to β annealed structure.
[b] Compared to bi-modal structure.
[c] Only applies to heavily stabilized alloys.

The effect of age-hardening on mechanical properties is discussed in Sect. 7.2.2. In Table 7.7 this effect is described by the row "Decreasing Age-Hardening", i.e. coarsening of the fine α platelets. The qualitative trends in mechanical properties shown in this row in Table 7.7 are valid to a varying degree for most of the different microstructural conditions of high strength β alloys discussed in Sect. 7.2.1. The effects of age-hardening on mechanical properties are very pronounced for β annealed microstructures. Since only the heavily stabilized β alloys are used in this β annealed condition the effect of age-hardening is especially important for these alloys.

The last row in Table 7.7 describes the tendency in properties for a reduction in β grain size in β annealed microstructures. This effect will be treated in Sect. 7.2.3 and is important for heavily stabilized β alloys.

7.2.1 Effect of Processing Route

7.2.1.1 Tensile Properties

The mechanical properties of high strength β alloys are dominated by the preferential plastic deformation along the continuous α layers at β grain boundaries. An example for this preferential plastic deformation is shown in Fig. 7.16. In this micrograph the tensile stress axis is horizontal and the large plastic displacement of the dark surface marker at the β grain boundary is clearly visible. This preferential plastic deformation along the continuous α layers can be treated in a very simplified way as a dislocation pile-up against the adjacent grain boundary triple point. The slip length is equal to the distance between grain boundary triple points [188]. The local stress concentration at the grain boundary triple point is causing crack nucleation and the relationship between the grain boundary length D and the tensile fracture properties can be described as σ_F (true fracture stress) proportional to $D^{-1/2}$ and ε_F (true fracture strain) proportional to D^{-1} [188]. This predicted effect of grain size on ductility and fracture stress can be seen in Table 7.8 by comparing the β annealed condition (large β grain size) with the bi-modal condition (small β grain size). Both of these conditions exhibit continuous α layers at β grain boundaries, as can be seen in the microstructures in Figs. 7.6a and 7.14a. The comparison in Table 7.8 uses the β-CEZ alloy as an example and the final two steps in the processing route (the intermediate annealing and the aging treatments) were identical for the different microstructures. The continuous α layers have no measurable effect on the yield stress because of their low volume fraction. This can be seen from Table 7.8 by comparing the yield stress values of the β annealed and bi-modal conditions. In summary, the unacceptably low ductility of high strength β alloys in the β annealed condition can be drastically improved by α+β processing to a bi-modal microstructure with a small β grain size, see also column "ε_F" in the qualitative summary in Table 7.7.

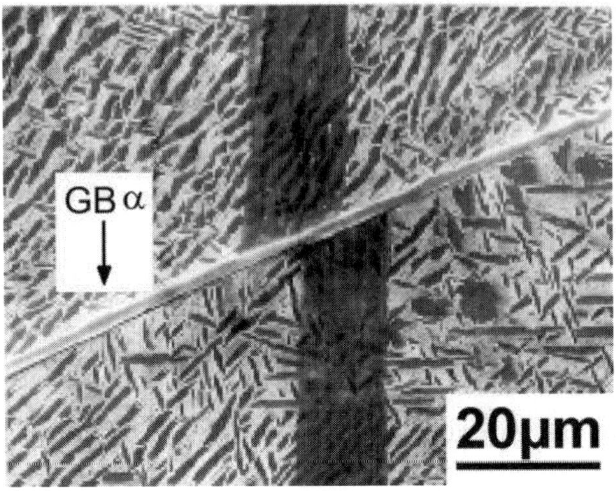

Fig. 7.16. Preferential plastic deformation (see large displacement of dark surface marker) along continuous α layers at β grain boundaries in high strength β titanium alloys, tensile stress axis is horizontal, SEM (courtesy J. C. Chesnutt)

Table 7.8. Tensile properties of β-CEZ

		$\sigma_{0.2}$ (MPa)	UTS (MPa)	σ_F (MPa)	T.E. (%)	RA (%)
β Annealed		1180	1280	1415	4	10
Bi-Modal		1200	1275	1660	13	34
β Processed	L	1190	1275	1480	10	16
	45°	1145	1200	1220	2	2
	ST	1185	1280	1410	6	10

In case of the β processed condition with unrecrystallized elongated β grains, the anisotropy of the tensile properties is demonstrated in Table 7.8 by test results in the longitudinal direction (L), the short transverse direction (ST), which is the through-thickness direction, and by tests under an inclined angle of 45° to the L and ST test directions (45°). Whereas the lower yield stress of the 45° test direction as compared to the other two directions can be explained by the crystallographic texture of the β matrix, the fracture properties are dominated by the angle of the tensile stress axis with the elongated β grain boundaries containing α layers. The effect of preferred plastic deformation along the α layers should be much more pronounced if the large, flat grain boundary planes were oriented in the maximum shear stress position, that is at an angle of 45° to the tensile stress axis as compared to being parallel (L test direction) or perpendicular (ST test direction). The resulting minimum values in fracture stress and ductility for the 45° test direction can be seen in Table 7.8. It also is obvious from Table 7.8 that the values in ST direction are lower than the ones in L direction, although they should be about equal from a pure shear stress argument along β grain boundaries assuming a very simplified grain shape. This effect can be more clearly seen by including 30° and 60° test directions and plotting the tensile ductility (RA values) as a function of that inclined angle (Fig. 7.17). For a discussion, the schematic drawing in Fig. 7.18 might be helpful showing a simplified sketch of the elongated β grains and all test directions. It is evident from Fig. 7.17 that the ductility as a function of inclined angle does not show a symmetrical relation. Instead, the ductility values are lower for the 60° and ST test directions as compared to the 30° and L test directions, respectively. The results can be explained by dividing the total plastic deformation up to fracture into a part occurring prior to crack nucleation and a part occurring during crack propagation. In the simple pile-up model for crack nucleation the short grain boundary length D_2 in Fig. 7.18 is the critical slip length for the L and ST directions whereas the large grain boundary length D_1 in Fig. 7.18 is the critical slip length for the 30°, 45°, and 60° directions explaining the minimum in ductility under 45° but otherwise predicting a fairly symmetrical dependence. An additional consideration is needed to account for the asymmetry observed in Fig. 7.17. Once a crack has been formed and propagates along the grain boundary the normal (mode I) stress acting on the crack becomes an important factor for crack propagation. This normal stress is much higher for the 60° and ST directions as compared to the 30° and L directions, respectively, explaining the asymmetrical shape of the curve in Fig. 7.17. It also should be mentioned, that similar results to those shown in Fig. 7.17 were obtained for high strength Al-alloys containing weak precipitate-free-zones (PFZ) along grain boundaries in a pancake shaped grain structure [189].

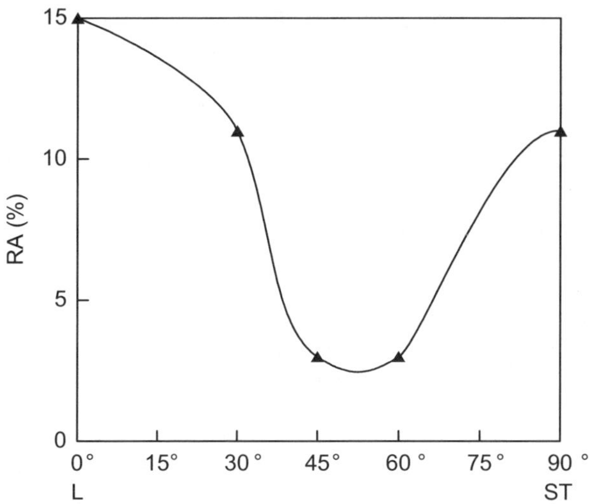

Fig. 7.17. Tensile ductility as a function of test direction of β processed material, Ti-6246

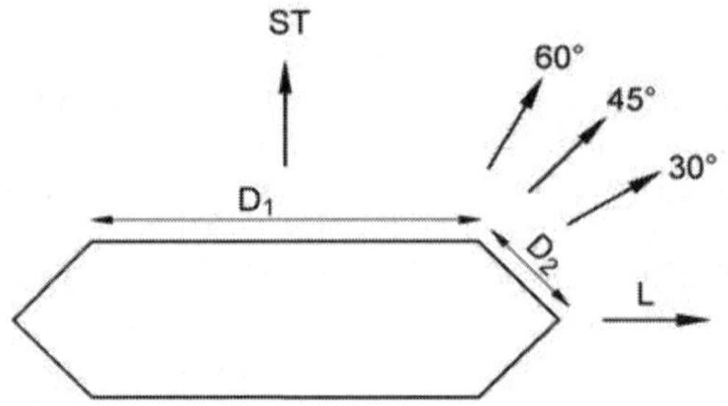

Fig. 7.18. Geometrical model for pancake shaped grains with GB α layers (schematically)

7.2.1.2 Fatigue Properties

In titanium alloys the HCF strength at 10^7 cycles is often stated as the resistance to fatigue crack nucleation. The fatigue results for β processed conditions of the alloys β-CEZ and Ti-6246 are shown in Figs. 7.19 and 7.20, respectively, for the three test directions L, ST, and in between these two directions (45°). The corresponding yield stress values of β-CEZ are shown in Table 7.8. The values for Ti-6246 follow the same trend. In case of Ti-6246 the material was upset forged to a symmetrical disk (pancake shaped grains) whereas the β-CEZ material was unidirectionally forged to a rectangular slab (cigar shaped grains). It can be seen from Figs. 7.19 and 7.20 that for both alloys the 45° test direction exhibited the lowest HCF strength. The HCF strength in ST test direction was higher than in L direction for both alloys. A possible reason for this behavior will be discussed later. For both alloys the fine grained bi-modal microstructure showed a higher HCF strength than the coarse grained β annealed structure. This is shown in Fig. 7.21 using the β-CEZ alloy as an example (for tensile properties see Table 7.8). It should be emphasized that for both alloys, the 45° test direction of the β processed condition had an even lower HCF strength than the coarse grained β annealed microstructure as demonstrated in Fig. 7.21.

Typical fatigue crack nucleation sites for the β annealed microstructure and the bi-modal microstructure are shown in Fig. 7.22. In case of the coarse grained β annealed structure, the fatigue cracks always nucleated on the continuous α layers at β grain boundaries (Fig. 7.22a). In the fine grained bi-modal structure the fatigue cracks nucleated either on the continuous α layers at β grain boundaries or on the largest intragranular α plates (see Fig. 7.22b). In this fine grained bi-modal microstructure, the largest α plates are in many cases as large as the β grain size. Typical examples for the fatigue crack nucleation sites of the β processed condition tested in L, 45°, and ST directions are shown in Fig. 7.23. For a better illustration, a schematic drawing of a pancake shaped grain structure is shown in Fig. 7.24 with the crack nucleation sites indicated. For specimens tested in 45° direction the cracks always nucleated at the intersection of the large, flat grain boundary planes with the specimen surface as schematically shown in Fig. 7.24. From the micrograph in Fig. 7.23b it can be seen that the cracks nucleated at the interface between the α layer and the β matrix and not inside the grain boundary α layer. Crack nucleation for the specimens tested in L and ST directions occurred also under an angle of about 45° to the stress axis but at the short segments of the elongated grains as indicated schematically in Fig. 7.24. However, it should be pointed out that especially for the L test direction cracks also nucleated at the interface of large α plates with the grain boundary segment involved being located close-by in the adjacent volume of material [190]. Examples for both cases are shown in Fig. 7.23a (nucleation at a large α plate) and in Fig. 7.23c (nucleation at a grain boundary α layer). Corresponding examples to Figs. 7.22 and 7.23 showing the fracture surfaces with the crack nucleation sites have been reported elsewhere [190].

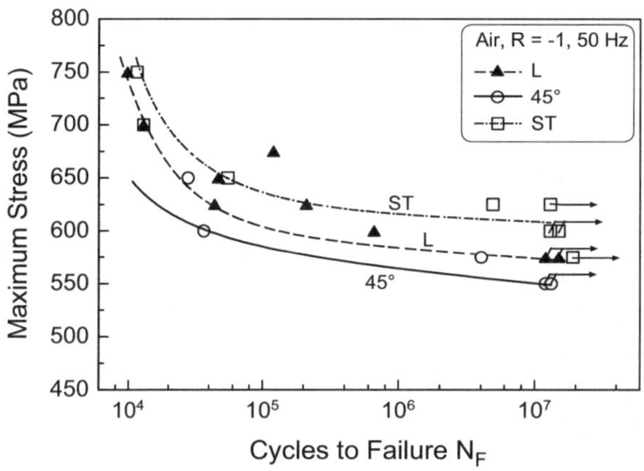

Fig. 7.19. S-N curves of β processed material, forged rectangular slab, β-CEZ

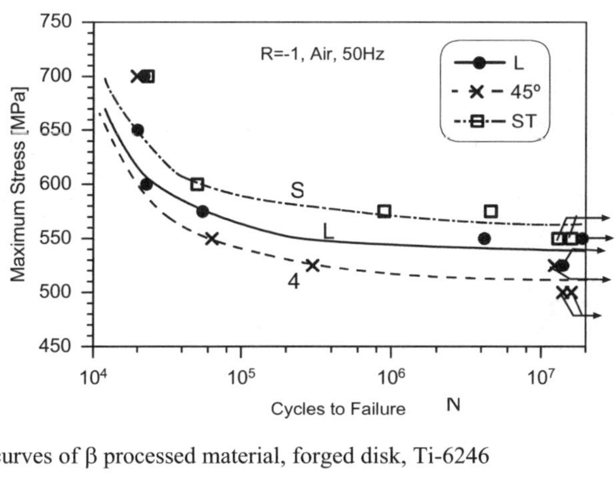

Fig. 7.20. S-N curves of β processed material, forged disk, Ti-6246

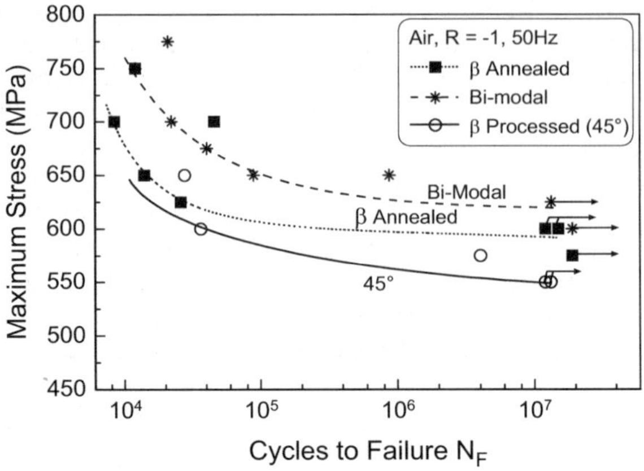

Fig. 7.21. S-N curves for different microstructures, β-CEZ

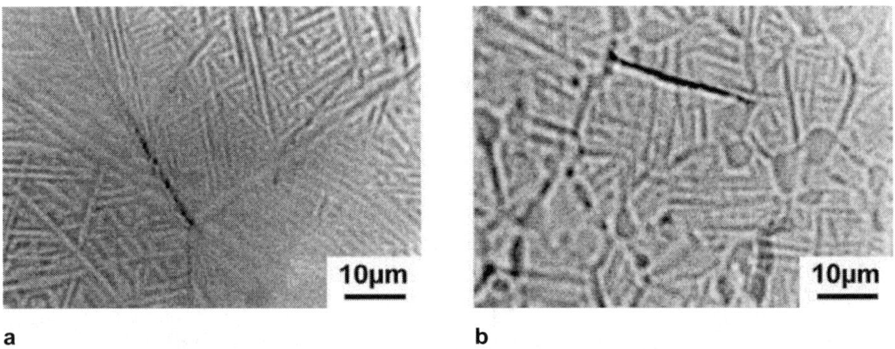

Fig. 7.22. Fatigue crack nucleation, β-CEZ, LM: (a) β annealed microstructure (b) Bi-modal microstructure

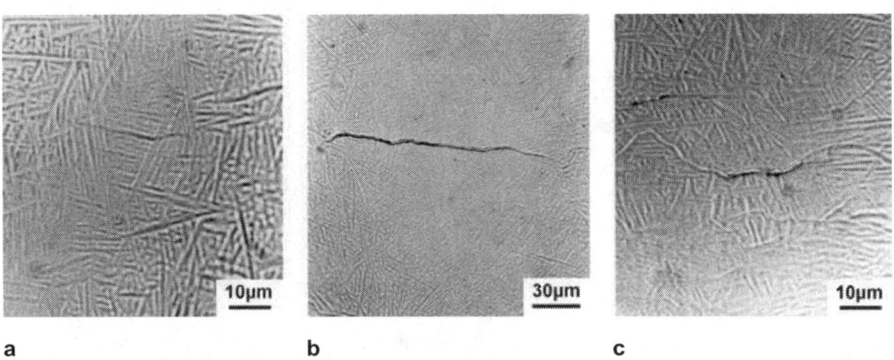

Fig. 7.23. Fatigue crack nucleation in β processed material for different test directions, β-CEZ, LM: (a) L (b) 45° (c) ST

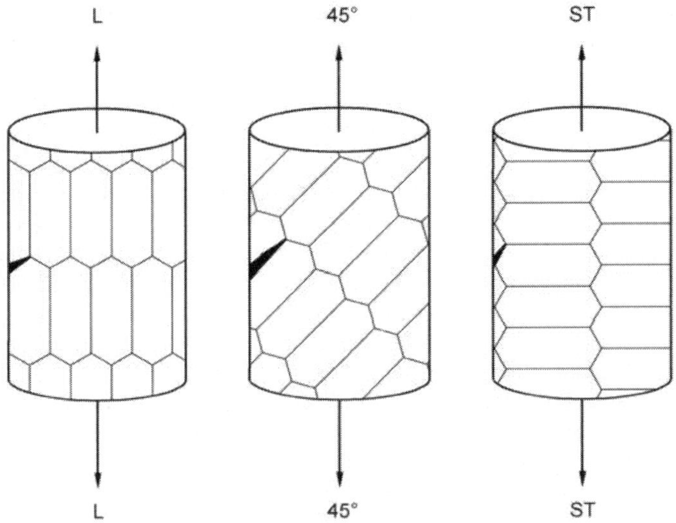

Fig. 7.24. Fatigue crack nucleation sites for pancake shaped grains with GB α layers (schematically)

A possible but very simple explanation for the higher HCF strength in ST test direction as compared to the L test direction observed in Figs. 7.19 (β-CEZ) and 7.20 (Ti-6246) might be that the two-dimensional schematic drawings in Figs. 7.18 and 7.24 are too oversimplified indicating that the critical slip length D_2 is equal for the two test directions. For both grain shapes (pancake and cigar shaped) a three-dimensional evaluation of the grain boundary segment length D_2 will result in shorter average D_2 values for the ST test direction as compared to the L test direction explaining the higher HCF strength in ST test direction. This is in disagreement with the well-known behavior of age-hardened aluminum alloys with PFZ and pancake shaped grains where the HCF strength is lowest in ST test direction. This effect in aluminum alloys is due to the additional presence of Fe and Si rich inclusions arranged in isolated arrays at grain boundaries. These arrays serve as crack nucleation sites. If the Fe, Si content is lowered drastically in those alloys, then also the ST test direction exhibits a higher HCF strength than the L test direction [191], in agreement with the results on high strength β alloys.

The propagation rates of the small, self initiated surface cracks (microcracks) for the different microstructures, see Figs. 7.22 and 7.23, are shown in Fig. 7.25 together with the results for macrocracks obtained using CT specimens. It can be seen that the microcracks exhibited the fastest propagation rate for the β annealed structure. In this microstructure the cracks, which nucleated along the continuous α layers at β grain boundaries, propagated very fast along these boundaries until they reached the first grain boundary triple point. Most of the cracks could not follow the grain boundary of the adjacent grain due to geometrical reasons and due to the small dimension of the plastic zone size at the crack tip. Instead, they deviated into the grain interior and the propagation rate slowed down. Even the fastest growing microcracks reached the slower crack propagation curve of the macrocracks (see Fig. 7.25) at a surface crack length 2c of about 1000 µm, i.e. when the crack front encountered approximately three β grains. The propagation rate of microcracks in the fine grained bi-modal microstructure was about one order of magnitude slower as compared to the coarse grained β annealed structure (Fig. 7.25). This is due to the small β grain size (about 30 µm) of the bi-modal structure slowing down the propagation rate even for very small crack sizes. Again, the microcrack propagation curve of the bi-modal structure merged with the macrocrack propagation curve at crack sizes of about a few grain diameters, in the present case 2c is approximately 100 µm. The microcrack propagation curve of the β processed condition with the cigar shaped grain structure tested in L direction is located in between the curves for the β annealed and bi-modal structures, as can be seen in Fig. 7.25. In this case, the small cracks propagated partially along the short grain boundary segments oriented approximately 45° to the stress direction and partially through the β matrix. This mixed propagation path explains the location of the propagation curve in Fig. 7.25. It should be mentioned, that the propagation curve for microcracks of this β processed condition when tested in the ST direction is located above the curve for the L direction but below the curve for the β annealed condition in Fig. 7.25 [192].

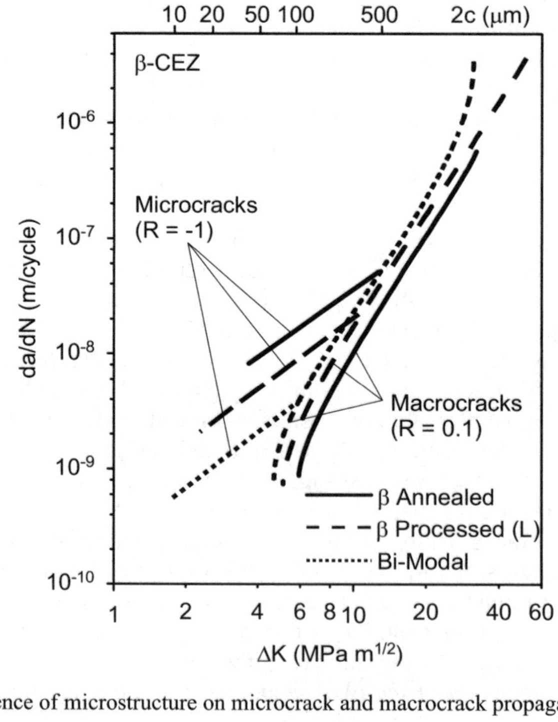

Fig. 7.25. Influence of microstructure on microcrack and macrocrack propagation, β-CEZ

The results of the fatigue crack propagation rate measurements for macrocracks (through-thickness cracks in CT specimens) show that the influence of the different microstructures tested (β annealed, β processed, and bi-modal) was rather small, see Fig. 7.25. This behavior is due to the small plastic zone size at the crack tip. Consequently, the cracks will propagate only within short segments of the total length of their crack front, which is the specimen thickness, along the soft zones of the microstructure (α layers at β grain boundaries). The remainder of the crack front is propagating through the β matrix containing the mixture of coarse α plates and small α platelets. Since the final annealing and aging treatments were kept constant for the different microstructures tested, the size and distribution of the α plates and α platelets within the β grains were similar for these microstructures, as seen by comparing Figs. 7.6, 7.8, and 7.14. The ranking of the three da/dN-ΔK curves for the macrocracks in Fig. 7.25 can be explained by considering the details of the crack front geometry for the three microstructures (Fig. 7.26). Due to the small β grain size of the bi-modal microstructure and due to the fact that the maximum size of the coarse α plates also is limited by the β grain size, the crack front profile was very smooth for this microstructure (Fig. 7.26b). This smoothness correlates with the fastest crack propagation rate among the three microstructures tested (Fig. 7.25). For the other two microstructures the crack front profile is more irregular, see Figs. 7.26a and c, leading to a reduction in the crack propagation rate. These deviations in crack path might result either from a local propagation along α layers at β grain boundaries or, more likely, from crack deflection at coarse α plates. On average, the coarse α plates were larger in the β annealed structure as compared to the β processed structure. This is because of the reduced grain width in the latter structure. This variation in coarse α plate size might explain the ranking of the da/dN-ΔK curves for these two microstructures in Fig. 7.25 (see also Table 7.7).

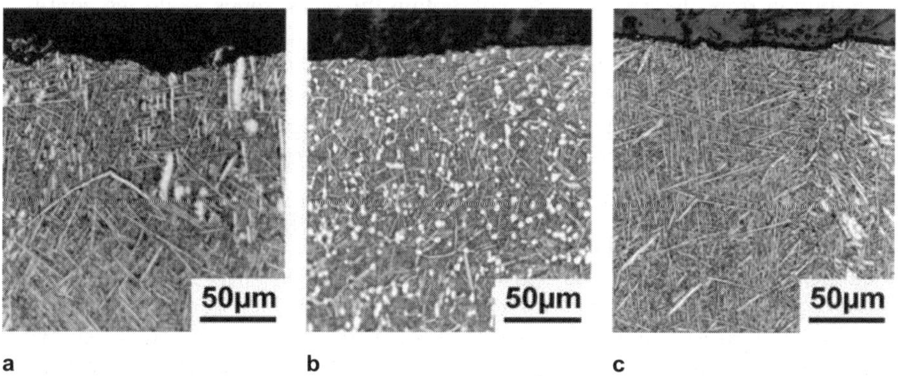

Fig. 7.26. Crack front profiles of macrocracks at about 10^{-9} m/cycle, β-CEZ, LM: (a) β annealed structure (b) Bi-modal structure (c) β processed structure (L)

7.2.1.3 Fracture Toughness

The fracture toughness values of the β annealed and bi-modal microstructures as well as those of the β processed condition for the three different testing directions (L, 45°, ST) are shown in Fig. 7.27 for Ti-6246. As can be seen, the fracture toughness of the β annealed condition was much higher than that of the bi-modal condition. The fracture surface at the transition from pre-cracking to unstable fracture is shown in Fig. 7.28 for the two microstructures. The β annealed microstructure clearly showed a ductile fracture mode along β grain boundaries (Fig. 7.28a). Apparently, the large crack tip plastic zone size, which forms during the onset of rapid crack extension, encountered whole β grains. In this case, the crack can follow the highly deformed region along α layers at β grain boundaries. This is in contrast to the situation present in fatigue crack propagation tests in the slow propagation regime with the small plastic zone size which was described above in Sect. 7.2.1.2. It should be mentioned, that these different crack propagation mechanisms for small and large plastic zone sizes are also observed for high strength aluminum alloys containing weak precipitate free zones at grain boundaries [193]. The rough crack front profile characteristic of monotonic loading crack extension in the β annealed condition correlates with the relatively high fracture toughness value. In contrast, the fine grained bi-modal microstructure exhibited a very smooth fracture surface topography (Fig. 7.28b) resulting in a much lower fracture toughness value than the β annealed structure (Fig. 7.27). This influence of crack front profile on crack propagation resistance was discussed in detail in Sect. 5.2.1. The crack extension mechanism leading to the high fracture toughness in β annealed structures is analogous to the one shown schematically in Fig. 5.24 if the fracture along slip bands is interchanged with fracture along α layers at β grain boundaries. Apparently, for β annealed structures the positive contribution to fracture toughness due to the rough crack front profile dominates over the negative contribution from the easier crack path within the weak zones along the continuous α layers at β grain boundaries, see (+) symbol for K_{Ic} in the first row in Table 7.7. In previous overview articles [194, 195], a (–) symbol was used in that position to point out that the fracture toughness would decline in the absence of any crack front profile effect, i.e. the intrinsic crack propagation resistance within the soft zones along the continuous α layers is lower than the crack propagation resistance of the matrix.

The fracture toughness of the β processed condition showed the highest value for the L test direction, followed by the 45° test direction, and the lowest fracture toughness value was observed for the ST direction (Fig. 7.27). The fracture surface for the three different test directions are shown in Fig. 7.29a-c. Comparison of these three SEM photos show that the specimen tested in L direction exhibits a very rough or irregular crack front profile (Fig. 7.29a). The crack tried to follow the soft grain boundary regions in this pancake shaped grain structure, but the grain boundaries are aligned mainly parallel to the applied stress direction, i.e. very unfavorable for crack propagation. In the example given here for the L test direction, the crack propagation direction was the ST direction. A similar crack growth retardation mechanism is present and a similar high fracture toughness is observed when the crack propagation direction is the LT direction [196]. In case of the other two test directions (45° and ST) the crack propagated nearly exclusively along α layers at β boundaries, as can be seen from Fig. 7.29b and c. Both fracture surfaces looked similar in the SEM photos but in case of the 45° test direction the crack propagated macroscopically at an angle of 45° to the applied stress direction, whereas in case of the ST test direction the crack propagated perpendicular to the applied stress direction. This can be seen more clearly in crack path profiles (Figs. 7.30b and c). The applied stress direction in the micrographs in Fig. 7.30 was vertical and the nominal crack propagation direction was horizontal. The macroscopic alignment of the crack plane at 45° to the applied stress direction in case of the 45° test direction (see Fig. 7.30b) is the main reason for the higher fracture toughness value measured as compared to the ST test direction (Fig. 7.27). Cracks propagating in this way are mixed mode I and II cracks and using for K the standard mode I solution to calculate fracture toughness doubles the value.

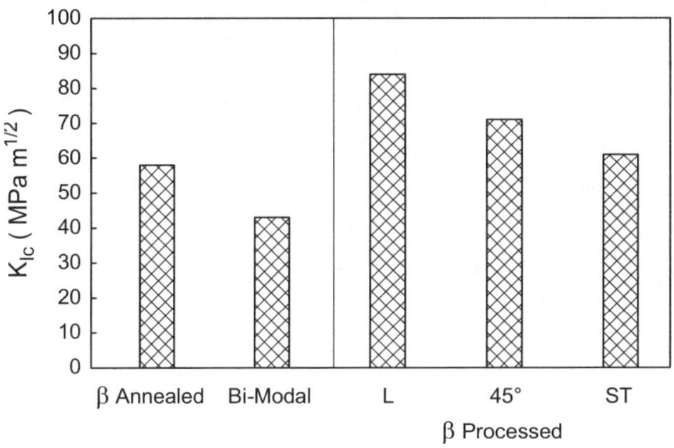

Fig. 7.27. Fracture toughness for different microstructures and for different test directions (β processed material), Ti-6246

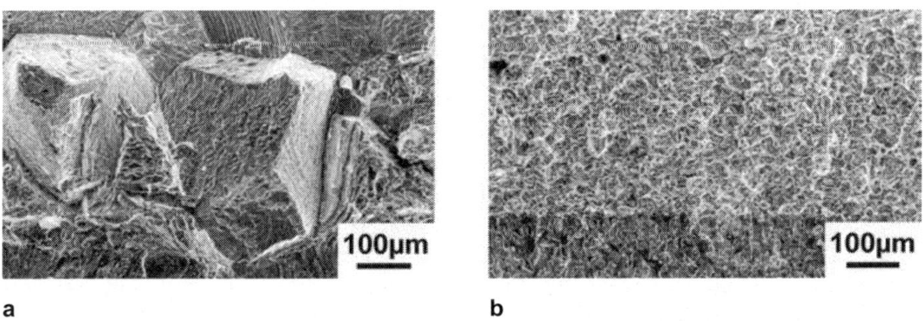

Fig. 7.28. Fracture surface of K_{Ic} specimens, Ti-6246, SEM: (a) β annealed structure (b) Bi-modal structure

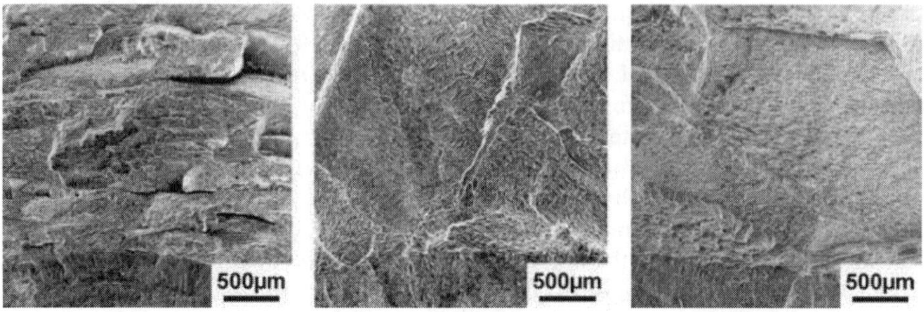

a b c

Fig. 7.29. Fracture surface of K_{Ic} specimens of β processed material for different test directions, Ti-6246, SEM: (a) L (b) 45° (c) ST

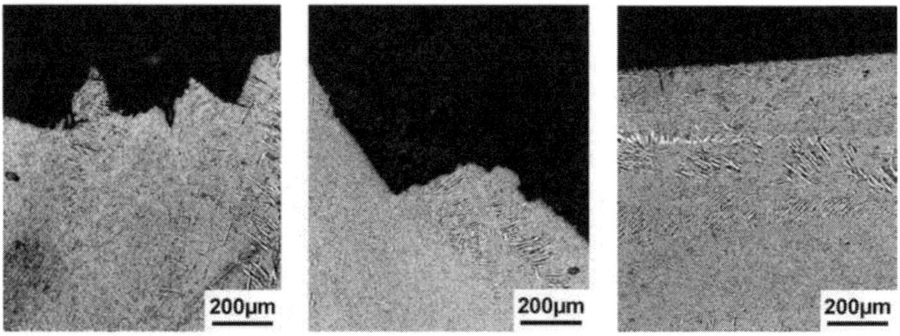

a b c

Fig. 7.30. Crack path profiles of K_{Ic} specimens of β processed material for different test directions, Ti-6246, LM: (a) L (b) 45° (c) ST

7.2.2 Effect of Age-Hardening

The yield stress in the case of incoherent α platelets is proportional to d^{-1} (Orowan relationship), where d is the interparticle distance. The maximum in the age-hardening curve, i.e. yield stress as a function of aging time at constant aging temperature, is reached when the equilibrium volume fraction for that aging temperature is precipitated. With increasing aging temperature the maximum in the age-hardening curve decreases (lower equilibrium volume fraction of α platelets) but the aging time to reach the maximum is shortened. In commercial practice, long aging times (above 24 h) are usually avoided. Therefore, a lower yield stress is usually achieved by increasing the aging temperature (step IV in Fig. 7.1 and Table 7.1 for heavily stabilized β alloys). As pointed out already in Sect. 7.1.1, for high strength β alloys overaging can also be achieved by lowering the annealing temperature in the intermediate step IVa or decreasing the cooling rate from that temperature (see Fig. 7.5 and Table 7.3), both leading to an increase in volume fraction of coarse α plates and thus leading to a decrease in volume fraction of α platelets.

For demonstration of the maximum possible effect of age-hardening on mechanical properties (tensile properties, fracture toughness, fatigue crack propagation of macrocracks), two extreme microstructures were produced for the β annealed condition of β-CEZ (A and C in Table 7.9). In the very fine β annealed microstructure A (Fig. 7.31) only small α platelets were present, i.e. the intermediate annealing step IVa in Fig. 7.5 was eliminated completely and a fast cooling rate of 600°C/min was used after the β annealing at 920°C (step III in Fig. 7.5). In the very coarse β annealed microstructure C (Fig. 7.32) only large α plates were present. In this case, a slow cooling rate of 1°C/min was used from the intermediate annealing temperature of 820°C. The β annealed microstructure B in Table 7.9 was achieved by a more commercial heat treatment practice using a cooling rate of 100°C/min after β annealing and after the intermediate annealing step. This microstructure can be seen in Fig. 7.6 and such microstructures using more commercial practice for heat treatments and processing were used in the comparison between the different processing routes in Sect. 7.2.1. The final heat treatment for all three microstructures A, B, and C was identical (8h 580°C). More details about the complete treatment for A, B, and C can be found elsewhere [192]. From Table 7.9 it can be seen that condition A failed in the "elastic" regime, i.e. the yield stress $\sigma_{0.2}$ was not reached. Apparently, the strength difference between the matrix and the soft zones along the continuous α layers at β grain boundaries was too large. The preferred plastic deformation within the soft zones was so concentrated that cracking occurred before any macroscopically measurable plastic deformation of the matrix occurred. Decreasing the yield stress level (conditions B and C) reduces the strength difference between matrix and soft zones to such a level that increasing plastic deformation of the matrix can occur before fracture within the soft zones takes place, i.e. increasing tensile ductility is observed (Table 7.9). All fracture surfaces looked similar to the fracture surface shown in Fig. 7.28a (unstable crack extension part of the fracture toughness specimen) exhibiting a ductile fracture mode along β grain boundaries and some portion of ductile transgranular fracture. The main difference was that the percentage of ductile transgranular fracture increased from about 10% for condition A to about 25% for condition C.

Part of the increase in fracture toughness from 33 MPa $m^{1/2}$ (condition A) to 73 MPa $m^{1/2}$ (condition C), see Table 7.9, can be attributed to the increase in plastic deformation of the matrix before the onset of unstable crack extension occurs within the soft zones along grain boundaries [197]. But as was shown in this recent investigation, an increase in the retarding effect of the rough crack front profile with decreasing strength level from A to C also contributes significantly to the observed increase in fracture toughness [197].

Although not determined experimentally up to now, the fatigue crack propagation rate of microcracks, which propagate in the early stages within the soft zones (see Sect. 7.2.1.2), should be faster for condition A as compared to condition C with the curve for the intermediate condition B, which can be seen in Fig. 7.25 (β annealed curve for microcracks), lying in between. This anticipated trend is also shown by the (+) symbol in Table 7.7.

The fatigue crack propagation rates of macrocracks for the two conditions A and C are shown in Fig. 7.33 [192]. Since macrocracks propagate through the matrix, as was discussed in detail in Sect. 7.2.1.2, the difference of the two curves in Fig. 7.33 probably represent the largest possible effect of α plate size variation on fatigue crack propagation rate in β titanium alloys. The difference in the threshold region is about 2 MPa $m^{1/2}$ (or a factor of 1.5). It should be mentioned that the curve for the intermediate condition B, which can be seen in Fig. 7.25 (β annealed curve for macrocracks), is located in between the two curves in Fig. 7.33 but lies very close to the curve for the coarse β annealed condition C. This shows that the smaller volume fraction of coarse α plates present in the intermediate condition B (see Fig. 7.6) causes enough local crack deflection for crack growth retardation. It also shows, that the presence (condition B) or absence (condition C) of additional small α platelets in between the coarse α plates has no significant influence on the fatigue crack propagation curves for these two conditions B and C.

Table 7.9. Effect of age-hardening on tensile properties and fracture toughness of β annealed microstructures, β-CEZ (see text for heat treatments A, B, C)

	$\sigma_{0.2}$ (MPa)	UTS (MPa)	σ_F (MPa)	T.E. (%)	RA (%)	K_{Ic} (MPa $m^{1/2}$)
A	$\sigma_0 = 1370$	1370	1370	0	0	33
B	1180	1280	1415	4	10	52
C	1030	1110	1315	12	17	73

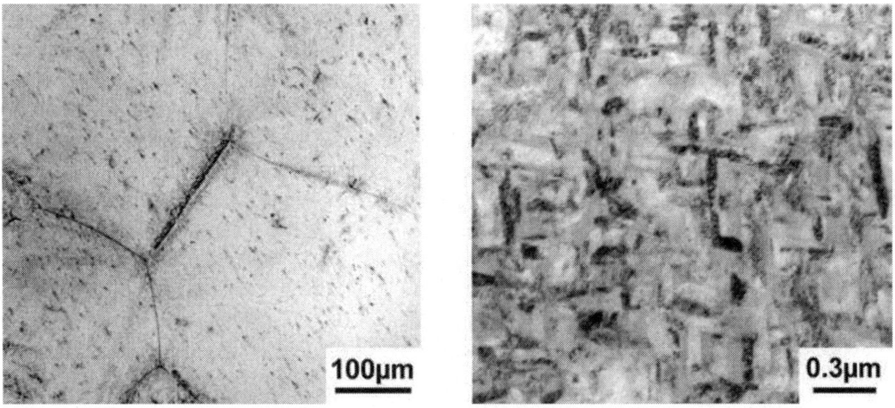

a b

Fig. 7.31. Fine β annealed microstructure (only α platelets), β-CEZ: (a) LM (b) TEM

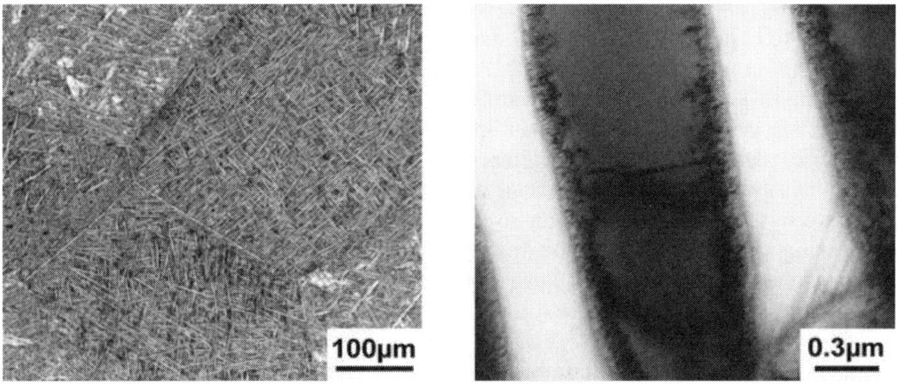

a b

Fig. 7.32. Coarse β annealed microstructure (only α plates), β-CEZ: (a) LM (b) TEM

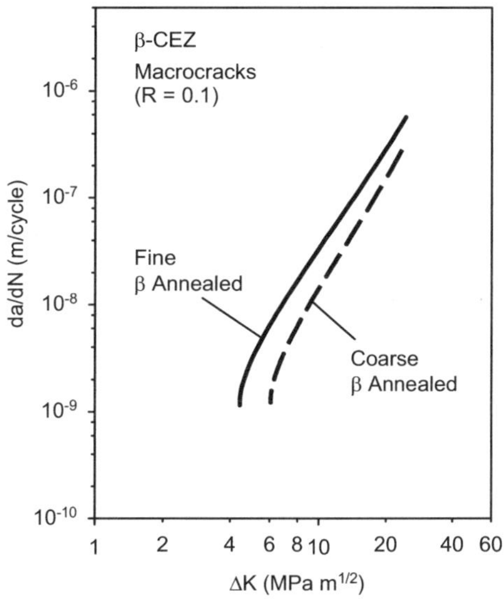

Fig. 7.33. Fatigue crack propagation for macrocracks of the β annealed microstructures shown in Figs. 7.31 and 7.32, β-CEZ

As in the case of microcrack propagation, S-N curves for conditions A and C have not been measured up to now. Therefore, the effect of decreasing age-hardening on HCF strength is illustrated on the example of the heavily stabilized β titanium alloy Beta 21S in two differently aged β annealed conditions. The two microstructures used for the comparison are shown in Fig. 7.2a ($\sigma_{0.2} = 1040$ MPa) and in Fig. 7.4b ($\sigma_{0.2} = 890$ MPa). The corresponding S-N curves are shown in Fig. 7.34 [184]. It can be seen that the condition with the aging treatment of

8h 593°C ($\sigma_{0.2}$ = 1040 MPa) exhibited a higher HCF strength (500 MPa) as compared to the condition with the two-stage aging treatment (8h 500°C + 24h 725°C) and the lower yield stress of 890 MPa which exhibited a HCF strength of only 450 MPa. In both β annealed conditions, crack nucleation occurred in the soft zones along grain boundaries, as outlined already in Sect. 7.2.1.2 (see Fig. 7.22a). Therefore, the HCF strength is basically declining with decreasing age-hardening, see (–) symbol in Table 7.7, but not proportional to the yield stress. The ratio of HCF strength to yield stress increased in the example shown in Fig. 7.34 with decreasing age-hardening demonstrating the effect of strength difference between age-hardened matrix and soft zones on fatigue crack nucleation. This tendency was also noticed in the publication showing the effect of pre-aging (see Fig. 7.3) on microstructure, yield stress, and HCF strength for the heavily stabilized β titanium alloy Beta C [57]. In combination with a later publication [198], it can be deduced that in case of the microstructures investigated for Beta C the ratio of HCF strength to yield stress decreased with increasing yield stress level from about 0.55 to about 0.45.

Although the β phase is less creep resistant than the α phase due to the much faster diffusion rates in the bcc lattice as compared to the hexagonal lattice (see Fig. 2.30, Sect. 2.9.1), some of the high strength β alloys, e.g. Ti-17 and Ti-6246, are used in aero-engines up to temperatures of about 400°C, as will be shown in Sect. 7.3. The reason for this relatively good creep resistance is the small interparticle distance of the non-deformable α platelets resulting from the high volume fraction of α platelets present in those alloys. Since the interparticle distance determines also the yield stress, the creep strength is lowered with decreasing age-hardening, see (–) symbol in Table 7.7. It can be further seen from Table 7.7, that all other microstructural features of β alloys discussed in Chap. 7 have no significant effect on creep strength, as shown by the (0) symbols in Table 7.7.

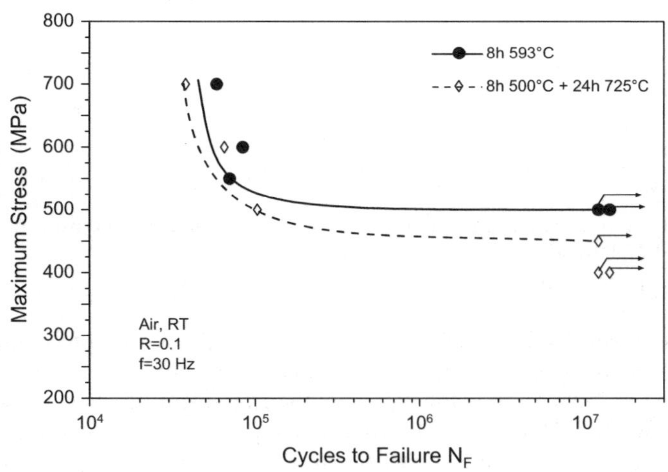

Fig. 7.34. S-N curves for differently aged β annealed microstructures of the heavily stabilized Beta 21S alloy

7.2.3 Effect of Beta Grain Size

The effect of β grain size on mechanical properties of β annealed microstructures has been treated in principle already in Sect. 7.2.1 by comparing the bi-modal microstructure (small β grain size) with the β annealed microstructure (large β grain size). A few more examples will be given in this section on β annealed microstructures of β-CEZ for which the β grain size is fairly large (400 μm) for conventionally β annealing practice (furnace heating). It has been already mentioned in the chapter on α+β titanium alloys, that fast heating rates by using induction heating can reduce the β grain size significantly, e.g. for β-CEZ a β grain size of 60 μm was obtained [199]. Although the β annealed condition of β-CEZ and other high strength β alloys is not used for application, the results presented in the following might serve as an example for grain size effects in heavily stabilized β titanium alloys.

The results of tensile tests on these two conditions of β-CEZ with 400 μm and 60 μm grain size are shown in Table 7.10 [199]. It can be seen that the β grain size had no effect on yield stress which is the expected result for microstructures hardened by incoherent particles. The presence of the soft zones also has no effect on yield stress because of the low volume fraction. But it can be seen that the tensile ductility is increased from 1% to 21% (RA values) by the reduction in β grain size from 400 μm to 60 μm. This is a significant improvement considering the fairly high yield stress of 1275 MPa. As already discussed in Sect. 7.2.1.1, this ductility improvement is due to the reduced slip length in the soft zones (grain boundary length).

Table 7.10. Effect of β grain size on tensile properties and fracture toughness of β annealed microstructures, β-CEZ

β Grain Size	$\sigma_{0.2}$ (MPa)	UTS (MPa)	σ_F (MPa)	T.E. (%)	RA (%)	K_{Ic} (MPa m$^{1/2}$)
400 μm	1275	1335	1350	1	1	39
60 μm	1275	1350	1590	8	21	35

The influence of β grain size on HCF strength of β annealed conditions is demonstrated in Fig. 7.35 for the two grain sizes of β-CEZ [199]. The resistance to fatigue crack nucleation is increased by the reduction of the slip length in the soft zones in analogy to the mechanism discussed for the tensile ductility. In the present example in Fig. 7.35, the HCF strength increased from about 650 MPa to about 700 MPa which amounts to an increase in the ratio of HCF strength to yield stress from 0.51 to 0.55.

The effect of β grain size on fatigue crack propagation of microcracks and macrocracks is shown in Fig. 7.36 for the two grain sizes of β-CEZ [199]. It can be seen that the microcracks, which showed after nucleation the typical behavior of propagating for the first few grains within the soft zones along β grain boundaries (see Sect. 7.2.1.2), exhibited in these early stages of growth a much faster propagation rate in the coarse grained material as compared to the fine grained material. In contrast, the macrocracks showed nearly the same propagation curves for both grain sizes with the tendency of a slightly faster rate for the fine grained material (Fig. 7.36). This insensitivity to grain size in β annealed microstructures is due to the fact that fatigue crack propagation of macrocracks occurs nearly exclusively through the matrix and only occasionally within the soft zones along grain boundaries, as already discussed in detail in Sect. 7.2.1.2. The few steps in the crack front profile (see for example Fig. 7.26a) result from propagation along grain boundaries and are higher for the coarse grained material and might be the reason for the tendency in ranking, see (–) symbol in Table 7.7.

The fracture toughness values for the two β annealed conditions with different β grain sizes are listed together with the tensile properties in Table 7.10. For both conditions crack propagation occurred nearly completely within the soft zones along β grain boundaries. The higher value of 39 MPa $m^{1/2}$ for the coarse grained material as compared to 35 MPa $m^{1/2}$ for the fine grained material shows that from the two contributions to fracture toughness (intrinsic resistance and crack front roughness) the crack front roughness contribution (discussed in detail in Sect. 7.2.1.3) dominates over the "ductility" term (discussed in detail in Sect. 7.2.2) when the β grain size is varied in β annealed microstructures.

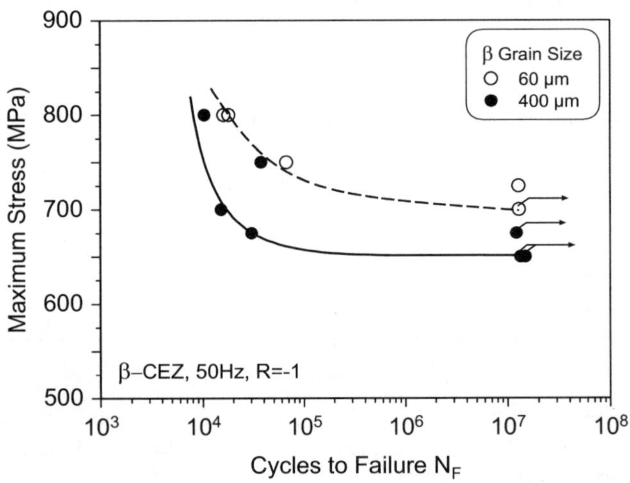

Fig. 7.35. Effect of grain size on HCF strength, β annealed microstructure, β-CEZ

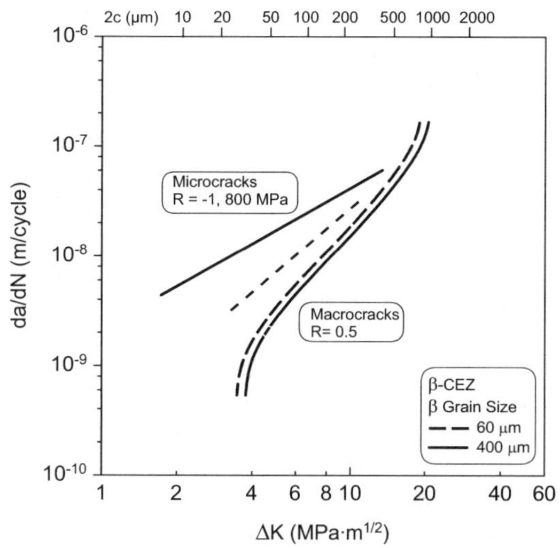

Fig. 7.36. Effect of grain size on fatigue crack propagation of microcracks and macrocracks, β annealed microstructure, β-CEZ

7.3 Properties and Application

The intention of this section on application of β titanium alloys is not to give a complete overview of all possible application areas for these alloys but rather to illustrate on some selected examples the relationship between processing, microstructure, properties, and application, like it was done in Sect. 5.3 for α+β titanium alloys (Ti-6Al-4V) and in Sect. 6.3 for high temperature application of α+β alloys. Excellent articles covering all aspects of application of β titanium alloys can be found elsewhere [161, 162, 200, 201].

In the area of aerospace application, there is the general tendency that β titanium alloys are increasing in application at the expense of the widely used Ti-6Al-4V alloy because of their higher strength level, i.e. the 4% application for β alloys shown in Fig. 1.3 for the USA market in 1998 is increasing steadily at the expense of the 56% application for the Ti-6Al-4V alloy. The Boeing 777 aircraft was the first commercial airplane for which the volume of β alloys outnumbered the volume of Ti-6Al-4V. The main reason was the application of the high strength β alloy Ti-10-2-3 in the landing gear structure (Fig. 7.37). Nearly all parts in the landing gear structure are Ti-10-2-3, except for the outer (and inner) cylinder and the axles, all being steel. The biggest single item was the truck beam with a length of about 3 m and a diameter of about 0.34 m. In the beginning the truck beam was fabricated by forging three pieces and welding them together by electron beam welding. Later, it was changed to a one piece forging. The basic requirement for the mechanical properties of the truck beam are high yield stress and good ductility with a minimum value for fracture toughness. The published processing practice for the Ti-10-2-3 alloy consists of β forging followed by α+β forging to plastic strains of about 15-25% [202]. The material is then heat treated similar to the heat treatment sequence shown in Fig. 7.13 for a bi-modal microstructure. In this case, the resulting mechanical properties of the truck beam should be similar to the properties of the bi-modal structure assuming that the 15-25% plastic deformation in the (α+β) phase field is high enough to form equiaxed primary α and small, equiaxed β grains similar to the microstructure shown in Fig. 7.14. Ti-10-2-3 recrystallizes more rapidly than most heavily stabilized β alloys because it contains no slow diffusing elements (Mo, Nb) and contains rapidly diffusing Fe (see Sect. 2.9.1). As demonstrated in Sect. 7.2.1, this bi-modal microstructure would exhibit high ductility at high strength levels but would have low fracture toughness values, i.e. the minimum value for fracture toughness of truck beam material would be very difficult to fulfill with a bi-modal type of microstructure even with small volume fractions of primary α resulting in a slightly larger β grain size. Based on the discussion in Sects. 7.1 and 7.2, the best microstructure for this type of application assuming the major service stresses are in longitudinal direction of the truck beam would then be an elongated β grain structure (cigar shaped). This microstructure would have in L direction besides fairly high ductility also a high fracture toughness (see Table 7.8 and Fig. 7.27). But, due to the large dimensions of the truck beam, it is very difficult to create an unrecrystallized, elongated β grain structure through normal β processing and prevent recrystallization during the processing procedure.

Fig. 7.37. Aircraft landing gear of the Boeing 777 using Ti-10-2-3 (courtesy R. R. Boyer, Boeing)

Smaller parts of this high strength β alloy Ti-10-2-3 are easier to forge and some examples of precision forgings are shown in Fig. 7.38. The parts shown in this figure are used in the Boeing 777 cargo handling system. The lower forging temperature, due to the low β transus temperature of only about 800°C (see Table 2.6), together with the better forgeability results in a lower price for these Ti-10-2-3 forgings as compared to Ti-6Al-4V forgings counting also the lower tooling costs [203].

Titanium nut clips are another interesting application (Fig. 7.39). These are small parts but they are used in the Boeing 777 in the order of 15 000-20 000 per airplane attaching the metallic floor structure to the composite floor beams [200]. These titanium clips replace cadmium plated steel and are much more corrosion resistant. The nut clips are made from thin sheets of the heavily stabilized β alloy Ti-15-3 by a simple sheet forming process and aging. Therefore, the processing route is similar to the β annealed processing route shown in Fig. 7.1.

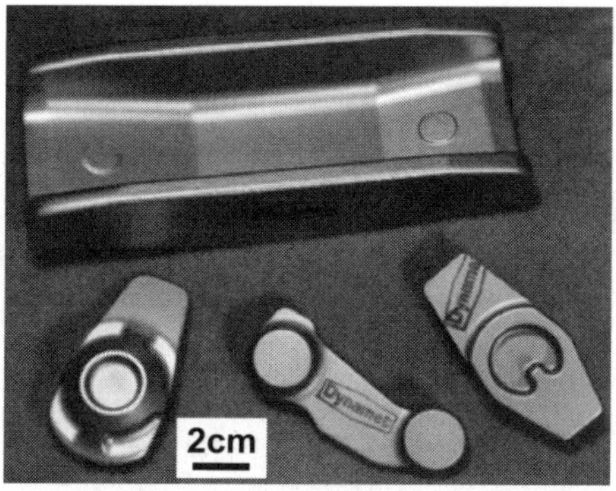

Fig. 7.38. Cargo handling fittings used for the Boeing 777, Ti-10-2-3 (courtesy R. R. Boyer, Boeing)

Fig. 7.39. Nut clips used in aircraft, heavily stabilized β alloy Ti-15-3 (courtesy R. R. Boyer, Boeing)

Springs made out of heavily stabilized β titanium alloys are used in a large variety of shapes and sizes in airplanes. The prime reason is the low modulus of elasticity of these alloys in combination with the relatively high yield stress. Some examples are shown in Fig. 7.40. To illustrate the size of the springs, the diameter of spring "a" in Fig. 7.40 is about 200 mm. Details about the specific use of the springs in Fig. 7.40 can be found elsewhere [200]. When springs are fabricated from strip, then usually the alloy Ti-15-3 is used (spring "a" in Fig. 7.40). If springs are made from wire (round or square type), then mostly the alloy Beta C is used (springs "b" and "c" in Fig. 7.40). The processing route for the springs is the route for β annealed microstructures shown in Fig. 7.1 which is normally used for heavily stabilized β alloys. If the requirement for a spring is mainly a very low modulus of elasticity and not so much high strength, as in case of the clock spring in Fig. 7.40 (spring "a"), then the material is used without aging, because with increasing volume fraction of α phase the modulus of elasticity is increased. Titanium springs have two major advantages. First, they can save as much as 70% in weight. Second, steel springs are susceptible to corrosion and require painting whereas titanium springs do not.

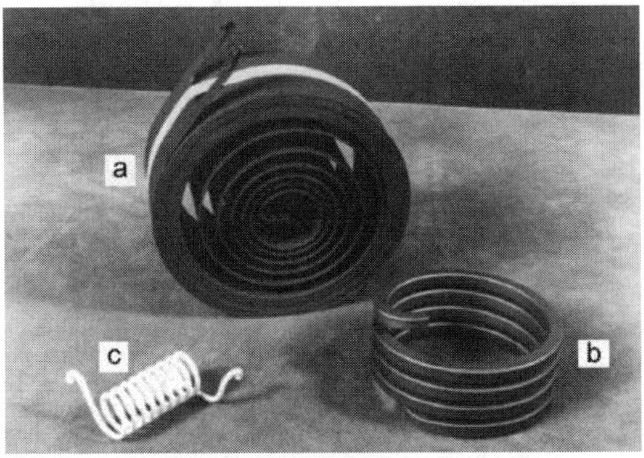

Fig. 7.40. Aircraft springs, heavily stabilized β alloys: (a) Ti-15-3 (b, c) Beta C (courtesy R. R. Boyer, Boeing)

Another application worth mentioning is the use of the heavily stabilized β alloy Beta 21S in the nacelle structure (exhaust plug, nozzle, and aft cowl) of all aero-engines used for the Boeing 777 aircraft [203]. Although the nacelle structure is, in principle, part of the aero-engine structure, the aircraft manufacturer often assumes for the nacelle structure the responsibility instead of the aero-engine manufacturer. For this application Beta 21S sheets are used because of the excellent oxidation resistance of this alloy which allows long time operating temperatures in the range of 480-565°C with short time periods up to 650°C [168]. The alloy Beta 21S containing 15% Mo and 2.7% Nb (see Table 2.6) was specially designed for high oxidation resistance and can be used for this application because the stresses acting on the nacelle structure during operation at high temperature are small. Beta 21S also has excellent resistance to embrittlement when exposed to hydraulic fluid at elevated temperatures. The Beta 21S sheets are basically applied in the β annealed condition (Fig. 7.1), but two different final aging treatments are used at Boeing [168]. These two different aging treatments result in two different microstructures, similar to those shown in Figs. 7.2a and 7.4a, with quite different yield stress levels. A major European effort was initiated recently to use Beta 21S sheets in the exhaust structure of helicopter engines in order to reduce effectively the noise level of civil helicopters.

Castings can also be made from heavily stabilized β alloys, such as Ti-15-3, Beta C, and Beta 21S [204]. As an example for such a casting, a brake torque tube used in fighter aircraft is shown in Fig. 7.41. The Ti-15-3 casting replaced in this case Ti-6Al-4V castings and the higher strength of the Ti-15-3 casting allowed a reduction in the volume of the torque tube which in turn allowed a higher volume of carbon in the brakes resulting in a longer brake life [168]. After the casting process the parts are hot isostatically pressed in the β phase field and subsequently aged. Therefore, the microstructure in these castings of heavily stabilized β alloys is essentially identical to the β annealed microstructure of wrought products. In the processing route (Fig. 7.1) only the deformation step II is missing for castings and the recrystallization treatment (step III in Fig. 7.1) is replaced by the hot isostatically pressing treatment in case of castings. The β grain size of castings will therefore be somewhat larger (depending on section size) as compared to wrought products resulting in a lower ductility (see Table 7.7). This factor limits the maximum useful section size that can be cast, but is alloy and casting process dependent. The yield stress of castings will be identical to wrought products as long as the cooling rate from the hot isostatically pressing temperature is fast enough to prevent the formation of coarse α plates during cooling. This last requirement is the main reason why high strength β titanium alloys are not used for castings.

Fig. 7.41. Cast brake torque tube for fighter aircraft, Ti-15-3 (courtesy R. R. Boyer, Boeing)

The high strength β titanium alloys Ti-17 and Ti-6246 are used in the compressor part of aero-engines replacing the Ti-6Al-4V alloy because of their higher basic strength. A good example is the large Ti-17 fan disk shown in Fig. 3.24 which is used in the GE-90 aero-engine. The fan disk shown in that figure has an outside diameter of about 800 mm and is manufactured from three separate forgings which are welded together using solid state inertia welding (see Sect. 3.6.2). The three pieces are β forged and conventionally heat treated, i.e. they received the two-stage heat treatment shown in Fig. 7.7 (steps IVa and IVb). Since the β grains in the fan disk forgings will be pancake shaped typical for β processed material (see Sect. 7.1.2) with the thin dimension parallel to the axial direction of the disk. The major applied stresses acting on the fan disk will be parallel to the pancake plane of the β grains, only the properties described in Sect. 7.2.1 as properties in L test direction of β processed material (see also Table 7.7) will be important.

As described already in Sect. 5.3, the LCF strength, i.e. a combination of fatigue crack nucleation resistance and microcrack propagation resistance, is the important design criteria for compressor disks. Both of these properties are good in L testing direction (Table 7.7) considering the anisotropy of the pancake shaped β grains. It should be pointed out, however, that those properties would be even better for the bi-modal microstructure and that the major drawback of the bi-modal microstructure, the low fracture toughness (see Fig. 7.27), is not important for compressor disks, because the fracture toughness has an insignificant effect on life time and no effect on inspection intervals, see discussion in Sect. 5.3. As also pointed out already in Sect. 5.3, for fatigue life calculations of disks based on fatigue crack propagation data it is important to use the microcrack propagation data because macrocrack propagation data would result in overestimation of cycles to failure (see Fig. 7.25).

As described at the end of Sect. 7.2.2, high strength β titanium alloys are also used in the hotter part of aero-engine compressors up to about 400°C. An example can be seen in Fig. 3.78 which is a two-stage HP compressor rotor made out of the high strength β titanium alloy Ti-17. The production route for this HP compressor rotor is the same as described above for the fan disk, i.e. the individual disks or stages are β processed, heat treated, inertia welded together, and finally machined. The high yield stress after aging of this high strength β alloy Ti-17 results in the required adequate creep strength in addition to the good LCF strength. LCF strength was discussed above for the fan disk. Creep strength is an additional requirement for elevated temperature application.

Another application using relatively large forgings of the high strength β alloy Ti-10-2-3 is the Super Lynx helicopter rotor head shown in Fig. 7.42. The three different forgings made out of Ti-10-2-3 are indicated in that figure. In this case, the HCF strength is the most important design criteria. This is a complete different requirement as compared to the large Ti-10-2-3 forging discussed earlier (high fracture toughness requirement for the truck beam forging in Fig. 7.37). For forging the helicopter rotor head parts, the final α+β forging apparently used (although not discussed in the open literature) leads to some kind of bi-modal microstructure (see Figs. 7.13 and 7.14) which is the correct microstructure to obtain a high HCF strength and which is even higher than for β processed material tested in L direction, see Sect. 7.2.1.2. On the example of the disk forging for the rotor head arms, the influence of various microstructural parameters on mechanical properties was measured and discussed in detail and the results can be found in [205]. Most of these results were the basis for an overview paper [206] on microstructure/property relationship in β titanium alloys containing also a summary table similar to Table 7.7.

Another area, for which heavily stabilized β titanium alloys are used, is the so-called downhole service area (oil and gas drilling and production, geothermal wells, etc.). The primary advantages of these β alloys over the Ti-6Al-4V alloy for this kind of application are the higher yield stress and the lower modulus of elasticity combined with equal or better corrosion resistance in aggressive environments. Particularly Beta C is used for these applications [207] with the usual β annealed and aged microstructure. It should be mentioned again, that for this β annealed microstructure the mechanical properties have to be adjusted either by the size of the equiaxed grains (Sect. 7.2.3) or by the aging treatment (Sect. 7.2.2).

Other application areas of β titanium alloys are in the biomedical field, in automobiles, and in the area of sporting goods. These applications will be covered in the relevant sections of Chap. 10.

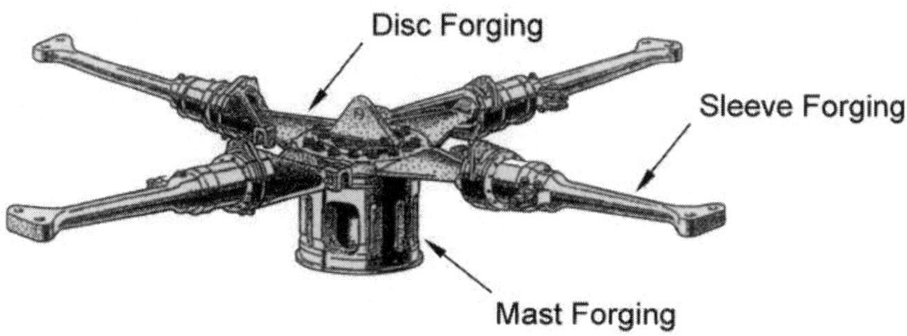

Fig. 7.42. Helicopter rotor using forgings of Ti-10-2-3 (courtesy G. Terlinde, Otto Fuchs Metallwerke)

8 Titanium Based Intermetallics

Intermetallic compounds[1], especially those formed between light elements such as Ti and Al are attractive, because of their low density and good elevated temperature strength. However, the formation of intermetallic compounds also usually reduces the symmetry of the parent metal lattice. In turn, this places additional restrictions on the available deformation modes. These restrictions usually are manifested as increased strength, at least at elevated temperatures, reduced ductility and fracture toughness. Historically, the issues associated with reduced ductility and fracture toughness have been viewed as outweighing the benefits of increased strength. Consequently, the use of intermetallic compounds in structural applications has been very limited.

Intermetallic compounds based on titanium and aluminum are light (low density), relatively stiff (high modulus), and have attractive high temperature mechanical properties (tensile and creep strength). Because of this promise, there has been a great deal of work on these compounds beginning in 1953 [208] and continuing up until now. Examination of the Ti-Al phase diagram, Fig. 2.11, shows there are three compounds in this system, Ti_3Al (α_2), TiAl (γ), and $TiAl_3$. Among these, only Ti_3Al and TiAl have been extensively studied. Ti_3Al has the hexagonal DO_{19} structure and TiAl has the tetragonal $L1_0$ structure. In both cases these compounds exhibit limited ductility at room temperature, particularly in the binary state. Alloying has been shown to improve ductility and fracture toughness of these compounds as will be discussed later in this chapter. Alloying also permits the formation of additional constituents, for example the disordered β phase (A2), the ordered β phase (B2), called β_2, and the ordered orthorhombic phase based on Ti_2AlNb, called the O phase.

In some regards, the effort to introduce these compounds for structural applications illustrates the difficulty associated with the introduction of any new class of material. That is, more than 40 years have passed and the titanium aluminides are still considered developmental materials. The primary barrier has been concern about brittleness, but newer alloyed versions of both Ti_3Al and TiAl have alleviated many of these concerns. The principal concern at present is cost and attainable performance improvement. This chapter will briefly describe the basic metallurgy of the titanium aluminides, outline some of the potential applications for them, and summarize the current barriers to their introduction into products. There are a number of fairly recent review articles that deal with the detailed aspects of γ alloys [209-221] and α_2 and orthorhombic alloys [222-227].

---

[1] Alloy compositions in Chap. 8 "Titanium Based Intermetallics" are given in at%.

8.1 Alloying and Microstructure

As already mentioned, there are two binary intermetallic compounds of interest in the Ti-Al system. The first compound is the hexagonal α_2 phase based on the composition Ti_3Al with a DO_{19} structure. The long-range order in this compound affects the deformation modes described in Chap. 2 for α titanium with the result that fewer modes are readily activated. In particular, twinning is not observed in this compound and $\vec{c}+\vec{a}$ slip is much less frequently observed. Further, the ordering makes $\vec{a}$ slip (1/3 $<11\bar{2}0>$ as referred to the disordered hexagonal lattice) on the basal plane more difficult because the movement of such dislocations creates disorder. It is possible for a second dislocation with the same Burgers vector to follow the first one and "repair" this disorder. Even though the stress to move these dislocation pairs is about half of that for movement of a single dislocation, the high anti-phase boundary energy in Ti_3Al still makes extensive basal slip by $\vec{a}$ dislocations difficult. This is because cross slip onto prism or pyramidal planes is inhibited. Consequently, when basal slip is observed, it is extremely planar as can be seen in Fig. 8.1. Marcinkowski [228] was first to examine the effect of $\vec{a}$ slip on the perfection of the DO_{19} superlattice. He pointed out that motion of 1/3 $<11\bar{2}0>$ dislocations (referred to the disordered lattice) on the basal plane always creates disorder across the slip plane that separates the sheared and unsheared regions of the lattice. Figure 8.2 shows the atomic arrangement in the basal planes of the DO_{19} lattice. Here the "small" atoms belong to the close-packed plane lying just above the plane of the diagram. It can be seen that basal slip with an $\vec{a}$ vector creates two wrong nearest neighbors. Further analysis shows that four wrong second nearest neighbors are also created. This is in contrast to shear on the prism planes. The atomic arrangement in the DO_{19} lattice results in two types of prism shear, as has been analyzed in detail by Blackburn [229]. These are labelled type 1 and type 2 in Fig. 8.2. It can be seen that type 1 prism shear creates no wrong nearest neighbors and, in fact, creates only a total of two wrong second nearest neighbors whereas type 2 prism shear creates two wrong nearest neighbors and four wrong second nearest neighbors. These are labelled as "easy" and "hard" prism shears in Fig. 8.2. The prototype system for the DO_{19} superlattice is Mg_3Cd. Here it has been shown that the ordered form of a Mg-25% Cd alloy is, surprisingly, more ductile than the disordered alloy. This is because the creation of "easy" prism slip planes favors prism slip in the ordered lattice, despite the large c/a ratio, which usually favors basal slip. Because the long-range order facilitates operation of additional deformation modes, Mg_3Cd is more ductile in the ordered condition than when it is disordered. In addition to $\vec{a}$ slip, hexagonal metals require operation of a slip mode that has a non-basal component to satisfy the Taylor-Von Mises criterion for general polycrystalline plasticity. The most common of non-basal deformation modes in α titanium is $\vec{c}+\vec{a}$ slip [33, 35]. In the DO_{19} lattice, the $\vec{c}+\vec{a}$ slip vector has an $\vec{a}$ component which creates disorder during motion. This makes the $\vec{c}+\vec{a}$ slip mode in the ordered DO_{19} α_2 phase difficult to activate in comparison to the disordered α phase. In general, the presence of long-range order also inhibits cross slip, which leads to planar slip as shown in Fig. 8.3.

In such circumstances, dislocation pile-ups occur at the head of the slip bands and the stresses associated with these pile-ups can initiate cleavage cracks such as shown in Fig. 8.4. Consequently, in polycrystalline form, the ordered α_2 phase exhibits limited ductility compared to the disordered α phase. Both the reduction in slip systems and the occurrence of planar slip appear to be largely responsible for the low ductility of the α_2 phase at room temperature and temperatures up to about 600°C.

Fig. 8.1. Micrograph showing basal slip bands and $\vec{a}$ dislocations on prism planes in Ti-25Al, TEM

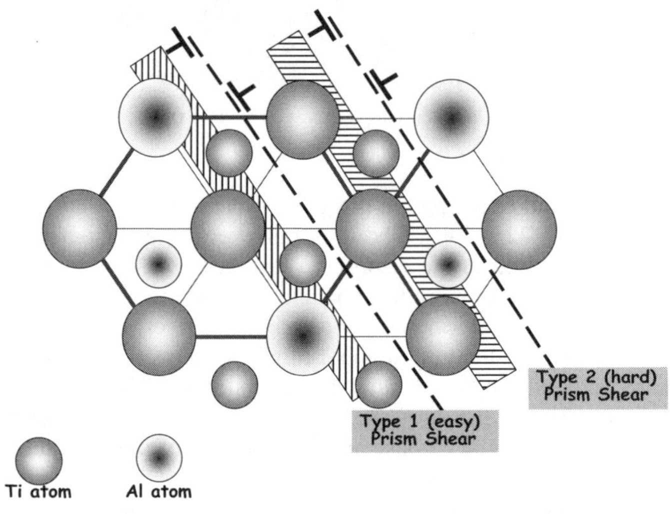

Fig. 8.2. Schematic drawing of the basal planes of the DO_{19} lattice indicating "easy" and "hard" slip on prism planes

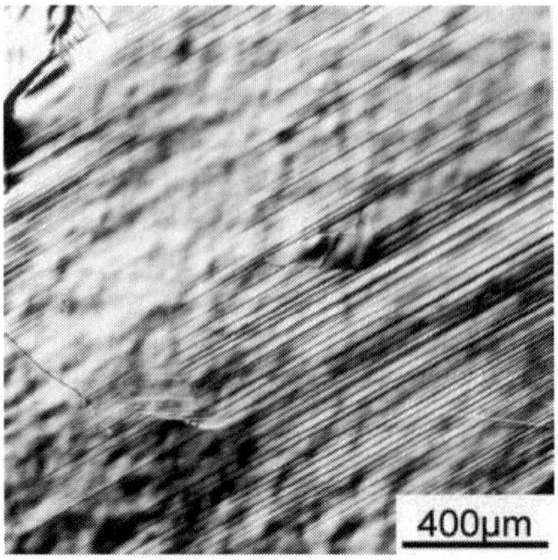

Fig. 8.3. Planar slip lines on the polished surface of Ti-25Al, LM

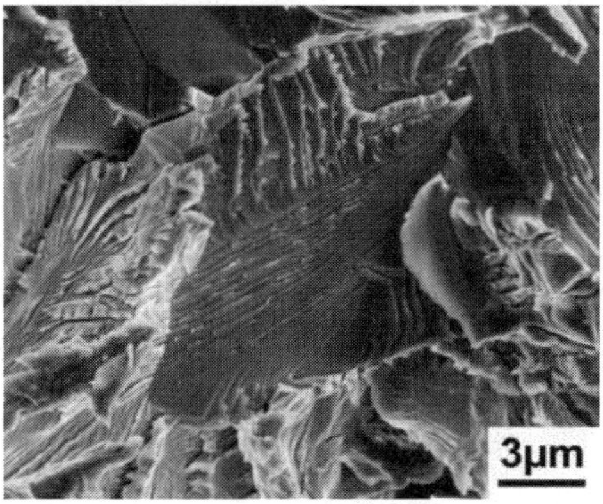

Fig. 8.4. Fracture surface of Ti-25Al showing cleavage cracks, SEM

The other compound of interest in the Ti-Al system is the γ phase (TiAl), which has the $L1_0$ structure [228, 230]. The long-range order in this structure creates a slight tetragonal distortion in the disordered face-centered cubic lattice (c/a=1.04) because of the different atomic radii of Ti and Al. The ordering also causes operation of four of the six possible ½ <110> slip vectors to create wrong next nearest neighbors. Consequently, the total number of slip systems that can operate in the γ phase also are divided into "easy" and "hard" systems. This point is illustrated in Fig. 8.5. As in the case of the α_2 phase, the presence of order decreases the plastic deformation capability of the γ phase and the ductility of polycrystalline γ is quite low up to about 750°C.

In light of the foregoing discussion of the crystal plasticity of the α_2 and γ phases, much of the recent effort to improve the ductility of these compounds has been devoted to alloying effects. In principle, there are four beneficial effects from alloying intermetallics as it concerns ductility. The first is to relax the restriction on slip modes already described. The second is to retard the kinetics of ordering or to change the degree of long-range order. The third is to alter the alloy constitution by introducing the β or β_2 phase as a ductile second phase that can arrest or hinder the growth of microcracks that form in the brittle major constituents, the α_2 and γ phases. The fourth is to alter the transformation behavior during heat treatment and processing to permit microstructural refinement and control. There is no compelling evidence that any of the alloying elements tried have had any significant effect on the intrinsic crystal plasticity of either the α_2 or the γ phases, although it has been reported that $\vec{c}+\vec{a}$ slip is more active in the O phase (described later) as compared to the α_2 phase [231]. Additions of Nb do alter the kinetics of the $\alpha \rightarrow \alpha_2$ ordering reaction, but do not significantly affect the equilibrium degree of order of the α_2 constituents. In alloys intended for high temperature service, kinetic effects are not very helpful since ordering can proceed in service. The role of alloying additions on constitution has proven to be effective for both α_2 and γ alloy classes. Whereas the role of alloying on transformation kinetics is useful for α_2 alloys, it has only limited applicability to γ alloys. One example of this in the case of α_2 alloys is the addition of Nb, Mo, and other β stabilizers to change the $\beta \rightarrow \alpha$ transformation kinetics. The most common alloying addition for both the DO_{19} and $L1_0$ compounds is Nb, but W, V, Mo, Mn, Si, and Cr also have been added to one or both of the compounds with clear benefit, at least to selected properties.

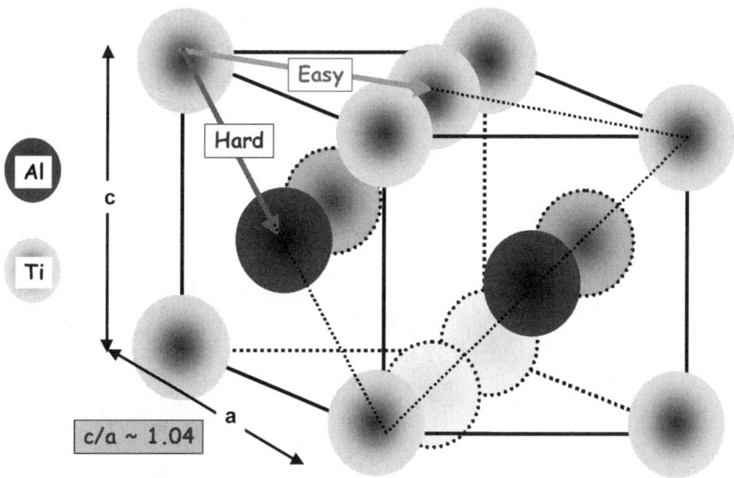

Fig. 8.5. Schematic drawing of the $L1_0$ structure cell showing "easy" and "hard" slip directions

The principal benefit of alloying, especially in the α_2 phase alloys, is to permit microstructural alteration and refinement by heat treatment [232]. Alloying additions such as Nb and/or W alter the $\beta \rightarrow \alpha$ reaction kinetics and permit the creation of a finer scale microstructure such as that shown in Fig. 8.6 [233]. The individual, differently oriented α_2 plates in this structure each belong to a different variant of the Burgers relation. This structure is often called Widmanstätten or "basket weave" structure. Consequently, this structure has the effect of reducing the slip length and the magnitude of the corresponding stresses associated with dislocation pile-ups. At Nb concentrations greater than about 8%, there is also a small volume fraction of Nb stabilized β_2 phase present. In this class of alloys, when there is a bcc constituent present, it is always the ordered β_2 phase, even though the terminology in the literature is somewhat inconsistent on this point. In other alloys, e.g. those containing W, the β_2 phase is present at W concentrations as low as 1%, presumably because of the low solubility of W in the α_2 phase. Figure 8.7 shows the β_2 phase in an alloy containing 2% W. This β_2 phase behaves as a ductile inclusion that permits the stress relaxation at the tips of microcracks that form in the brittle α_2 phase. Other alloys contain sufficient Nb to stabilize a small volume fraction of bcc phase. Consequently, the room temperature ductility of Ti-24Al-11Nb is typically 4-5% as opposed to essentially 0% for the unalloyed binary Ti_3Al. This beneficial effect of Nb additions on ductility also increases the density of the alloy and reduces the creep strength, presumably due to the presence of the Nb stabilized β_2 phase. Because of the range of possible alloying effects on microstructure of α_2 alloys, there have been numerous efforts to develop α_2 phase alloys. The alloy Ti-24Al-11Nb was an early composition that was made in significant quantities and was used for component fabrication demonstrations [234]. The density corrected strength of this and similar ternary alloys was marginal compared to conventional titanium alloys or to Ni base alloys. Consequently, other more complex compositions were explored. These alloy development efforts culminated in the definition of several other compositions, but the one that was most widely endorsed as a second-generation α_2 phase alloy was Ti-25Al-10Nb-3V-1Mo [232]. This alloy is sometimes known as "Super α_2".

Fig. 8.6. Grain boundary α_2 and α_2 plates in the matrix, Ti-24Al-11Nb, LM

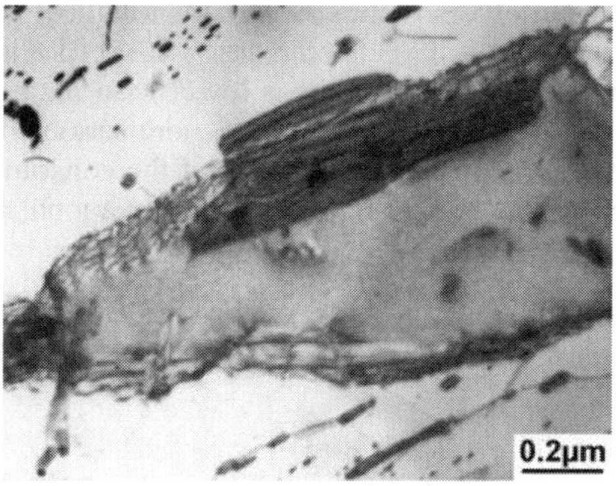

Fig. 8.7. Small β_2 precipitates in α_2 alloy containing 2% W, TEM

At Nb concentrations of about 15% or higher, the high temperature bcc phase can be retained as a metastable phase on quenching but exhibits an ordered B2 (CsCl) structure [235]. The ordered β_2 phase has much more limited ductility than the disordered bcc β phase. For example, an alloy containing 25% Al and 15% Nb that has been rapidly quenched from the β phase field consists of 100% β_2 phase and is extremely brittle. This is most likely because the metastable β phase can undergo a series of rapid decomposition reactions [235, 236]. These reactions both strengthen and embrittle the β_2 matrix. This effect, coupled with the density increase due to the Nb, defines the upper concentration limits of Nb that can be usefully added to α_2 phase alloys for the purpose of ductility improvement. This is at least true for applications where the alloys could be potentially subjected to rapid cooling (e.g. during joining) that would result in metastable retained β_2 phase. Alloys of this composition (Ti-25Al-15Nb) quenched from very high temperatures exhibit many anti-phase boundaries (APBs) in the β_2 phase, suggesting that the ordering reaction has occurred during quenching. This same alloy quenched from high in the (α_2+β_2) phase field shows essentially no APBs, but diffraction shows that the bcc phase has the B2 structure. This is consistent with the existence of an order/disorder transformation in the bcc phase at an intermediate temperature. The exact temperature of the $\beta \rightarrow \beta_2$ ordering transition appears to depend on both the Al and Nb concentration in the alloy [236]. This dependence has not been systematically determined.

Alloyed versions of the α_2 alloys containing ≥10% Nb (or the equivalent amount of other β stabilizers) and the so-called orthorhombic alloys exhibit a response to thermo-mechanical processing that is similar to that of the conventional titanium alloys described in Chaps. 5 and 7 [237, 238]. In this class of alloys based on the α_2 phase, the entire range of microstructures ranging from fully lamellar to bi-modal to equiaxed can be produced with the proper processing and heat treatment [237]. Examples of the range of attainable microstructures in Ti-25Al-10Nb-3V-1Mo (Super α_2) are shown in Fig. 8.8. The morphology of the α_2 phase affects properties in a similar manner as observed in conventional titanium alloys. Since the α_2 phase alloys are mainly of interest for elevated temperature applications, the fully lamellar structure is perhaps the most interesting because it has the best creep strength. As it is also the case with $\alpha+\beta$ alloys, the tensile ductility of the fully lamellar microstructures is lower than for bi-modal or equiaxed ones. In the case of the α_2 phase alloys, this is a more serious limitation because of the intrinsically low room temperature ductility of the constituents in these alloys, presumably because of the long-range order in hexagonal α_2 and the bcc β_2 phases.

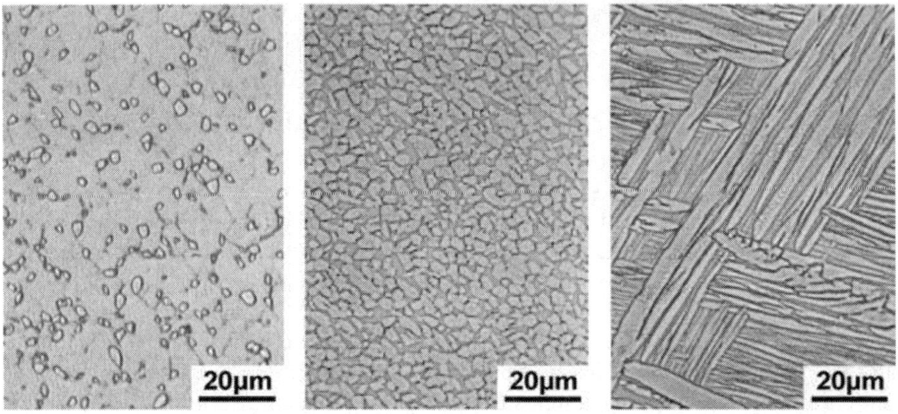

a b c

Fig. 8.8. Microstructures of Super α_2 created by different thermo-mechanical treatments, final heat treatment 2h 830°C, LM: (a) Bi-modal (b) Equiaxed (c) Lamellar [237]

There is an additional class of intermetallic alloys that contain 25% Al and 12.5-24% Nb and which form an ordered orthorhombic phase (the O phase) based on the composition Ti_2AlNb [239]. The orthorhombic phase forms at higher Nb concentrations than those corresponding to the stability range of the α_2 phase. The structure of this phase along with the orthorhombic cell (heavy lines) is shown in Fig. 8.9. Because the atomic radii of Nb and Al are different, this structure has unequal axis lengths along the rows that consist of Ti and Nb and those that consist of Ti and Al. Consequently, the hexagonal α_2 structure is distorted and must be represented as an orthorhombic cell. These so-called orthorhombic alloys have considerably improved room temperature ductility and better density corrected strength than the Super α_2 alloys, as will be discussed later. One mode of formation of the O phase is in Nb rich regions that are the result of solute partitioning due to Nb rejection from the α_2 phase during slow cooling or during aging. It has been suggested that the orthorhombic phase forms as a result of a miscibility gap in the Ti_3Al-Nb system [240]. This explains the formation of the Nb rich regions that subsequently order to form the Ti_2AlNb orthorhombic phase. The phase equilibria in alloys containing 20-30% Al and 15-30% Nb have been studied and described. It has been shown that $O+\beta_2$, $O+\alpha_2$, and $O+\alpha_2+\beta_2$ constitutions are possible, but that $O+\beta_2$ exists over much of the composition range cited earlier [240].

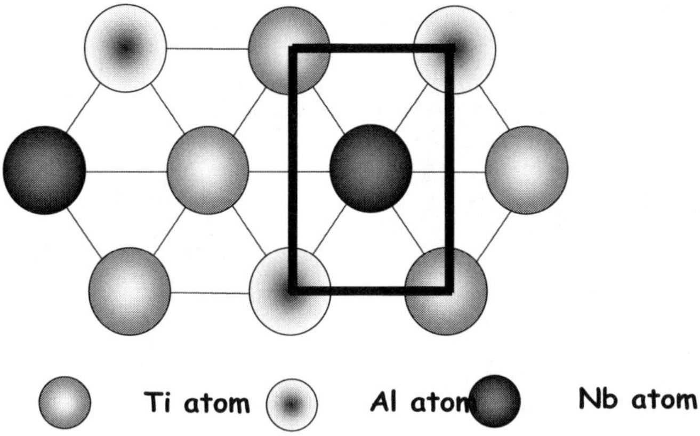

Fig. 8.9. Schematic drawing of orthorhombic cell (dark lines) and its relation to DO_{19} structure

The γ alloys have always been the more promising class of titanium aluminides in terms of their high temperature behavior [241]. They also present the biggest challenge in terms of low ductility and fracture toughness at low temperatures. Alloying has proven beneficial to a degree in raising the room temperature ductility, but the values are still low (about 2% elongation). The most important composition effect in γ alloys is the effect of Al concentration on constitution. This is mainly because Nb, Mo, W, and V are relatively insoluble in the γ phase. Alloys with 45-48% Al lie in the two phase (α_2+γ) field at room temperature. Cooling from the single phase α region into the (α_2+γ) phase field results in a fully lamellar structure as shown in Fig. 8.10a [242, 243]. If the cooling rate is slow enough, some γ phase also is heterogeneously nucleated at the α grain boundary triple points. It also is possible to have an equiaxed α_2+γ structure in these alloys (Fig. 8.10b) or to have a bi-modal structure consisting of equiaxed γ grains and lamellar grains. The lamellar structure consists of α_2 laths interspersed between γ laths. These laths have an orientation relation characterized by parallel close-packed planes and parallel close-packed directions [242], i.e. $(0002)_\alpha \parallel (111)_\gamma$ and $[11\bar{2}0]_\alpha \parallel [1\bar{1}0]_\gamma$. Ignoring the small tetragonality of the γ phase, this is the same orientation relation observed in many hcp to fcc transformations, e.g. the allotropic transformation of Co. The γ laths also belong to more than one variant of the α_2/γ orientation relation and the adjacent laths often share a common $\{111\}_\gamma$ plane, making them twin related. The crystallography of the α_2+γ lath structure makes slip transmission across the α_2/γ lath boundaries favorable when "easy" slip systems are aligned. Fig. 8.11 shows the α_2+γ lath structure and Fig. 8.12 shows a selected area electron diffraction pattern consistent with the orientation relation described earlier. If the two phase α_2+γ alloys are quenched from a very high temperature (>1300°C) the α_2 phase contains anti-phase boundaries, consistent with the occurrence of ordering during cooling. Fig. 8.13 shows the APBs in the α_2 laths. This lath structure has important consequences on the ductility as will be discussed later.

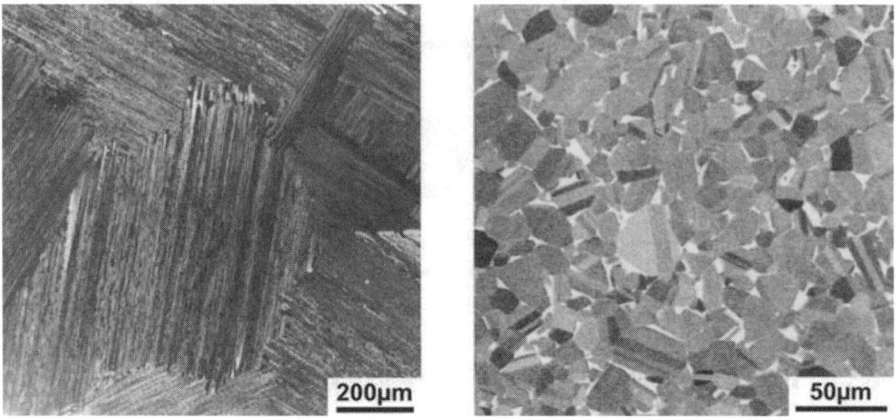

a b

Fig. 8.10. Microstructures of Ti-48Al-2Cr, SEM: (a) Fully lamellar (b) Equiaxed [243]

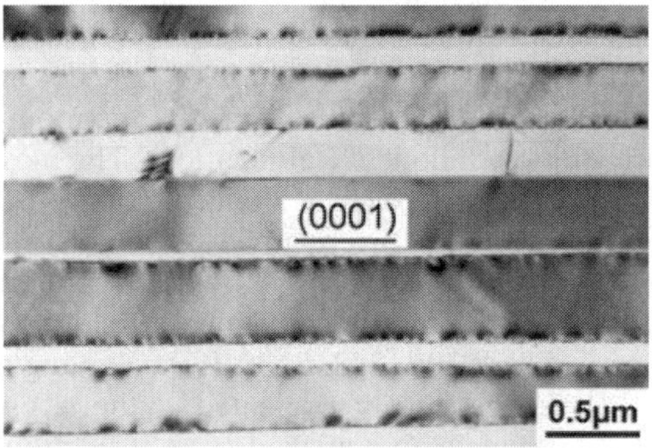

Fig. 8.11. Lamellar α_2+γ structure in Ti-44Al, TEM

Fig. 8.12. Selected area diffraction pattern taken from lamellar α_2+γ structure with two twin related γ variants in α_2 matrix, TEM

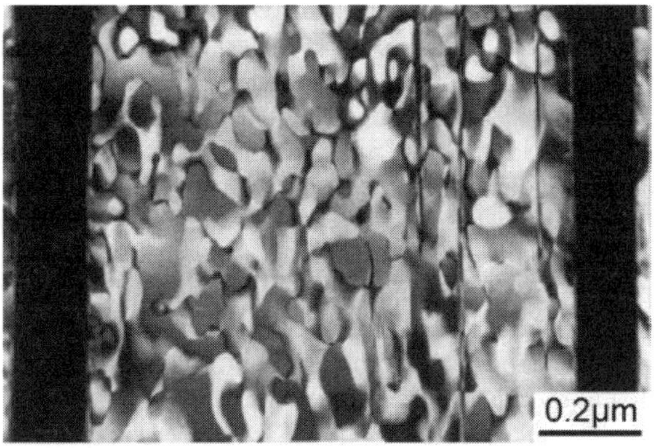

Fig. 8.13. Dark field electron micrograph showing APBs in the α_2 phase of Ti-44Al quenched from (α+γ) phase field, TEM

The addition of Cr and Nb is also beneficial to the low temperature ductility of the γ phase. An early alloy that has received a lot of study is Ti-48Al-2Cr-2Nb. In this alloy the intrinsic effects of Cr and Nb on ductility are not clear. It is clear that this composition opens the heat treatment window for creating the fully lamellar structure by shifting the α_2+γ phase boundary to the right at higher temperatures. Another γ alloy that has received attention is the alloy Ti-47Al-5Nb-1W because it has good creep strength [244]. As in the case of the α_2 alloys, the W has very limited solubility in the γ phase also. Consequently, this alloy and others containing W additions have small β_2 phase particles present in the microstructure. An example of these β_2 phase particles is shown in Fig. 8.14. These β_2 particles are far too coarse and too widely spaced to cause precipitation or dispersion hardening. They do tend to cause the microstructure to be finer and this leads to higher strengths up to about 700°C.

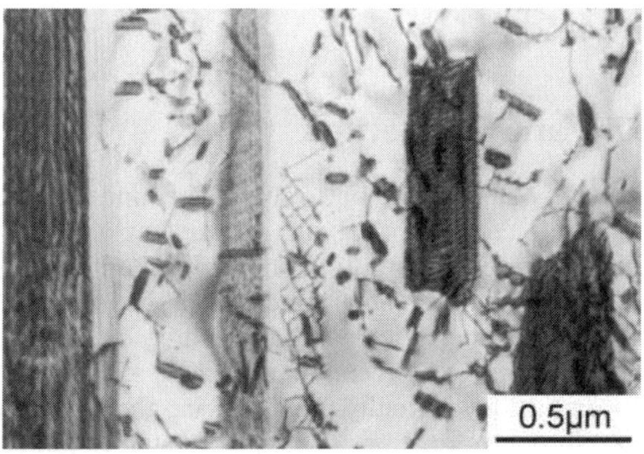

Fig. 8.14. Small β_2 precipitates in γ alloy Ti-49Al containing 2% W, TEM

It also is important that any alloying addition does not adversely affect the stability of the surface oxide, because these materials are principally attractive for use at elevated temperatures in the uncoated condition. An early generation γ phase alloy was Ti-48Al-1V and it became clear that the V addition was beneficial to the low temperature ductility. It also became clear later that V is harmful to elevated temperature oxidation and hot corrosion resistance. This caused a diminished interest in this alloy. The additions of Nb and Cr are beneficial to the oxidation behavior and this relationship is shown in Fig. 8.15 [245]. The reason for this improvement appears to be the stabilization of the Al_2O_3 scale when these elements are present. It is anticipated that the use of γ phase alloys in the uncoated condition is highly preferred both from a cost and a reliability standpoint, which is another reason that alloying of γ phase is desirable.

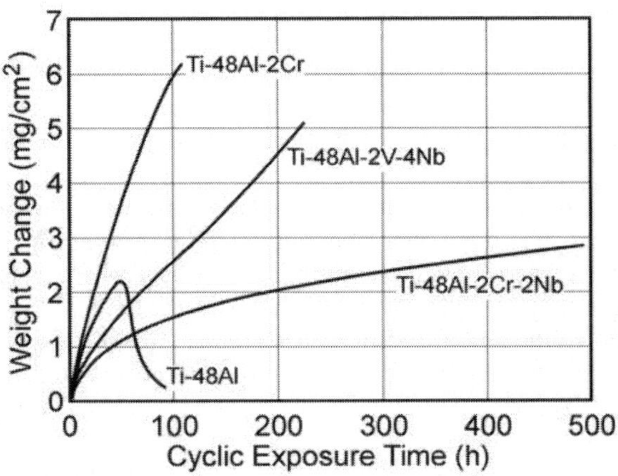

Fig. 8.15. Plot of weight change during cyclic thermal exposure in air at 850°C of four γ alloys [245]

8.2 Microstructure and Properties

It is important to clearly state at the outset of this section that there are currently no major or mass production applications of titanium aluminides. Further, the interest in applying alloys belonging to the α_2 and orthorhombic class has waned considerably due to the environmental cracking problem to be discussed later. There is continuing interest in γ alloys and guarded optimism regarding their ultimate application. In contrast to conventional titanium alloys, this lack of real service experience increases the difficulty in identifying the most important issues in the area of structure/property relations. This makes studies aimed at property optimization through incremental changes in composition or processing somewhat premature since no "hard" goals can be defined.

In principle, titanium aluminides have attraction because they extend the temperature capability of titanium based alloys, although at somewhat increased risk due to the lower ductility of this class of material. Still, they fill an important gap in lightweight materials capability in the temperature range from about 500 to 700°C. This section describes some of the approaches that have been taken to understand and improve various properties such as strength and ductility.

8.2.1 Alpha 2 and Orthorhombic Alloys

The alloyed versions of the α_2 and orthorhombic class of alloys can exhibit quite high strengths, particularly in the temperature range 550-650°C where conventional titanium alloys begin to rapidly weaken with increasing temperature. This is in part because the long-range order imparts a different temperature dependence to the flow stress of the ordered α_2, O, and β_2 constituents and in part because the microstructures are quite fine resulting in significant boundary strengthening contribution to the flow stress. In some regards, the latter strengthening contribution is similar to that observed in the aged β alloys as discussed in Chap. 7. It is difficult to directly compare α_2 or orthorhombic alloys with conventional titanium alloys. This is because the differences in composition cause large differences in transformation kinetics, making creation of microstructures of comparable length scales difficult to produce, even in the laboratory. A more pragmatic approach is to compare microstructure and properties at a constant cooling rate since this is representative of the way these alloys would be treated in production applications.

Table 8.1 shows some tensile properties of Ti-25Al-10Nb-3V-1Mo (Super α_2) in the coarse lamellar, equiaxed, and bi-modal conditions. The coarse lamellar and equiaxed conditions were created at very low cooling rates for the purpose of assessing some fundamental points of structure/property relations [237]. These microstructures would not be encountered in actual practice. The data in Table 8.1 compare the properties of Super α_2 to those of the conventional high temperature alloy IMI 834. In IMI 834 at 600°C, the yield and tensile strengths are much lower as compared to Super α_2. Comparing the slowly cooled, fully lamellar conditions of the two alloys, the Super α_2 is also stronger, but the strength differences are much smaller. This suggests that the principal strengthening contribution in Super α_2 is boundary strengthening. Examination of the size scale of the lamellar regions in the bi-modal microstructure in the two alloys shows that the Super α_2 has a much finer transformed structure than IMI 834. This is shown in Fig. 8.16. Cooling rate has a large affect on the scale of the transformed β regions of the Super α_2 alloy. This has a big effect on the strength of this alloy in the bi-modal condition as shown in Table 8.2 [246].

The tensile ductility of the Super α_2 is somewhat lower than that of IMI 834, but, with the exception of the room temperature ductility of the coarse lamellar structure, falls in the acceptable range (Table 8.1). In contrast, the room temperature toughness of the lamellar structure is roughly twice that of any of the bi-modal conditions [246]. This suggests that crack initiation is relatively easy in the lamellar structure but the crack extension energy is higher than for the bi-modal structures. This is consistent with the considerably rougher and more irregular fracture path in the lamellar structure compared to the bi-modal conditions, again, which is similar to the conventional $\alpha+\beta$ alloys discussed in Chap. 5.

Unlike the conventional high temperature alloys, the creep strength in Super α_2 is better for the fine bi-modal structures than for the fully lamellar structures. For example, Fig. 6.7 illustrates this point for IMI 834. Further, the creep strength of Super α_2 improves as the scale of the transformed regions in the structure decreases, which is in direct opposition to the behavior of conventional alloys such as IMI 834 in the fast cooling rate regime, as also seen in Fig. 6.7. While the reasons for this different creep behavior are not understood in detail, it is suggested that the long range order in the α_2 and the ordered β_2 structure provide good intrinsic elevated temperature strength.

Table 8.1. Tensile properties at room temperature and 600°C of Super α_2 and for comparison of IMI 834 (bi-modal microstructure) [237]

Alloy and Microstructure	Test Temp.	$\sigma_{0.2}$ (MPa)	UTS (MPa)	Elong. (%)	RA (%)
Super α_2 (Bi-Modal)	RT	1010	1235	5.2	8.6
Super α_2 (Lamellar)	RT	585	615	0.3	1.3
Super α_2 (Equiaxed)	RT	835	900	1.8	4.6
IMI 834 (Bi-Modal)	RT	950	1055	11.8	17.0
Super α_2 (Bi-Modal)	600°C	790	1045	8.0	19.4
Super α_2 (Lamellar)	600°C	370	560	6.3	16.7
Super α_2 (Equiaxed)	600°C	575	740	10.5	23.2
IMI 834 (Bi-Modal)	600°C	545	700	13.8	45.0

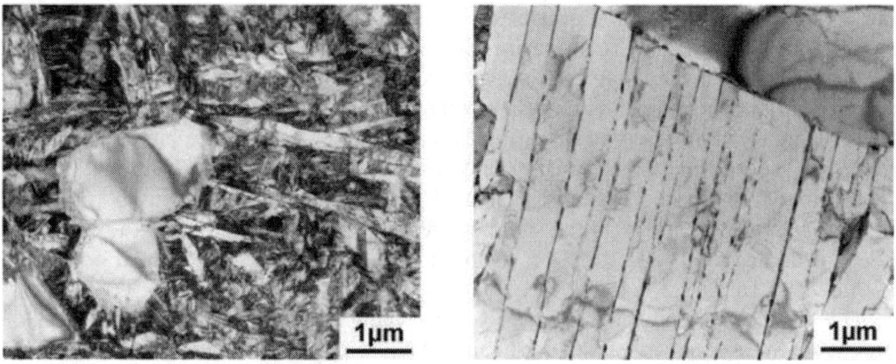

a **b**

Fig. 8.16. Micrographs comparing the lamellar regions of bi-modal microstructures, TEM: (a) Super α_2 (b) IMI 834 [237]

Table 8.2. Tensile properties at room temperature and 600°C of Super α_2 with bi-modal microstructure cooled at different rates [246]

Cooling Rate (°C/min)	Test Temp.	$\sigma_{0.2}$ (MPa)	UTS (MPa)	Elong. (%)	RA (%)
1100	RT	1130	1335	6.1	9.2
450	RT	1010	1235	5.1	8.6
100	RT	935	1180	6.4	10.9
25	RT	765	930	2.8	8.6
1100	600°C	890	1110	5.9	17.2
450	600°C	790	1045	7.8	19.3
100	600°C	690	965	7.0	21.3
25	600°C	550	825	11.2	20.9

The higher Nb content alloys based on the Ti_3Al composition form the O phase as has been previously discussed. These alloys have more attractive mechanical properties than the DO_{19} based α_2 and Super α_2 materials, largely because they have finer microstructures after comparable cooling rates. For example, Ti-22Al-27Nb has a higher room temperature strength (1075 MPa as compared to approximately 1000 MPa) than the comparable microstructural condition in Super α_2. This class of alloys is the only one that has density normalized strength that exceeds the values readily obtainable in wrought Ni base alloys such as IN718, as can be seen in the density normalized yield stress versus temperature plot, Fig. 8.17 [247]. As also can be seen in Fig. 8.17, the orthorhombic alloys are stronger up to about 750°C than either the first generation α_2 alloy (Ti-24Al-11Nb) or Super α_2. Table 8.3 contains tensile properties for three alloys of the orthorhombic class. Two of these alloys contain O+β_2 phases and the third contains O+α_2 phases. These data show that the O+α_2 is considerably weaker than the O+β_2 and that the tensile ductility also is less, particularly at room temperature [248]. The properties cited here for the Ti-22Al-27Nb alloy are the result of very limited microstructural optimization. It has been shown [248] that this alloy responds to different heat treatments, but the detailed knowledge necessary to permit optimization is not available. There also has been some question about the microstructural stability of the orthorhombic alloys during prolonged elevated temperature exposure. It has been shown that the strength and ductility remain essentially constant after 100 h exposure at temperature up to 540°C, begin to decrease after 100 h at 650°C and decrease significantly after 100 h at 760°C [248].

The creep strength of the orthorhombic alloys is slightly better than that of the α_2 and Super α_2 materials, which are slightly better than the strongest conventional high temperature titanium alloys such as IMI 834 [237]. Further the orthorhombic alloys have better room temperature ductility and fracture toughness than Super α_2. In the case of fatigue life, there is not a lot of data for the orthorhombic alloys. However, it is reasonable to suggest that the fatigue strength at 10^7 cycles scales with the yield strength. This has been demonstrated to be the case for the

Super α_2 materials [249] and can also be seen from the S-N curves in Fig. 8.18 for the three microstructural conditions discussed earlier (see Fig. 8.8 and Table 8.1) [237]. Fig. 8.18 also includes a S-N curve for IMI 834. These data are consistent with the generally accepted relation between high cycle fatigue strength and yield strength of titanium alloys. Since the orthorhombic alloys are stronger than the Super α_2 materials, it is likely that these alloys will also have better fatigue strength.

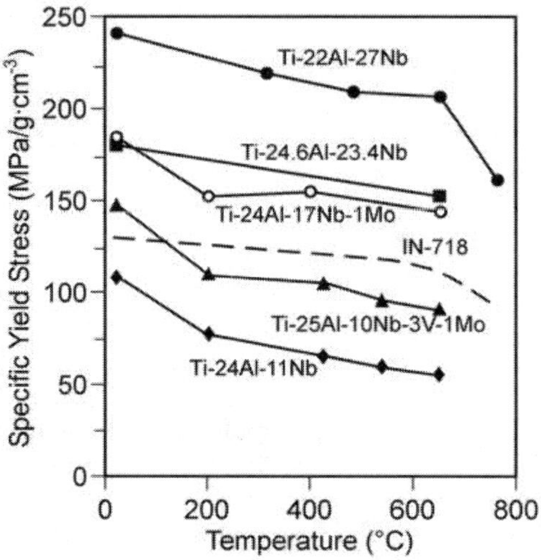

Fig. 8.17. Density normalized yield stress versus temperature curves for two α_2 and three α_2 orthorhombic alloys and for the Ni-alloy IN718 [247]

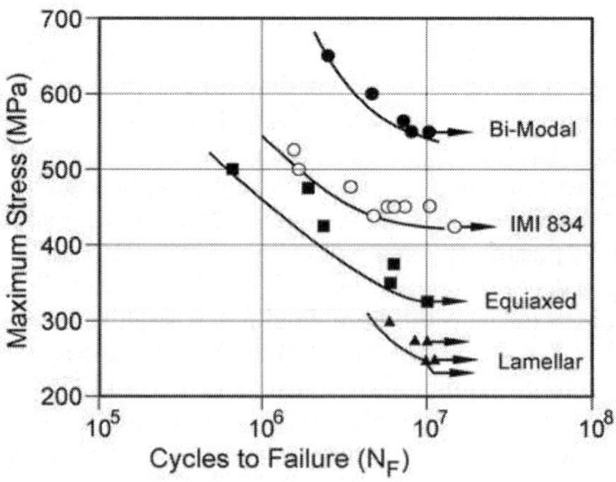

Fig. 8.18. S-N curves (R = 0.1) at 600°C of Super α_2 and for comparison of IMI 834 (bi-modal microstructure) [237]

Table 8.3. Tensile properties at room temperature and 650°C of orthorhombic alloys [248]

Alloy and Condition	Annealing Temp. (°C)	Test Temp.	Test Environm.	$\sigma_{0.2}$ (MPa)	UTS (MPa)	Elong. (%)
25Al-21Nb ($O+\alpha_2$)	1050+815	RT	VAC	845	880	0.4
25Al-21Nb ($O+\alpha_2$)	1175+760	RT	VAC	-	925	0.1
22Al-25Nb ($O+\beta_2$)	1000+815	RT	VAC	1245	1415	4.6
22Al-25Nb ($O+\beta_2$)	1125+815	RT	VAC	1135	1175	0.9
22Al-27Nb ($O+\beta_2$)	815	RT	VAC	1295	1415	3.6
22Al-27Nb ($O+\beta_2$)	1000+760	RT	AIR	1040	1120	2.8
22Al-27Nb ($O+\beta_2$)	1000+760 (1000h 650)	RT	AIR	1085	1145	2.6
25Al-21Nb ($O+\alpha_2$)	1050+815	650°C	VAC	680	935	18.1
25Al-21Nb ($O+\alpha_2$)	1175+760	650°C	VAC	730	945	2.5
22Al-25Nb ($O+\beta_2$)	1000+815	650°C	VAC	1005	1110	9.9
22Al-25Nb ($O+\beta_2$)	1125+815	650°C	VAC	880	1015	3.1
22Al-27Nb ($O+\beta_2$)	815	650°C	VAC	1120	1275	8.5
22Al-27Nb ($O+\beta_2$)	1000+760	650°C	AIR	800	940	13.4
22Al-27Nb ($O+\beta_2$)	1000+760 (1000h 650)	650°C	AIR	800	945	10.7

In the case of both creep and fatigue, it is important to emphasize that little effort has been devoted toward optimization of these properties. Therefore it is reasonable to suggest that there is at least opportunity for limited improvement of these properties for applications where they are limiting. It appears that the creep strength of these alloys may be the property that is least improved compared to IMI 834. Therefore, caution should be exercised if these materials are being considered for applications at temperatures that exceed the capability of conventional high temperature titanium alloys (>625-650°C) and that are mainly creep limited.

There is one other major technical obstacle to the use of the α_2 or orthorhombic alloys for critical applications. This is the severe loss of tensile ductility of these alloys when tested at elevated temperatures >600°C in air under normal strain rates (10^{-4} s^{-1}) [250]. For example, the Super α_2 material in the bi-modal condition exhibits an approximate 65% reduction in tensile ductility between tests conducted in air and vacuum at 650°C [250]. In this same study, the ductility differences between air and vacuum essentially vanished when the tests were conducted at $8x10^0$ s^{-1}. It was suggested [250] that this environmental effect is the result of accelerated transport of hydrogen by moving dislocations during the tensile tests. Both hydrogen and oxygen are known to embrittle titanium based alloys, but the detailed mechanisms in this case have not been identified. In another study the fracture toughness and fatigue crack growth of bi-modal Super α_2 were investigated. Little effect of environment on fracture toughness was seen but an acceleration of more than a factor two in fatigue crack growth rate was reported [251]. When the strain rates at the crack tip during fracture toughness and fatigue crack growth testing are considered together with the time of exposure to the environment of any crack tip volume element, these results appear consistent.

8.2.2 Gamma Alloys

Gamma alloys have been studied in three product forms: castings, forgings or other wrought products, and thin sheet. Consequently there are several alloy compositions for which property data have been generated. In this section a general summary of the trends in properties as a function of microstructure and composition will be provided. For a more comprehensive assessment of the details, the reader is referred to the (numerous) conference proceedings that address this subject [220, 252, 253].

The product form of γ alloys that is closest to acceptance is castings as will be described in the next section. There are several alloy compositions intended for use as castings that have been extensively evaluated, but there are more data for Ti-48Al-2Cr-2Nb and Ti-47Al-2Mn-2Nb+0.8vol%B than for other alloys. Moreover, these two cast alloys adequately demonstrate the behavior of cast γ alloys so they are considered representative for the purpose of discussing cast products. The principal effect of boron additions is to form small TiB_2 particles that refine the lamellar colony size. Smaller colonies lead to better room temperature ductility, but also can reduce the creep strength.

Wrought processing of γ alloys has been shown to be possible, but the lower ductility as compared to conventional titanium alloys is a perpetual restraint and will ultimately be an economic hardship if this class of materials is processed this way for production applications. The benefit of wrought processing is that greater microstructural flexibility can be achieved by combining thermal processing and recrystallization, just as in the other classes of titanium based alloys mentioned earlier. If the alloys containing boron are included, then control over recrystallization is comparable to that in conventional alloys even though the low ductility limits the amount of work that can be done and higher working temperatures are required. The most promising aspect of wrought processed γ alloys is the improved ductility that can be realized as a result of finer bi-modal or equiaxed microstructures.

The processing of thin sheet of γ alloys has been demonstrated [254]. This is a significant accomplishment considering the limited ductility of γ alloys and the plane strain conditions that accompany rolling. The thin sheet has a very fine grain size and the equiaxed structure of the alloy Ti-48Al-2Cr exhibits superplastic behavior at temperatures above 950°C. This creates a number of interesting possibilities for fabricated structures made from this thin γ sheet.

The γ alloys have considerably lower room temperature strength than the orthorhombic and α_2 alloys. Yield strength values at room temperature typically range from 375 MPa to 650 MPa, although higher values have been reported [255]. The corresponding range of tensile ductilities is 0.5-3% elongation. These alloys are used in three basic microstructural conditions: fully lamellar, equiaxed, and bi-modal. Two of these were shown in Fig. 8.10 [243]. The scale of these microstructures is quite different as would be expected by analogy to other titanium base alloy systems (conventional alloys, orthorhombic and α_2 alloys). The microstructural condition has a major influence on the balance of properties. The fully lamellar microstructure materials have very low room temperature strength and ductility but good creep resistance. The bi-modal microstructure materials have better strength and ductility but poorer creep strength. The fatigue strength scales with tensile properties, the ultimate strength being in this case a better normalizing parameter as shown in Fig. 8.19 [256]. The fatigue crack growth rates are also slower in the fully lamellar structures, which is consistent with microstructural scale effects seen in $\alpha+\beta$ titanium alloys as described in Chap. 5. The detailed mechanisms that control the variation in these properties are not clear for γ alloys, but there is no reason to believe that the qualitative correlations between microstructural length scale, constituent morphology, and strength should be different for these alloys than for conventional titanium alloys or for orthorhombic and α_2 alloys.

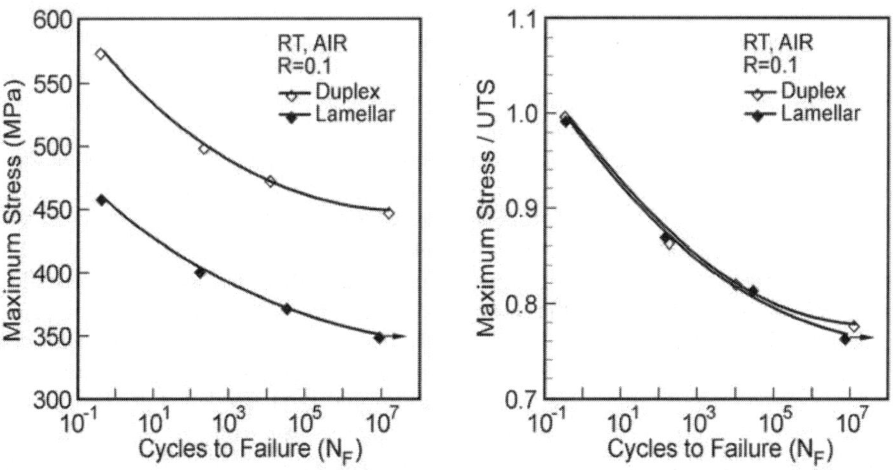

a b

Fig. 8.19. S-N curves for a γ alloy in the duplex (bi-modal) and fully lamellar conditions: (a) Applied stress versus life (b) Applied stress normalized by UTS versus life [256]

8.3 Applications

It was stated at the outset that there are no large scale production applications for titanium aluminides, although a few niche applications exist. For example, there are about 10 000 cast γ turbocharger rotors in service and γ exhaust valves are being used in some classes of competition auto racing engines. Using mid-2001 as a reference point, it is correct that there is a strong interest in γ alloys and that the interest in the orthorhombic and α_2 alloys as attractive high temperature materials has essentially disappeared. The latter loss of interest is the direct result of the environmental susceptibility that has been described in the previous section. Therefore, this section will describe promising potential applications of γ alloys, including some of the reasons why these alloys are attractive for such applications. Even then, the discussion is with the clear understanding that the applications described are potential rather than current actual ones.

Perhaps the single most attractive current application for γ alloys is for low pressure turbine (LPT) blades in aero-engines. For LPT blades, γ alloys would replace conventionally cast Ni base blades made from superalloys such as Rene 77. The maximum service temperature for these LPT blades is about 750°C and the γ alloys have adequate creep strength up to this temperature. The γ alloys also have adequate surface stability at these temperatures and so should retain their strength for extended service periods without embrittlement. Since these blades are rotating components, the reduced mass translates into lower loads on the LPT disk and the Ni alloy disk can be reduced in mass while maintaining constant levels of operating stress. In a large aero-engine such as the Boeing 777 class engine shown earlier in Fig. 1.6, two LPT stages of γ alloys allow about 100 kg of weight to be removed from the disks and blades compared to an all Ni base alloy construction. In the weight critical aero-engine industry, this is considered almost unheard of for a single material change. Thus it is extremely attractive. Further, these LPT blades would be cast, not forged, which is the potentially more economical component fabrication method for γ alloys. The technical feasibility of using cast γ alloy LPT blades has been demonstrated during an extensive factory engine test several years ago. In this test a last stage LPT rotor with cast Ti-48Al-2Cr-2Nb γ alloy blades was run for more than 1600 rejected takeoff cycles, which is a very demanding test. The rotor was also disassembled midway through the test. There was no damage observed in the blades and the test was considered fully satisfactory. The rotor from this test is shown in Fig. 8.20, while an example of individual LPT blades is shown in Fig. 8.21. Given the demonstration of technical feasibility for γ alloy LPT blades, what stands in the way of introducing them into commercial service? There are types of additional data needed, but there is no reason to believe any of these would technically prevent introduction. Included are fretting behavior between the Ni alloy disk and the γ alloy blades, surface stability of the γ alloy after extended service, a more extensive high cycle fatigue database for high mean stresses (i.e. a Goodman diagram), greater certainty that there is no susceptibility to hot salt stress corrosion cracking, and better definition of the range of machining parameters that can be safely used without introducing damage. The real issue, however, is cost. There has been several million dollars spent on γ alloy casting technology in the past 5-7 years. Consequently, considerable progress has been made in improving the yield and net shape process capability. However, there are several important business related or "cultural" issues involving the traditional supply chain participants. Among these are pricing, displacement of existing products by new, higher risk ones, and uncertainty of market size. These issues are not algorithmic in nature because of the difficulty associated with quantifying some of the variables needed to do a numerical calculation. Consequently, these issues create both real and perceived risks and remain as significant barriers. Resolving these is a commercial issue and further discussion of this is not consistent with the purpose of this book.

Fig. 8.20. Low pressure turbine (LPT) rotor from a 747 class engine containing cast γ blades after running in an extensive factory engine test (courtesy GE Aircraft Engines)

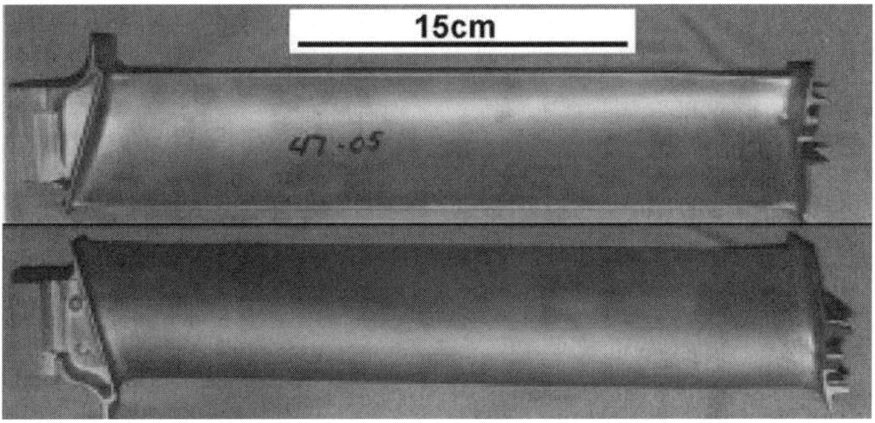

Fig. 8.21. Low pressure turbine (LPT) γ blades in the as-cast condition ready for machining (courtesy GE Aircraft Engines)

Another frequently discussed potential application for γ alloys are aero-engine high pressure compressor (HPC) blades. Many of the same benefits that pertain to LPT blades also are relevant to HPC blades, but many of the issues just described also pertain. Perhaps the greatest technical barrier is the relatively poor impact damage resistance of γ alloys because of their low ductility and low yield strength. Thus, susceptibility to impact damage is a concern. Further, HPC blades have greater exposure to foreign objects in an aero-engine because there is no equivalent of the combustor to "filter" particles as they come through the engine entrained in the gas stream. The other barrier to γ alloy HPC blades is cost. The cross section of HPC blades is much thinner than that of LPT blades and casting them will be considerably more challenging than casting LPT blades. This is because it will be more difficult to fill the much thinner leading and trailing edges of the HPC blade during casting. Forging is an alternative to casting HPC blades, but there are clear data that show forged γ alloy articles will be more expensive than cast ones. Thus forging does not "solve" the cost issue for γ alloy HPC blades.

There are several other potential structural applications for γ alloys, but these also vary in the time frame that they might actually emerge in real production applications. Further, for conventional titanium and nickel alloys, structural castings are cost competitive with machined forgings because they are very near net shape. As such they are typically irregular and unsymmetrical in shape and have large variations in cross sectional area. Consequently, complete filling of the shell is difficult and shrinkage, tears, and other casting defects are common in the as-cast condition. As a consequence, conventional Ni and Ti alloy structural castings have been successful largely because they can be extensively repaired after casting by welding (see Sect. 3.5.1). Welding of γ alloys has been proven feasible, but is difficult because it requires a significant preheat and slow post-weld cool, both of which add to the cost. The low ductility of γ alloys also increases the difficulty and therefore the cost of producing weld filler wire, which is essential for weld repairs. As in the case of LPT blades, there has been significant progress made in improving the casting technology for γ alloys. The most attractive potential static structure applications of γ alloys is where the low density and adequate strength or temperature capability allows them to displace Ni base castings or forgings. One example is the exhaust nozzle for a supersonic aircraft. Here, the significant weight reduction potential is the most significant reason to use γ alloys. This is particularly attractive because the exhaust nozzle in a large supersonic aircraft is located well behind the center of gravity and weight is extremely critical. The size of the exhaust nozzle parts depends on the nozzle design and the size of the aircraft, but can be quite large. This presents a casting technology challenge, but the ability to cast large parts has been demonstrated. An example of cast exhaust nozzle parts is shown in Fig. 8.22. When the time is right to build a large supersonic aircraft, the use of γ alloys in large structural castings such as this is a virtual certainty. Related applications such as single stage to orbit space planes will also be weight critical and γ alloy structural castings also have promise for these future applications. Lightweight, high section modulus structures are made from γ alloy sheet using hot formed stringers that have been laser welded to a flat sheet. An example of such a structure is shown in Fig. 8.23 [254]. This capability of fabricating γ alloy sheet into structural components that can operate up to about 750°C can have large benefits for hypersonic air vehicles.

There also has been interest in γ alloys for exhaust valves in internal combustion engines. Here the attraction is weight and temperature capability. A low weight valve train is believed to be capable of increasing the fuel economy of a mid-sized American car by as much as 5 miles per gallon which translates to reducing the fuel consumption for example from 8 liter/100 km to about 7 liter/100 km. Cost is so much more important for automotive applications, than it is for aerospace applications that it appears unlikely that this application of γ alloys will materialize in production autos unless there is another mandated improvement in fuel economy. Even in Europe where fuel is much more expensive than in the US, the economics of voluntary introduction of such high cost materials as γ alloys seems unlikely. Here the γ alloy valves are competing with steel and the cost increase is comparatively large. This only serves to decrease the likelihood of a voluntary γ alloy introduction.

The near term prospects for production introduction of γ alloys in any component are relatively small, which is unfortunate. In new materials such as the γ alloys, there is a clear economy of scale if several applications emerge and if these applications use the same or similar alloys. This is because significant levels of use generate an economically attractive stream of revert that further reduces the cost of the input material. For introduction of new classes of materials, it has been historically seen that there is a usage threshold. Once this threshold is exceeded there often is a nonlinear expansion of material usage. As will be discussed in Chap. 9, the barriers that currently prevent the use of titanium matrix composites are perhaps the best contemporary example of this point.

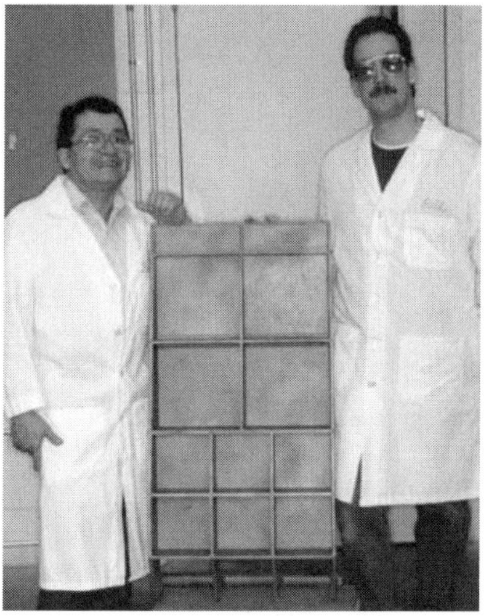

Fig. 8.22. Large cast γ exhaust nozzle flap for a supersonic transport engine (courtesy GE Aircraft Engines)

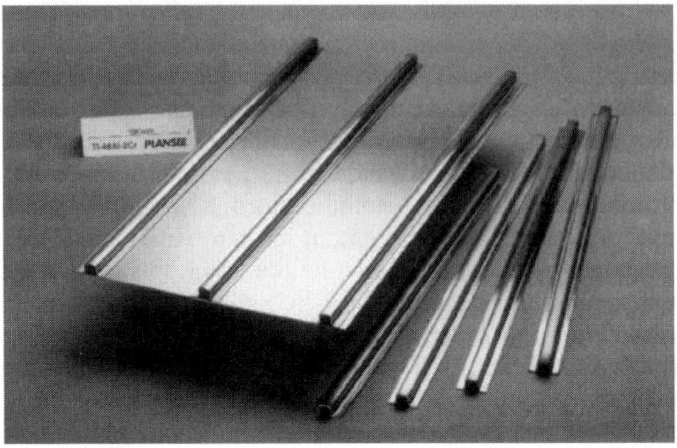

Fig. 8.23. High stiffness structural panel made from γ alloy sheet and hot formed stringers (also shown separately) [254]

9 Titanium Matrix Composites

Titanium matrix composites (TMCs) consist of a titanium matrix containing continuous reinforcing fibers. Development of these materials began more than 20 years ago when the primary reinforcing fiber being considered was boron. Since then, TMCs have evolved and have been improved with the availability of SiC fibers. The principal attractions of TMCs are strength and stiffness. On a density-corrected basis, continuous fiber (SiC) reinforced TMCs have about twice the ultimate strength and the stiffness of conventional titanium alloys, measured parallel to the fiber direction. In principle, this makes them among the most structurally efficient engineering materials known. In practice, it is often difficult to fully capitalize on the unidirectional capability of TMCs in a component, because off-axis loads are usually present. This lessens the impact of TMCs. Moreover, as is often the case, there are many other aspects of a successful material introduction than just one or two material properties (in this case UTS and E). Included are the additional considerations of reproducibility and variability of the properties, the cost and availability of the material, and the cost of finished components made from the material. There also is the question of design methods when the material is significantly different from those that it might replace. Because of the fiber reinforcement, TMCs are extremely anisotropic, which creates a challenge to maximize the benefits of the longitudinal properties while minimizing the penalties associated with the lower transverse properties. When this design situation is achieved, TMCs have much to offer.

This chapter will describe the characteristics of TMCs, the methods used to produce them, some potential applications, and some of the barriers that still exist to their widespread use. The organization of this chapter differs slightly from the previous chapters, each dedicated to a titanium alloy class, because a significant part of understanding TMCs is related to their processing, i.e. the methods by which the fibers are incorporated into the materials and the processes used to make components from the TMC once the material has been made. There is no direct equivalent to this situation, neither for monolithic titanium alloys nor for intermetallics.

9.1 Processing

Titanium matrix composites typically contain 35-40 vol.% reinforcing fiber. Early attempts (circa 1970) to produce TMCs used boron reinforcing fibers coated with SiC, the so-called Borosic™ fiber [257]. These fibers were extremely expensive, and, as it became clear that titanium Borosic composites were not going to be cost effective, most of the work on TMCs was discontinued for a number of years. Today the preferred (and really only available) reinforcement for TMCs is a SiC fiber that is grown by chemical vapor deposition by decomposing silane and methane on a hot filament. In the past, the most common filaments were made from either amorphous carbon or tungsten. In either case the filament was incorporated into the fiber at its center. The filament appears as a circular core at the center of the SiC fibers in micrographs showing the cross section of the composite containing the fibers, or of just the individual fiber. TMCs have been primarily produced using fibers from three manufacturers. One of the earliest fibers, SCS-6™, was produced on a carbon core by Textron in the US. A second fiber, grown on a W wire core, was developed by British Petroleum (BP) and was produced in the UK by DERA and in the US by ARC (Trimarc). It is called Sigma™ (Sigma Monofilament MMC, BP Information Sheet, November 1991). Additional information on these fibers and the composites that incorporate them may be found elsewhere [258]. More recently ARC has produced the Trimarc-2 fiber, which is grown on a carbon core. This fiber and composites containing it are now being produced commercially by FMW Composite Systems, Inc. [259]. An example of the microstructure of a TMC with all fibers having a unidirectional orientation normal to the plane of polish is shown in Fig. 9.1. These SiC fibers are typically 140 μm in diameter and are placed in the composite as a monofilament. There have been several processing methods developed to produce the titanium matrix containing continuous SiC fibers. Each of these methods has advantages and disadvantages. These method include the following: foil-fiber-foil [260, 261], physical vapor deposition (PVD) [262], spray-wind-spray [263], and powder cloth [264]. The foil-fiber-foil process creates a sandwich using multiple (two or more) layers of titanium alloy foil with the fibers in between them. This "sandwich" is then hot pressed to bond the titanium foil layers together and to bond the fibers to the titanium. In the PVD process, the titanium alloy matrix material is deposited directly on the fibers by physical vapor deposition, creating a cylinder of titanium alloy with a fiber at its center. These cylinders are laid side by side and hot pressed together to create the composite. The spray-wind-spray process employs plasma spraying of titanium alloy starting with powder to lay down a layer of matrix alloy on a mandrel that is rotated under the plasma torch. The fiber is placed on this matrix layer by winding. Then another layer of matrix is sprayed on top of it to create the composite. In the powder cloth method, the fiber array is surrounded by a slurry of titanium alloy powder mixed with an organic binder to create a green preform compact. This preform is sintered to consolidate the matrix and drive off the binder, creating the composite.

Irrespective of the method used, it is important that the fibers in the final composite do not touch one another because fibers do not bond to each other during consolidation [262, 265]. Fiber-to-fiber touching can greatly reduce the TMC strength. Consequently, materials made with such fabrication inconsistencies exhibit unacceptably large variation in properties. An early example of TMCs produced before a method for preventing lateral fiber movement and contact was developed is shown in Fig. 9.2. TMCs that contain fibers in contact with each other can have large (up to 50%) reductions in tensile and fatigue strength. Ensuring that fibers do not touch requires considerable care and poses a practical challenge during commercial scale production of large quantities of TMCs. This requirement also translates into a high production cost because of the difficulty in automating the fiber placement process. In TMCs, the elementary product that is equivalent to a mill product of conventional alloys (sheet, plate, bar, or billet) is a monotape. Monotape is a thin sheet of titanium matrix containing a single layer (monolayer) of unidirectional fibers located at its midplane in the thickness direction (hence monotape). A cross section of a monotape is shown in Fig. 9.3. Here the core and the SiC fiber can be readily seen. Careful examination of Fig. 9.3 also shows the thin dark layer of amorphous carbon present at the fiber/matrix interface to minimize the formation of reaction products. The monotape is rolled or coiled as shown in Fig. 9.4. These coils comprise the "raw material" for making articles from TMCs. To produce TMC components, this monotape is unrolled and laid up much in the same way that prepreg is laid up in the processing of continuous carbon fiber polymer matrix composites. After lay up, the monotapes are consolidated by diffusion bonding, typically in a vacuum hot press. Since the fibers in the monotape have unidirectional orientation, biaxial stiffening and strengthening is achieved with cross plies by orienting each monotape in the predetermined direction to create a multi-layer composite with the desired properties. One of the potential benefits of TMCs is the ability to tailor the properties in different in-plane directions through control of the fiber orientation in each ply of the TMC. Clearly, the properties in the direction perpendicular to the plane of the fibers will always be lower. As a consequence, properties of TMCs in the thickness direction are quite low. Fortunately, for many applications through-thickness loads are small, and this is seldom a limiting factor.

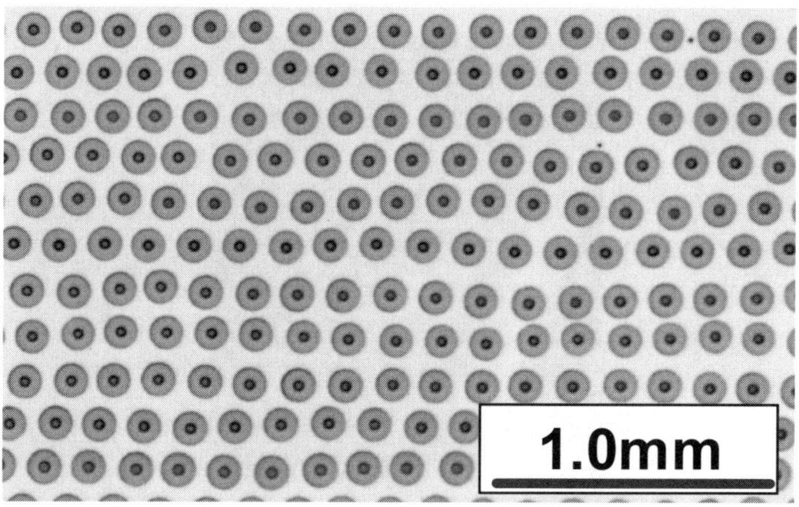

Fig. 9.1. Cross section of a TMC with all plies in the same orientation, LM (courtesy J. Jackson, GE Aircraft Engines)

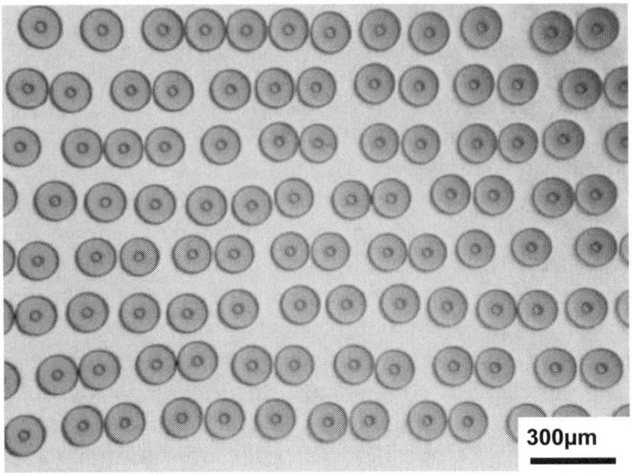

Fig. 9.2. Cross section of an early generation foil-fiber-foil TMC with fibers touching, LM (courtesy J. Jackson, GE Aircraft Engines)

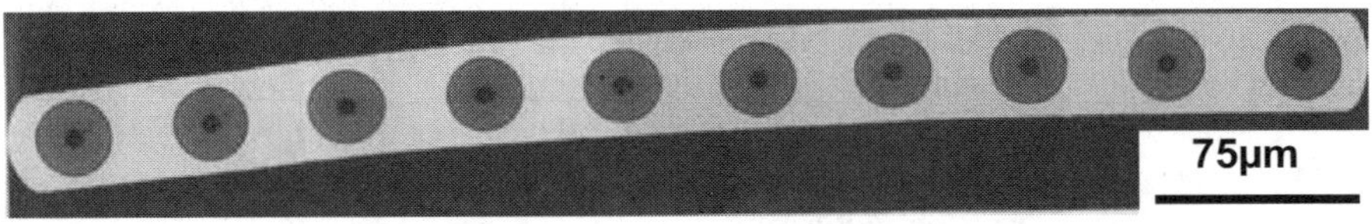

Fig. 9.3. Cross section of a monotape, LM (courtesy L. Johnson, GE Aircraft Engines)

Fig. 9.4. Roll of TMC monotape used as individual plies in articles with multidirectional fiber architectures (courtesy L. Johnson, GE Aircraft Engines)

Another issue that has limited the rate of development of TMCs is the reaction between the SiC fiber and the titanium alloy matrix during monotape production and during diffusion bonding of the monotape to produce the composite. If the reaction product is a brittle phase, then the transverse strength drops precipitously. Going back to the Borosic fiber era, the SiC coating was put on the boron fibers to avoid the formation of TiB_2 at the fiber/matrix interface [266]. This was generally successful, but there was some matrix reaction with the fibers if the bonding times were long or the bonding temperatures were high. In this case the reaction product typically was TiC. The SiC fibers used today have an amorphous C coating on them to minimize the extent of fiber/matrix interaction and to provide some compliance during thermal cycling to minimize interface damage. This coating also provides a weak fiber/matrix interface which deflects cracks growing perpendicular to the fibers. The Sigma™ fiber also has a layer of TiB_2 on top of the amorphous carbon to act as a diffusion barrier [267, 268]. This approach has proven quite effective. The other processing issue that has been mentioned earlier is the movement of the fibers so that they touch. As mentioned earlier, any point of fiber contact is an intrinsic flaw since the fibers do not bond to one another. The consequence of such flaws is to significantly reduce properties such as fatigue life and fracture strain. This will be further discussed in the section on properties. In the foil-fiber-foil method, lateral motion of the fibers during monotape production has been minimized by weaving fine titanium or molybdenum wire between the fibers perpendicular to their axis to prevent their motion [269]. The molybdenum wire was used mainly with higher temperature matrix materials, such as the α_2 alloy Ti-24Al-11Nb. A schematic of this is shown in Fig. 9.5 [270]. Clearly, this weaving operation is labor intensive and adds to the cost of producing these composites. Another disadvantage of the use of Mo wires to minimize lateral fiber motion is that these wires are prone to severe oxidation attack if they are exposed to the atmosphere and the TMC is used at elevated temperatures. Such exposure can occur either by matrix cracking or by exposure during machining of the composite matrix at attachments or other features (e.g. bolt holes). An example of the defects that are created by oxidation of the Mo wires is shown in Fig. 9.6 [270]. For this reason the Mo wire is no longer used in high temperature TMCs.

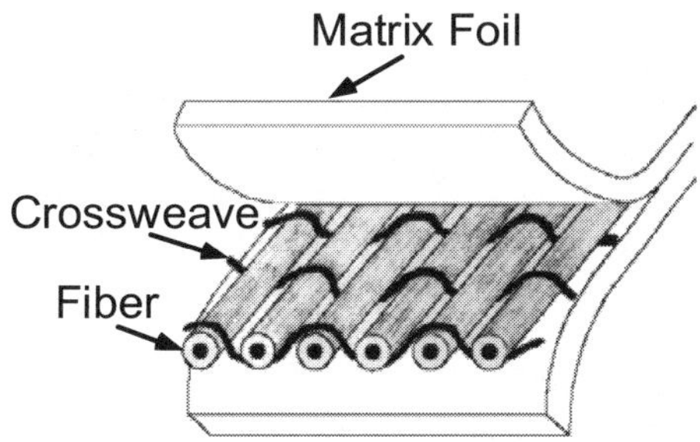

Fig. 9.5. Schematic showing the use of metal crossweave wires to ensure that fibers do not touch in the TMC [270]

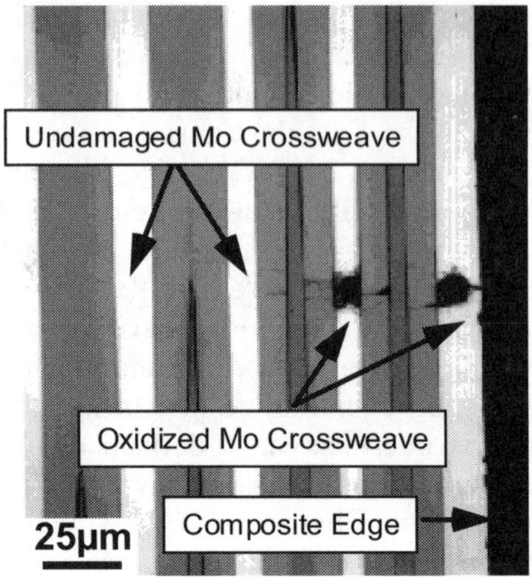

Fig. 9.6. Cross section of the TMC shown schematically in Fig. 9.5 after exposure at 815°C for 2 h in air, LM [270]

One of the reasons that the PVD process has attraction is because it eliminates the need for fiber weaving operation. Another issue with the foil-fiber-foil process is the cost of producing foil from titanium alloys. It is possible to make 125 μm thick foil from alloys such as Ti-6Al-4V [271, 272] using a Sendzimir mill as described in Chap. 3 and illustrated in Fig. 3.21, but this foil is very costly. The cost and availability of foil has prompted the use of β titanium alloys for the matrix of TMCs, since these alloys are more readily cold rolled into foil. Another advantage of using β alloys is the increased solubility of C and Si in the β phase, which reduces the risk of forming a brittle interface reaction product as mentioned earlier. The β alloy Beta 21S is perhaps the most commonly used β alloy. The composition of this alloy is included in Table 2.6. The Nb addition in Beta 21S imparts improved oxidation resistance to the alloy, which helps during processing and should be beneficial during elevated temperature service.

9.2 Properties

As mentioned in the introduction to this chapter, the potential for substantial improvement in properties of titanium based materials through the introduction of SiC fibers is the motivation for making TMCs. This section will examine the degree to which these properties can be improved, as well as the limitations to such improvements.

9.2.1 Tensile Properties

The static strength and stiffness of TMCs depends on the fiber modulus, volume fraction, and orientation of the fibers and of the matrix. The modulus of continuous, unidirectional fiber composites can be calculated with the rule of mixtures. In practice unidirectionally loaded components are infrequently found. Consequently, unidirectional fiber arrangements are seldom employed because the greatest benefit of this fiber architecture is realized in unidirectional loading. The effect of the presence of fibers and of fiber architecture on modulus of TMCs with several representative fiber architectures is shown in Fig. 9.7 as a function of temperature and for two different matrix alloys [273]. This figure includes data for TMCs with two different matrix alloys, Ti-24Al-11Nb and Beta 21S. This figure also shows the modulus of two conventional, unreinforced titanium alloys for comparison. From these data the increase in modulus of TMCs when the fiber axes are parallel to the loading direction is very clear. It also is clear that this improvement falls off sharply when the loading direction is not parallel to the fiber axis (see the curves labeled 90, 0/90, and 0/±45/90). The low off-axis stiffness and strength are direct consequences of the weak fiber/matrix interface because this inhibits load transfer to the fibers. Paradoxically, this weak interface is necessary to achieve crack deflection at fiber/matrix interfaces to obtain acceptable fracture resistance.

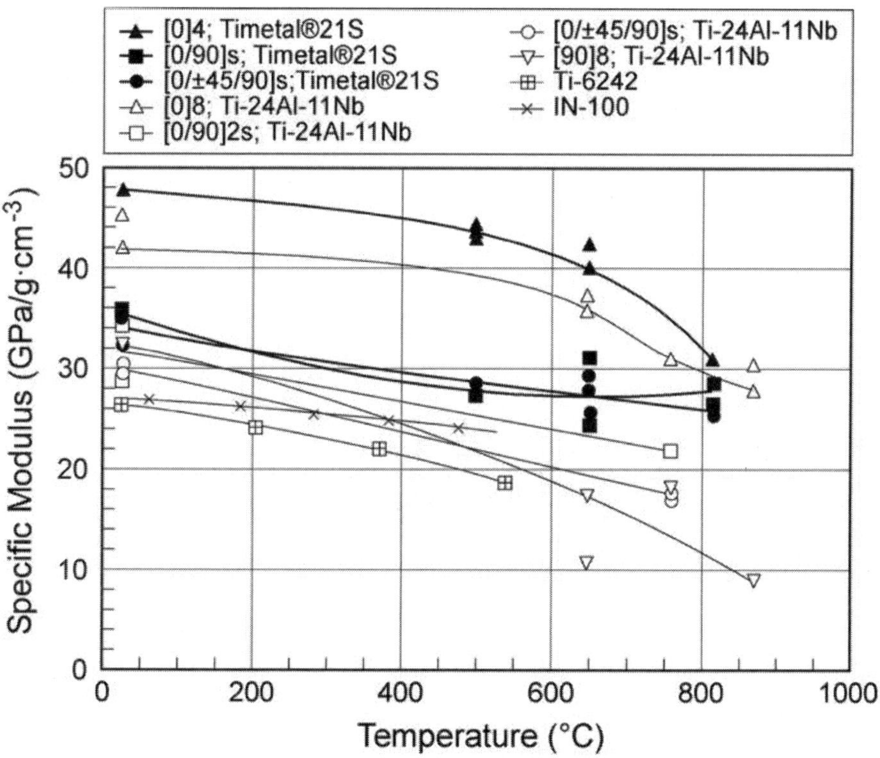

Fig. 9.7. Density normalized modulus of TMCs and two unreinforced metallic alloys (Ti-6242 and IN-100) as a function of temperature [273]

The effect of fiber orientation on ultimate strength is also very pronounced. Data similar to those shown in Fig. 9.7, but for ultimate strength, are shown in Fig. 9.8 [273]. The use of ultimate tensile strength is appropriate because the plastic strain to failure of the TMCs is so small that determination of the 0.2% yield stress is unreliable. The data in Fig. 9.8 show the effect of the higher room temperature strength of Beta 21S matrix and also the benefit of the higher temperature capability above about 650°C of the Ti-24Al-11Nb matrix. The curves for both of the unreinforced matrix alloys makes this point quite clear. As in the modulus case, the strengthening due to the fibers is much less pronounced when the load is not parallel to the fiber axis due to the separation of the weak fiber/matrix interface. This weak fiber/matrix interface bond leads to early failure when a normal stress is applied across it.

As a minimum, this strong fiber orientation effect creates the need for thoughtful design methods to maximize the loads parallel to the fiber direction and to minimize loads in other directions. In the absence of these design methods, the benefit realized from the use of composites relative to their cost will be unattractive.

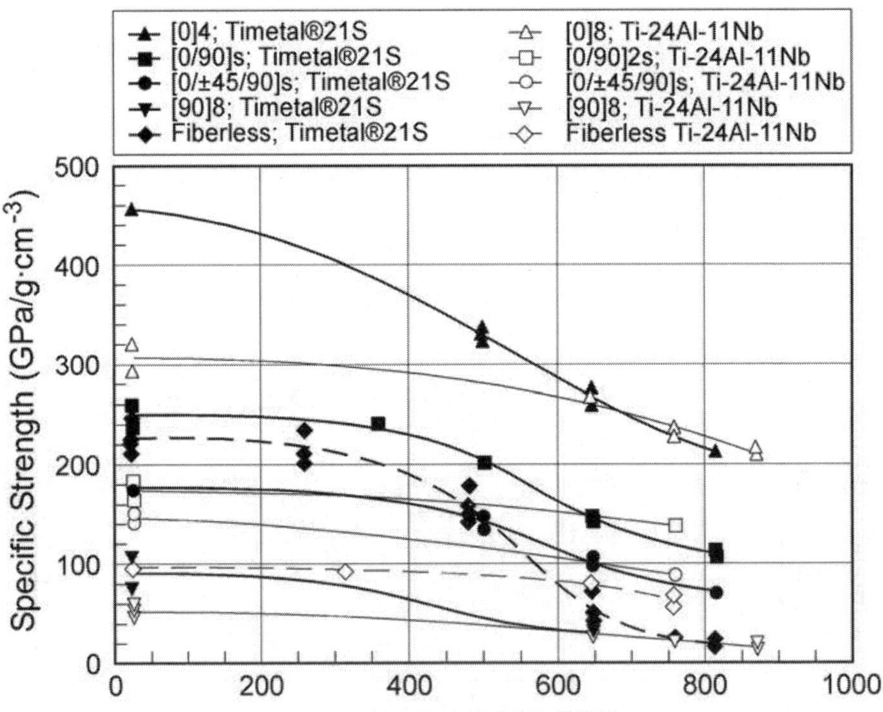

Fig. 9.8. Density normalized UTS of TMCs with different matrix alloys and fiber architecture as a function of temperature (also shown are the curves for the unreinforced matrix alloys) [273]

9.2.2 Fatigue Properties

The fatigue strength at 650°C of TMCs with a Ti-24Al-11Nb matrix is about 50% better than the unreinforced matrix at 10^3 cycles but the advantage decreases at longer lifetimes (10^6 cycles) to about 15%. These data are shown in Fig. 9.9 [273]. For the TMC, the load axis was parallel to the fiber axis. The shapes of the curves in Fig. 9.9 for the TMC with a Ti-24Al-11Nb matrix and the unreinforced matrix merit some discussion because the curves are not parallel. They actually appear to intersect between 5×10^6 and 10^7 cycles. This suggests that crack nucleation resistance in both materials is about the same, but that the propagation of microcracks is faster in the unreinforced material. This is consistent with the role of fibers in bridging the cracks and retarding their growth under a constant driving force (ΔK). At present, there is no direct evidence of the rate of microcrack propagation in TMCs. However, Fig. 9.10 [273], clearly shows that the presence of fibers retards the rate of macrocrack propagation. The curves in Fig. 9.9 for Beta 21S matrix composites also show the pronounced benefit of fiber reinforcement when the load is applied parallel to the fiber axis. In this case there are no data for the unreinforced matrix at 650°C, partly because this is a relatively high temperature for a β alloy. The curves for the Beta 21S matrix composites when the fiber axis and the load axis are not always aligned, as in a (0/90) lay up, or when the fiber axis architecture has been altered to approximate an isotropic material such as a (0/±45/90) lay up are quite instructive. In both cases, (0/90) and (0/±45/90), the fatigue strength is much lower than for the unidirectional (0) material. This indicates that the off-axis plies do not show the benefits from the presence of the fibers and that these plies dominate the overall behavior of the composite. One other piece of data that demonstrates the effects of a weak fiber/matrix interface is the reduction in elastic modulus during fatigue cycling. This is shown in Fig. 9.11 [274] from which it can be seen that the stiffness of the TMC in fatigue drops well below the initial stiffness. The initial stiffness was measured while the weak fiber/matrix was still intact. When the applied stress exceeded this bond strength, fiber/matrix interface separation occurred, and thereafter the lower composite stiffness reflects the loss of load transfer to the fibers.

As is the case for statically loaded structures, it is clear that the fatigue performance benefits that can be derived from the use of TMCs will depend stringently on the design practice employed. If a component is designed so there are essentially no loads in any direction other than the fiber axis, the structural efficiency of TMCs is significantly better than that of monolithic materials. Given the cost of TMCs, such design practice will be essential if the benefits that are realized from the use of TMCs will be great enough to justify their associated costs.

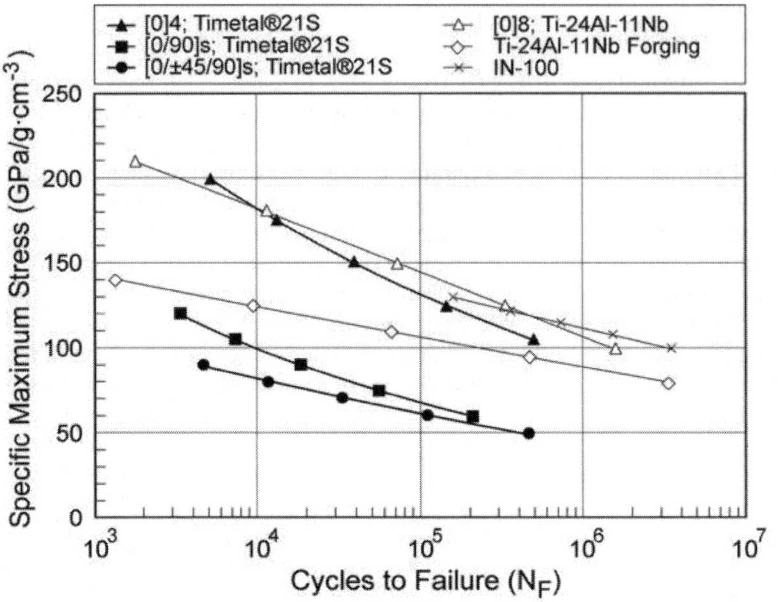

Fig. 9.9. Density normalized maximum stress (R = 0) vs. cycles to failure for TMCs and two unreinforced metallic alloys [273]

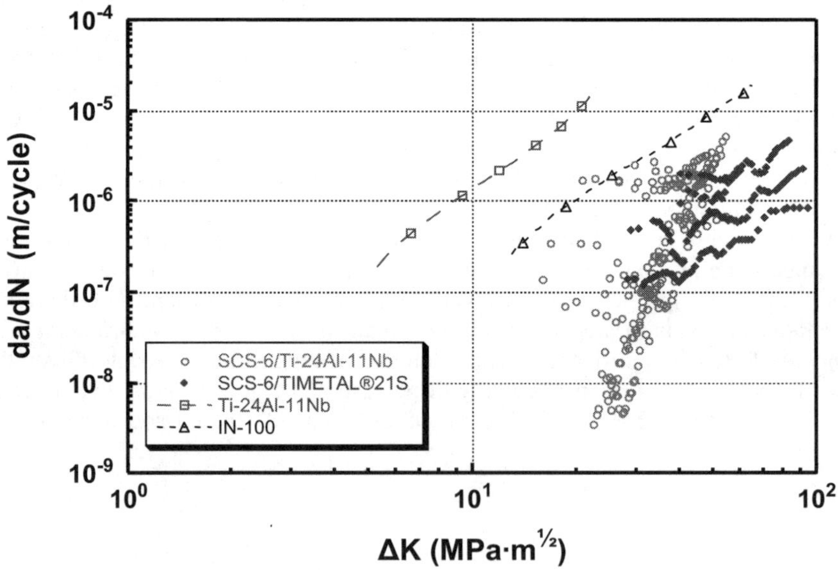

Fig. 9.10. Crack propagation curves for macrocracks in two TMCs and two unreinforced metallic alloys [273]

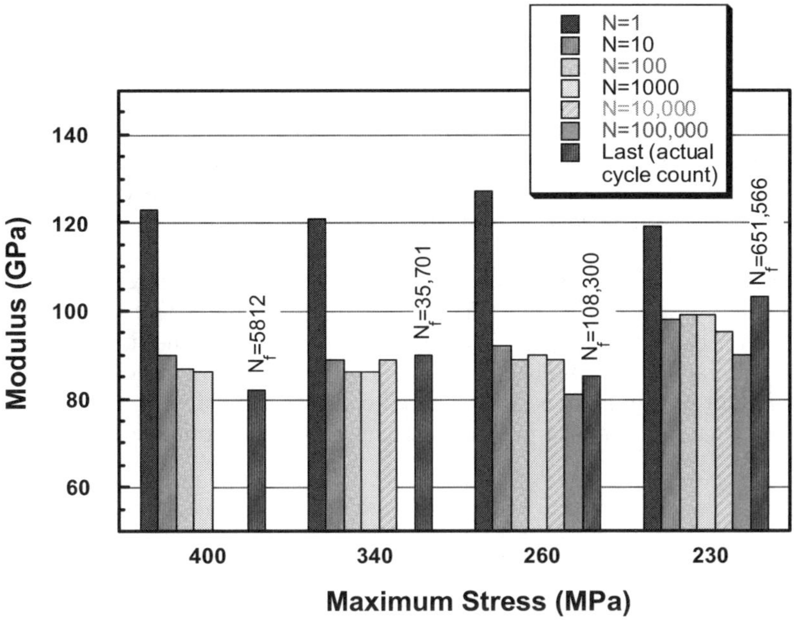

Fig. 9.11. Apparent modulus (compliance) of a Beta 21S matrix composite (0, ±45, 90) after cycling at four different maximum stress values [274]

9.2.3 Creep Properties

The performance of TMCs during sustained loading at elevated temperature is as dependent on the fiber orientation as are the other properties already discussed. The creep rupture behavior of Beta 21S and Ti-24Al-11Nb matrix composites is shown in Fig. 9.12 [273] along with properties of the unreinforced matrix alloys and two additional unreinforced high creep strength materials. All of the data in Fig. 9.12 are density corrected. From these data, it can be seen that the TMCs have very good rupture properties when the load is applied parallel to the axis of the fibers. The rupture behavior of the Beta 21S matrix composites loaded perpendicular to the fibers is slightly worse that that of the unreinforced matrix, due to the weak fiber/matrix interface. Also shown in this figure are data for (0/90) and (0/±45/90) fiber architectures. In each of these cases the rupture behavior is consistent with a rule of mixtures using the volume fraction of fiber aligned with the load axis.

There are two additional concerns regarding the creep behavior of TMCs. One is the time dependent degradation of the matrix due to oxidation and, of greater concern, the degradation of the fiber/matrix interface. The very low creep resistance of TMCs when fiber and matrix become disbonded has already been demonstrated. The occurrence of a matrix crack, its subsequent oxidation, and the degradation of the fiber/matrix interface are shown in Fig. 9.13. The second concern is associated with the application of creep data obtained from laboratory coupons to the design of actual components. Unlike monolithic metals, fiber reinforced TMCs have a time dependent load transfer to the fibers in a displacement controlled situation because of the creep strain that is accumulated in the matrix. Most laboratory test coupons are gripped at their ends and the compressive clamping loads from the grips maximize the transfer of load to the fibers by constraining the matrix. Consequently, the data obtained from these coupon tests are nonconservative. The degree to which this concern exists depends on the component configuration. For example, in a circular ring principally loaded in the circumferential direction, the concern is small. On the other hand, for a free-end component such as an airfoil, the error can be quite large. The relative insensitivity of the rupture behavior of TMC laboratory test coupons with matrix alloys that have large differences in creep strength is mainly caused by the end constraint just described. In practice, great care must be exercised in designing components to optimize the transfer of creep loads.

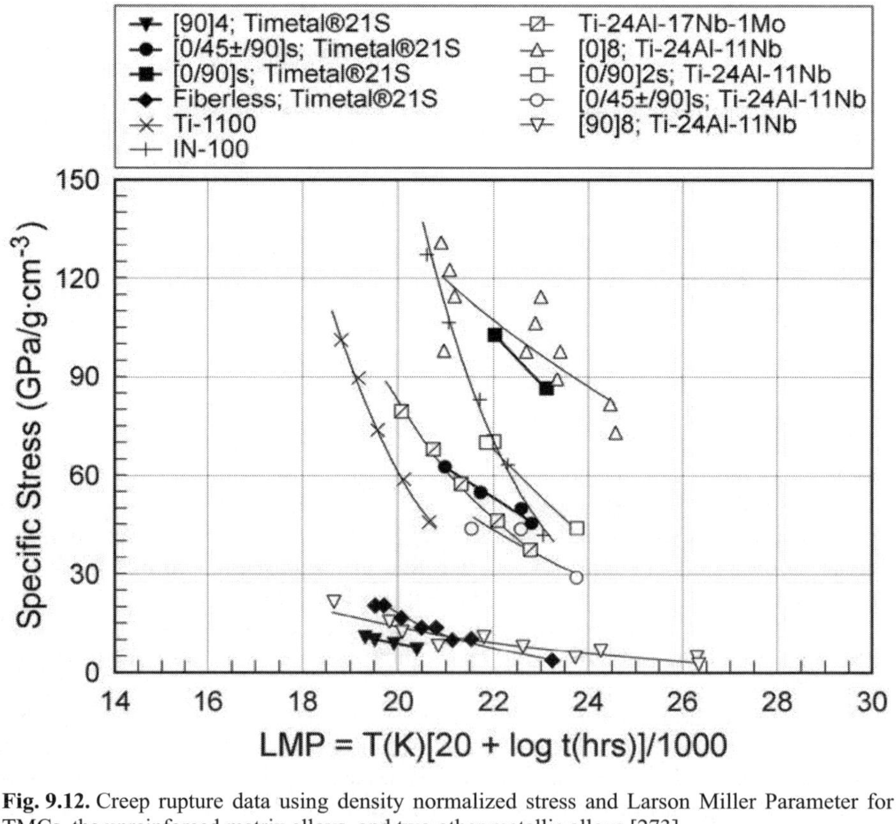

Fig. 9.12. Creep rupture data using density normalized stress and Larson Miller Parameter for TMCs, the unreinforced matrix alloys, and two other metallic alloys [273]

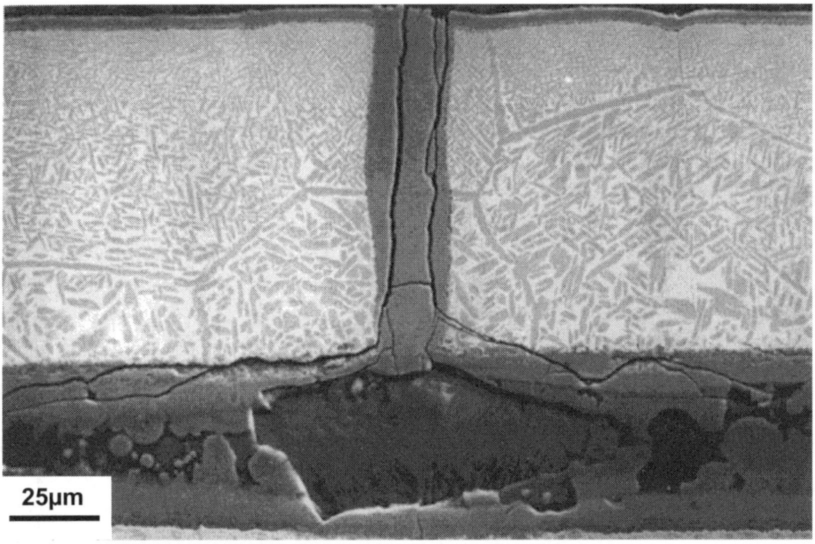

Fig. 9.13. Heavily oxidized matrix crack and degraded fiber/matrix interface in a Beta 21S matrix composite, SEM [273]

9.3 Applications

As in the case of titanium based intermetallics, there are essentially no production applications of TMCs. This is because the production of these materials on a commercial scale has not really been accomplished and, consequently, the cost of these materials remains very high. There is broad agreement that a "trigger price" for introduction is about 1000 US $ per kilogram stated in 1998 US dollars. The cost-volume relationship has been modeled and the volume sensitivity of the cost of TMCs is very high at low volumes (<5000 kg/yr). This relationship is shown in Fig. 9.14. The difference between business model and cost model in Fig. 9.14 is that the business model includes the cost of the plant and equipment required to produce the composite material. At sufficient volume, this cost contribution is diluted to a vanishingly small amount and the two curves converge. Clearly it is necessary for TMCs to be introduced into several production applications to realize lower cost per kg of the materials. Use of approximately 5000 kg per year is needed to stabilize and sustain the TMC supplier base.

There are several potential applications for TMCs that benefit from the properties described in the previous section. Perhaps the highest risk but highest payoff application is for reinforced rings for rotating parts of an aircraft engine. The high strength and stiffness of TMCs would allow elimination of the bores of the disk in rotating stages such as those shown in the cross section of Fig. 3.55. Elimination of the disk bores would allow the diameter of the rotating stages to be significantly reduced with attendant weight reduction. Smaller diameter engines (at constant thrust) also would create opportunities for new aircraft configurations. This could be especially beneficial for military aircraft where the engine is often built into the fuselage. Examples of TMC rings are shown in Fig. 9.15. These rings would be embedded into a titanium rotor forging to create a rotating stage. The principal stresses in a rotor are oriented in the circumferential direction, so the characteristics of TMCs are well suited for this application. Still, it is essential that care be exercised to maintain very low stress levels in the radial direction because of the low properties of TMCs in this direction. Care also must be exercised in the design of attachments (flanges, etc.) to avoid complex stress states in which at least one component exceeds the critically low property values in directions other that parallel to the fibers of the highly anisotropic TMCs. The fibers in a ring are wound continuously, so the free end question discussed earlier is not an issue.

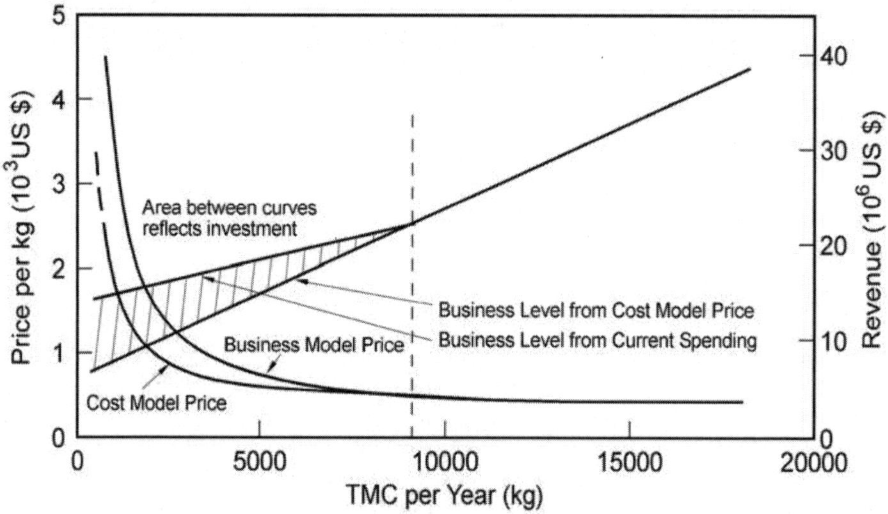

Fig. 9.14. Cost vs. volume curves derived from a cost model and a business model (courtesy L. Johnson, GE Aircraft Engines)

If rotors using TMC rings are incorporated into a design, the annular region along the engine centerline also decreases in size. In this case, the diameter of the shaft that connects the rotating fan and compressor stages with the turbine stages can become limiting. Fortuitously, a TMC shaft can have the torque capacity to minimize this constraint. The principal stresses in a shaft are torsional, so a shaft that is made by winding the fibers at a 45° angle to the shaft axis can transmit very high torque. In this case also it is essential to minimize stresses in other directions than the fiber orientation at the ends where the shaft is attached to the rotating stages. A demonstration shaft is shown in Fig. 9.16. In the photo, the large ends necessary to keep the stresses low at the attachment points can be seen.

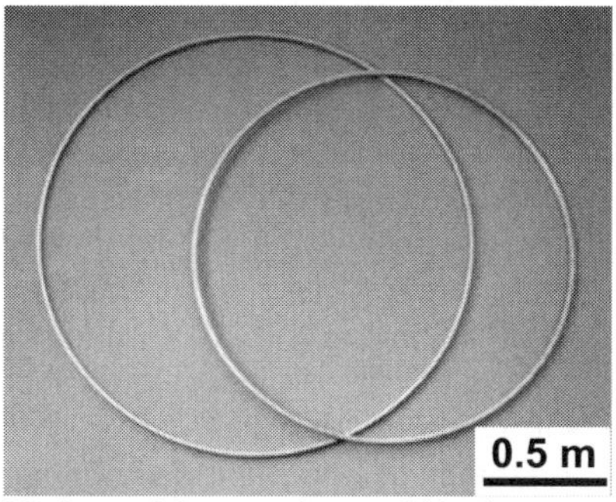

Fig. 9.15. Photo of two TMC rings of the type that could be incorporated into a rotating stage of an aircraft engine (courtesy L. Johnson, GE Aircraft Engines)

Fig. 9.16. Photo of a TMC shaft for an aircraft engine (courtesy L. Johnson, GE Aircraft Engines)

An early, but important, small volume production application of TMCs was established in 1999. This if for the augmenter actuator links in a large (F-16 class) military engine. These links are approximately 500 mm in length. One of these links is shown in Fig. 9.17 (upper photo), together with the Ni-base superalloy link (lower photo) it replaces. This application has low risk but high payoff because the augmenter links are located in the extreme aft position of the aircraft. This location makes them weight critical. The TMC link weigth is roughly 50% of that of the superalloy link, which makes the increased cost acceptable. An alternate means of utilizing MMCs at lower total cost is through the use of selective reinforcement. In selective reinforcement the MMC is preferentially located in regions of the component where the properties will create the greatest benefit. For example, in a bending-stiffness-limited application only the outer portions of the component would be MMC. Selective reinforcement is most attractive for stiffness-limited applications where the average section stresses are quite low ($<0.5\ \sigma_{0.2}$). This is because the region where the MMC and the unreinforced alloy are joined will be subject to large shear stresses. One means of introducing selective reinforcement is a process called "bicasting" [275]. In this process MMC inserts are placed in the desired position in the investment casting shell, and the unreinforced metal is cast around it to form a monolithic part.

As of 2002, there are a number of important developments in the commercialization of TMCs. The approximate cost for TMC rod has dropped below 2200 US $ per kg, and there are excellent prospects for substantial increases in levels of production of TMC [259]. In addition to the continued manufacture of the augmentor links just mentioned, a joint development agreement has been signed between ARC and BF Goodrich, Landing Systems Division, to produce commercial landing gear components for a number of large commercial aircraft, such as the Boeing 747 and 777 models [276], and NASA is considering the use of TMCs to satisfy specialized needs in the Space Shuttle and the Hubble Telescope. In the final analysis, the commercial viability of TMCs will depend on successful applications such as these and others.

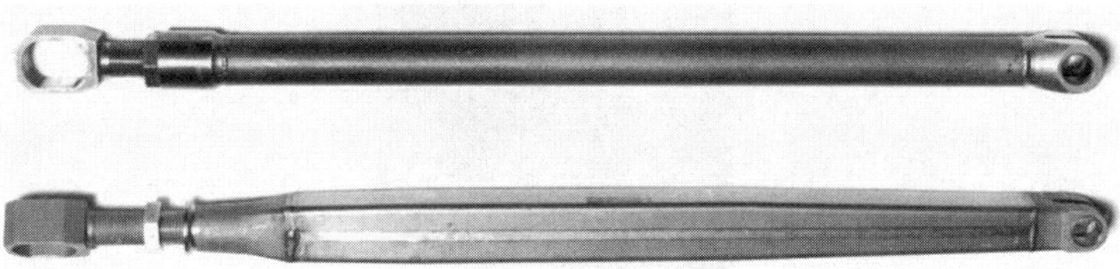

Fig. 9.17. Photo of two augmenter links, the top is TMC and the bottom is a Ni-base alloy (courtesy L. Johnson, GE Aircraft Engines)

10 Special Properties and Applications of Titanium

This final chapter contains short sections describing a range of titanium application areas which do not logically fit into Chaps. 4 through 9. Some of these areas are not alloy class specific, which makes it difficult to include them in these earlier chapters.

The first group of application areas is related to at least one "nontraditional" property or characteristic of titanium. Included are superconductivity, shape memory effect, burn resistance, and hydrogen absorption and de-absorption. Rather than distribute the discussion of individual alloys that have these characteristics into chapters organized by alloy class, it is possibly more efficient to group these according to a specific property or characteristic.

There also are application requirements for titanium, most of which are emerging from a market standpoint, that can be satisfied by materials from more than one class of alloys. Here it also is more efficient to discuss these by application area rather than by alloy class. Included in this group are biomedical devices, automotive components, sports equipment, and appearance related uses (jewelry and architectural creations). Some of these latter application classes are driven by combinations of unusual properties. Other applications are viable because of the appeal that titanium currently enjoys in the eyes of the consumer, but which are not always strictly supportable on the basis of cost-effectiveness.

10.1 Superconductivity

Superconductivity was first observed by Onnes in 1911 [277] when he discovered that, when cooled to a low enough temperature (about 4 K), a wire of mercury lost all of its electrical resistivity and became capable of conducting essentially infinite current. The change in resistance with temperature is shown in Fig. 10.1. Onnes called this loss of resistance phenomenon superconductivity. This remarkable characteristic was subsequently demonstrated for other pure metals including lead, tin, and indium. In each case, the temperature associated with the transition from normal to superconducting behavior is different, but in all cases it is low, on the order of a few Kelvin. The superconducting phenomenon has been extensively studied and discussed in great detail in the physics literature, but is less familiar to materials engineers. Therefore, a brief summary of the essential features associated with superconducting behavior will be reviewed here.

As mentioned earlier, the superconducting phenomenon is one that occurs in a number of metallic materials at very low temperatures when electrical resistivity decreases rapidly with temperature and essentially becomes zero at a critical temperature, T_c. For pure metals in the superconducting state, the internal magnetic field associated with the current being conducted penetrates the material at relatively low fields and destroys the superconducting characteristics. This transition occurs at a critical field H_c and also at a critical current I_c in the presence of an external field, including that from adjacent conductors in an electromagnet. These latter two critical parameters, H_c and I_c, make it impossible to achieve high field electromagnets using superconductors, even though superconductors, in principle, are capable of carrying very high currents with attendant high magnetic fields. In 1933, an additional effect associated with superconductivity was discovered, namely that materials in the superconducting condition can exclude the magnetic field from the interior of the superconductor and, consequently, exhibit very strong diamagnetic behavior (magnetic susceptibility about 1). Until 1933 it was anticipated that the field would be trapped inside the conductor by the surface eddy currents created as a consequence of Faraday's law. This turned out to be incorrect and the expulsion of the magnetic field at a critical field occurs abruptly and is completely reversible. This exclusion of the magnetic field to create a perfect diamagnetic condition is known as the Meissner effect (also called the Meissner-Ochsenfeld effect) [278]. The observed reversibility also permitted the application of thermodynamics to the superconducting transition because the superconducting condition could be treated as a material state.

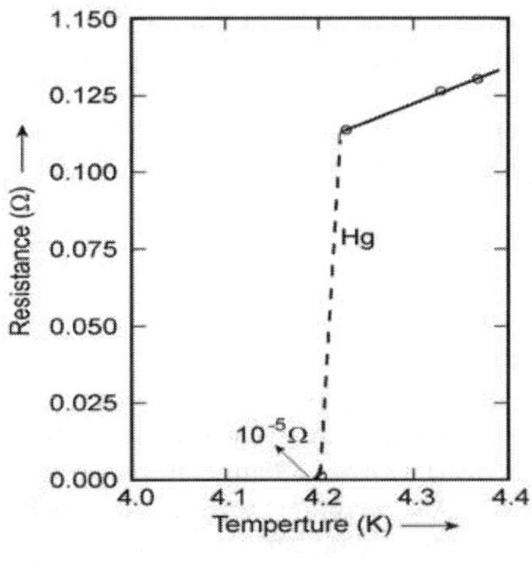

Fig. 10.1. Plot of electrical resistance vs. T for Hg showing the virtual disappearance of resistance at about 4.2 K

Most pure metals lose their superconducting character at a modest critical field, H_c, and concurrently lose their perfect diamagnetism. At $H>H_c$ they no longer have high diamagnetism and are not superconductors. These metals are called Type I superconductors. This behavior is schematically summarized in the magnetization applied field curves, Fig. 10.2. Figure 10.2a shows the relation for a conductor carrying infinite current without flux expulsion. Figure 10.2b shows the effect of flux expulsion which leads to Type I superconductivity. At applied fields less than H_c, the internal field, B, is zero and the material is superconducting.

In addition to pure metals, some alloys also exhibit a transition from normal electrical conductivity to a superconducting condition at low temperatures. In a number of these alloys the transition is not sharp as in a Type I superconductor and some of the superconducting behavior is retained up to a higher applied field. This behavior is the result of a mixed state of superconducting and normal material. This behavior is summarized in Fig. 10.3 from which it can be seen that the critical field, H_c, is replaced by two critical field values, H_{c1} and H_{c2}. Superconductors that exhibit this behavior are called Type II superconductors. Type II superconductivity was originally reported in Russia in the 1930's by Schibnikow [279] for several lead alloys but this was not widely noticed. Consequently, widespread recognition of Type II superconducting behavior did not occur until the 1960's. It is the mixed state that exists between H_{c1} and H_{c2} that enables the high magnetic fields envisaged by Onnes to be realized.

Fig. 10.2. Magnetization (M) – applied field (H) relationships: (a) Infinite current (b) Type I superconductor with flux expulsion (Meissner effect)

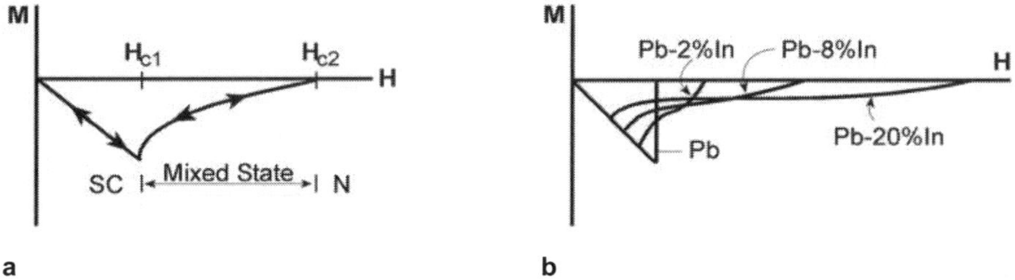

Fig. 10.3. Magnetization as a function of applied field for a Type II superconductor, showing the mixed state between H_{c1} and H_{c2}: (a) General schematic (b) Actual curves for Pb-In alloys

The mixed state in Type II superconductors allows partial penetration of the external magnetic field but this penetration is not uniform throughout the volume of the material. As mentioned earlier, it is this state that allows high magnetic fields to be obtained with the consequent emergence of strong electromagnets used in particle accelerators, medical imaging devices based on nuclear magnetic resonance (now called magnetic resonance imaging or MRI), energy storage devices, and experimental magnetic levitation trains. Today, most all of these devices are based on superconducting magnets whose coils are made from filaments of Ti-Nb alloys and pure Nb and/or Cu. The major role of Ti-Nb alloys in this application of superconductivity is the reason that this subject is treated here.

In the mixed state the material contains an array of linear features called fluxoids embedded in a superconducting matrix. Each fluxoid can be thought of as a cylindrical region which has material at its center that exhibits normal conductivity (as opposed to superconductivity) and which has been penetrated by the external magnetic field. Surrounding these fluxoids in the superconducting matrix are vortices of supercurrent which confine the magnetic field to the fluxoids, leaving the remaining matrix superconducting. The fluxoids repel one another and their axes become distributed as a regular two dimensional array, or lattice, where the fluxoid radii define the plane of the array. Numerous experimental techniques have been used to image the fluxoid lattice, thereby demonstrating its existence. In the mixed state, Type II superconductors maintain most of their volume in the thermodynamically favorable superconducting state up to a critical current (I_c) above which complete penetration of the magnetic field occurs and superconductivity is lost. There is a relation between the density and spatial distribution of fluxoid pinning sites and the critical current at any external field. Therefore, the key to achieving high critical currents at a given applied field is the microstructure of the superconducting material. Currently, the most popular low temperature, high current superconductors are made from Nb-47% Ti alloys. The Ti-Nb system is a β isomorphous one where the two elements are continuously soluble in each other above the β transus temperature. Nb is a relatively weak β stabilizer. Consequently, the transus of Nb-47Ti is about 450°C. An array of Nb-47%Ti wire (strands) is shown in Fig. 10.4 and at a higher magnification in Fig. 10.5. Each wire has been specially processed to create highly elongated α phase precipitates on a very fine scale. These α phase precipitates become the fluxoids. Because of the composition difference between the α precipitates and the β matrix due to alloy element partitioning, these sites are immobile and automatically satisfy the requirement of pinning the fluxoids. The processing scheme of the superconducting wire is basically one of initially swaging followed by cold drawing, then aging at 350-375°C to precipitate a uniform distribution of α phase precipitates and redrawing to produce a highly directional distribution of these precipitates. As discussed earlier in this book (Chaps. 2 and 7), the cold work creates a high density of nucleation sites for the α phase precipitates. The equilibrium fluxoid spacing for maximum critical current performance depends on the external field and the optimum spacing decreases with increasing field. For example at a field of 5 Tesla, the optimum spacing is about 20 nm.

Clearly, in the production of Ti-Nb alloy wire, it is important to begin with chemically homogeneous ingots, which for Ti-Nb alloys is challenging because of the difference in melting temperature of the two elements. While the processing of the Nb-47Ti alloy superconducting strands requires a great deal of care, the processing of this alloy is possible because it is quite ductile in the all β condition and even is relatively ductile after forming the α phase precipitates. In part, this is because the precipitate volume fraction is only about 25%. This practical benefit has made Nb-47Ti the alloy of choice for superconducting magnet applications. By comparison, there is an intermetallic compound, Nb_3Sn, that actually has higher critical current values, but it is so difficult to process into wire and wind into magnets that it has been of very limited practical use. Currently the ability to produce relatively compact 5-10 tesla superconducting magnets has enabled the development and production of medical imaging devices and particle accelerators that would not otherwise be possible.

Recently, there have been attempts to artificially introduce pinning centers into the superconducting material by mechanically assembling Ti-Nb inside annular Nb regions and drawing this assembly down to the dimensions of the desired superconducting strands. Materials made by this method have been designated Artificial Pinning Center (APC) materials. While some encouraging results have been obtained from laboratory produced materials, there also has been a number of disappointing results reported, presumably due to poor processing technique. It appears that production of the APC materials will require even more care than the Ti-Nb alloys produced by conventional processing routes. This additional care will, as a minimum, translate into higher cost.

There also are ongoing efforts to use tantalum to replace some or all of the Nb in the superconducting wire, but these are only laboratory scale studies. It is clear that the availability of superconducting electromagnets made from Nb-47Ti wire has enabled major instrumentation advances both in imaging and in particle accelerator technology.

There is a new class of high temperature superconductors based on complex oxides, one of which is Y-Ba-Cu-O. These superconductors have T_c values above 90K, which means they can be operated in liquid nitrogen instead of liquid helium. While these materials may someday supplant Ti-Nb superconductors, this seems unlikely until higher critical current and critical field values are achieved and the processing challenges posed by the brittle nature of the complex oxide are overcome.

Fig. 10.4. SEM photo of superconductor showing Nb-47Ti strands assembled in a Cu matrix (courtesy E. W. Collings, The Ohio State University)

Fig. 10.5. Higher magnification SEM photo of Fig. 10.4 showing individual Nb-47Ti wires (courtesy E. W. Collings, The Ohio State University)

10.2 Burn Resistance

When Ti combines with oxygen it creates a strong exothermic reaction. This reaction is not self sustaining in air at atmospheric pressure and oxygen partial pressures as is the case for Mg which readily burns in air, once ignited. However, under elevated temperature and pressure and under high mass flow conditions, such as those which exist in the compressor of a jet engine, titanium ignition can be self sustaining. There have been numerous examples of titanium fires in jet engines over the course of the past 20 years. The causative event is typically a hard rub of a rotating compressor airfoil on a titanium static member (vane or casing) or the impact of a liberated compressor airfoil against a titanium casing. Once ignited, the fire spreads to other titanium components and causes widespread damage to the engine. Figure 10.6 shows the compressor section of a military engine after a titanium fire. It can be seen that the casing has been completely breached by the fire [280]. In some cases the fire can even spread to the airframe leading to loss of the aircraft and significant danger to the crew. This is particularly true for military aircraft where the engine is contained inside the fuselage and in close proximity to the crew, the fuselage, and the wing carry-through structure. Titanium fires are of sufficient concern that significant engine design changes have been made to minimize the risk of fire. Among these measures are elimination of the use of titanium compressor casings in favor of steel in many engines or coating the titanium casings with Viton™, which is an adherent elastomer that minimizes the exposure of fresh titanium to the high pressure, high flow rate gas stream in the compressor.

In recent years, there also have been several programs to develop new alloys that are less prone to titanium fires. These efforts have focused on identifying alloy compositions that are less susceptible to burning. In conjunction with these programs there have been development projects to identify laboratory test methods that permit the qualitative ranking of alloys with respect to their propensity to burn. This section will describe the phenomenology of titanium fires and then discuss the physical metallurgy rationale for developing so-called burn resistant titanium alloys. In this regard it is important to recognize that all metals will burn under the proper conditions of pressure, flow rate, and temperature, so burn resistant is at best a relative term.

Fig. 10.6. Photo of a military aircraft engine compressor section after a titanium fire has breached the casing [280]

10.2.1 Phenomenology of Titanium Fires

The process of titanium ignition and subsequent damage creation in an aircraft engine can be thought of as occurring in four stages.

Stage 1: The first stage is the ignition of the titanium. This is defined as sustained combustion which depends on a combination of the following factors: the local amount of fuel (titanium); the rate and level of energy input from external sources; the enthalpy of the oxidation reaction; the occurrence of local melting which absorbs energy; the availability of oxidizer (in this case high pressure air); any solid/liquid chemical reactions that may occur; and geometry that alters the surface to volume ratio of the titanium "fuel".

Stage 2: The second stage is spreading of the titanium fire to other downstream, combustible components. This depends on a number of factors including the following: the amount and temperature of combustion products; the presence of other, downstream combustible components (e.g. titanium); the volume normalized heat content of the combustion products; the oxygen supply; the possibility of "thermite" reactions; and the travel distance of the combustion products.

Stage 3: This stage involves liberation of the combustion products from the containment vessel (usually the compressor case). The factors that determine the ease or difficulty of occurrence of stage 3 include: the containment vessel material characteristics such as heat capacity, thermal conductivity, melting temperature, and strength as a function of temperature. In addition, all of the factors that are important in stage 2 are also important here. Finally, since the incidence of stage 3 depends on rupture of the containment vessel, the pressure inside the vessel is important.

Stage 4: This stage involves secondary damage to other components as the result of a titanium fire. In this case, the factors that matter include the relative location of the critical components; the volume normalized heat content of the containment vessel; the heat capacity of the critical components including their contents; the combustibility of the critical components, and all the factors important to stage 2.

Recognizing these stages of a titanium fire and the factors that affect the propensity to incur such an event including its severity allows this possibility to be minimized through design. Clearly, the intrinsic flammability of a given titanium alloy cannot be altered, but the probability of occurrence of a titanium fire can be reduced by utilizing designs that take the factors mentioned earlier into account. Alloy composition also affects the probability of a fire and its severity as will be addressed in the next section.

10.2.2 Alloy Selection for Fire Risk Mitigation

Over the past 10-12 years there have been several significant efforts devoted to development of fire resistant titanium alloys. There are now several proprietary alloys that appear to be less susceptible to fire than the existing alloys commonly used in jet engines. Included among these alloys would be Alloy C (Ti-35V-15Cr), BurTi (Ti-25V-15Cr-2Al-0.2C), and Ti40 (Ti-25V-15Cr-xSi). These alloys have been developed in the US, the UK, and China respectively. There are basically two tests to assess the burn resistance of a titanium alloy. The first (type 1) is to impinge a drop of molten metal on the alloy and observe the reaction. The second (type 2) is to ignite the end of a rod of the alloy under controlled conditions of oxygen fugacity, which can be flow rate, pressure, or oxygen partial pressure in a closed chamber. Here the measurement is the amount of titanium alloy rod consumed before the fire extinguishes. Based on the four stages discussed earlier, the first test is more closely related to stage 2 whereas the second test is related to stage 1.

An analysis [281] of the thermodynamics of titanium fires suggests the following reactions are important in understanding the relation between alloy composition and flammability effects.

$$\mathrm{Ti\ (l) + O_2\ (g) = TiO_2\ (l)}$$

$$\mathrm{4/3\ Al\ (l) + O_2\ (g) = 2/3\ Al_2O_3\ (s)}$$

$$\mathrm{4/3\ V\ (l) + O_2\ (g) = 2/3\ V_2O_3\ (s)}$$

$$\mathrm{4/3\ Cr\ (l) + O_2\ (g) = 2/3\ Cr_2O_3\ (s)}$$

Here g, l, and s denote gas, liquid, and solid, respectively. The free energy for each of these reactions, stated in kJ/mol, is −517, −632, −410, and −360, respectively. Therefore, the composition trends toward less titanium, no Al, and more Cr or V are consistent with reduced heat release during local ignition in stage 1. Test of ignition conditions in a type 2 test show that the pressure and temperature for ignition shift to higher values when Ti-6Al-4V is compared to a burn resistant alloy. A specific comparison is shown for Alloy C and Ti-6Al-4V in Fig. 10.7 [64, 282]. Thus a compressor case made of a burn resistant titanium alloy would reduce the risk of fire compared to a Ti-6Al-4V casing or reduce the engine weight if it were substituted for a steel casing. Both of these results are attractive but the associated cost of the burn resistant alloys is quite high.

Burn resistant titanium alloys are not widely used, mainly because of their cost. The decision to incorporate a burn resistant alloy into an engine depends on the design philosophy and the extent to which the weight reduction is needed. It is more likely that this type of alloys will see application in military engines. This is because these engines are often more weight critical and the trade of cost for weight is typically more favorable [283].

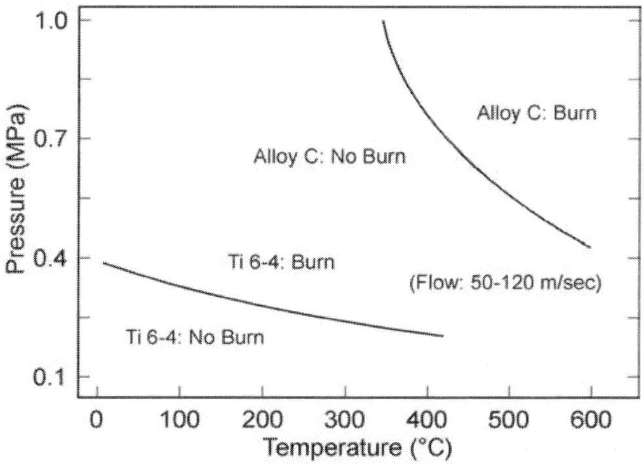

Fig. 10.7. Diagram showing the improvement in temperature pressure combinations for combustion in a burn resistant alloy compared to Ti-6Al-4V

10.3 Hydrogen Storage

Titanium has an affinity for absorbing hydrogen as was discussed earlier. In α alloys, this affinity is manifested as a relatively low solubility and the accompanying formation of the titanium hydride phase, TiH_2. This makes α alloys largely unsuitable for hydrogen storage because it limits the amount of hydrogen that can be stored per unit volume of titanium and the uptake and evolution of hydrogen is more difficult than for materials where the hydrogen is present in solid solution. Nevertheless, titanium based hydrogen storage systems remain attractive because of the thermodynamic feasibility of using titanium as a storage host and because titanium also is relatively lightweight. Consequently, there has been a modest continuing effort to develop a titanium based hydrogen storage medium and, until recently, these efforts have been focused on the intermetallic compounds TiFe and TiMn. These compounds have an additional advantage of being very brittle and therefore easy to crush. Since the rate of uptake and release of hydrogen is the combined result of a surface reaction and bulk diffusion, granular media are attractive because they have a high ratio of surface to volume and the maximum diffusion distances are limited to roughly half the maximum particle size.

Hydrogen powered vehicles are environmentally attractive because they are zero emission entities. In fact, there has been a successful proof of concept demonstration using a few hydrogen powered vehicles for residential mail delivery by the United States Post Office in the early 1980's timeframe. These vehicles burned hydrogen as fuel in an internal combustion engine without the benefit of a great deal of optimization. Unfortunately, economic reality set in after this demonstration and interest in this technology has waned considerably. This reality is strongly influenced by the relatively low cost of oil and currently by the minimal economic penalty for emissions of No_x and CO_2 from autos and trucks. Also, the cost of even a minimal hydrogen distribution system would be quite high. There is renewed interest in hydrogen as a fuel because of the resurgence of activity around devices called fuel cells that make electricity by combining oxygen and hydrogen. Experimental fuel cell powered vehicles are currently in operation but economic barriers to their broad acceptance are still quite high. These vehicles are electric powered but require no gasoline engine as is the case for the hybrid vehicles currently in production. A viable fuel cell based energy source requires readily available hydrogen. In principle this hydrogen could be derived from gasoline or other hydrocarbon fuels such as methanol or diesel. In practice, extraction of hydrogen from a hydrocarbon source requires a catalytic re-forming device that cracks the hydrocarbon molecules to release the hydrogen and other byproducts. Here again, this re-forming process has been demonstrated but the catalysts involved typically contain significant concentrations of noble metals, mainly Pt and Pd. These catalysts have obvious associated cost issues. Work to find more affordable and effective catalysts is continuing, but no practical solution is in sight at present. Consequently, broadbased implementation of fuel cells as a significant source of electricity still appears to be a way off. However, fuel cells are compact, efficient, and emission free. They have long been used in spacecraft because of their compactness and simplicity. There may be other niche markets for fuel cells. If this possibility materializes, a safe, reliable (if not affordable) source of hydrogen will be required. Titanium based hydrogen devices may well have appeal in this case. Meanwhile, as the market economics for hydrogen as a fuel source sorts itself out, there are important technical issues to be addressed regarding hydrogen storage. Consequently, there is basic research being conducted as witnessed by a collection of hydrogen storage articles in a recent journal [284-286]. These recent studies have concentrated on the Ti-V-Cr-Al class of alloys. They are mainly concerned with the quantity of hydrogen that can be absorbed and the kinetics of absorption and desorption as a function of the host composition, processing history, and service history, meaning the number of times the host has been hydrogen charged and subsequently depleted. One of the findings is that the capacity to absorb hydrogen changes after a number of use cycles. This has been related to a change in microstructure of the host but many questions remain unanswered. Considering the well-known effect of hydrogen on the structure of conventional titanium alloys such as Ti-6Al-4V, this is not surprising. Considering the cost of a titanium based hydrogen storage device, this life limiting problem must be solved before any significant market acceptance can be expected. Nevertheless, titanium based hydrogen storage remains an attractive option, at least until the economical catalyst question is answered.

10.4 Shape Memory Effect

Shape memory alloys have the unique characteristic of being able to recover their initial geometric shape after they have been plastically deformed. This recovery after deformation at a lower temperature occurs on reheating to a few hundred degrees Celsius. This reversibility of geometric shape is referred to as the Shape Memory Effect (SME). Alloys based on the nominal composition of the compound TiNi have been recognized to exhibit the shape memory effect for many years. One of the original shape memory alloys dates back to 1950 and was based on AuCd [287]. The temperature range over which this strain reversal is possible is very sensitive to the alloy composition but SME occurring from –150°C to +150°C is readily achievable [288]. Under the right conditions of composition and temperature, this class of materials also exhibits the capability to deform elastically over large strain ranges. This latter characteristic is known as superelastic behavior. At present, the TiNi class of SME alloys is among the most widely used. For this reason, a short discussion of shape memory alloys and applications of titanium based shape memory alloys is included in this chapter. The discussion begins with a brief description of the mechanism that causes the shape memory effect in TiNi and concludes with a few examples of shape memory alloy applications.

10.4.1 Phenomenology of the Shape Memory Effect

The compound TiNi has a bcc B2 (CsCl prototype) structure which transforms martensitically to a monoclinic B19 (AuCd prototype) structure when cooled below the martensite start (M_s) temperature, as shown schematically in Fig. 10.8. The martensite is internally twinned. The twins represent the mode of lattice invariant deformation using the standard terminology from the crystallographic theory of martensite. Part B of Fig. 10.9 shows these twins. Upon deformation, the martensite untwins, as shown in part C of Fig. 10.9, leaving a shape change equivalent to the volume fraction of twins multiplied by the twinning shear. In

TiNi this reversible shape change can be equivalent to as much as an 8% shear strain. If the material containing the deformed (untwinned) martensite is reheated to above the austenite start temperature (A_s) shown in Fig. 10.8, the martensite becomes unstable and reverts to the parent B2 phase. When this occurs, the original shape of the starting piece is recovered, as shown in part A of Fig. 10.9. This reversible martensitic phase tranformation is the basis for the shape memory effect, not only in compounds based on TiNi but also in a variety of other alloys based on Cu, Au, Ag, and Fe. The shape memory effect is described in much greater detail elsewhere [289].

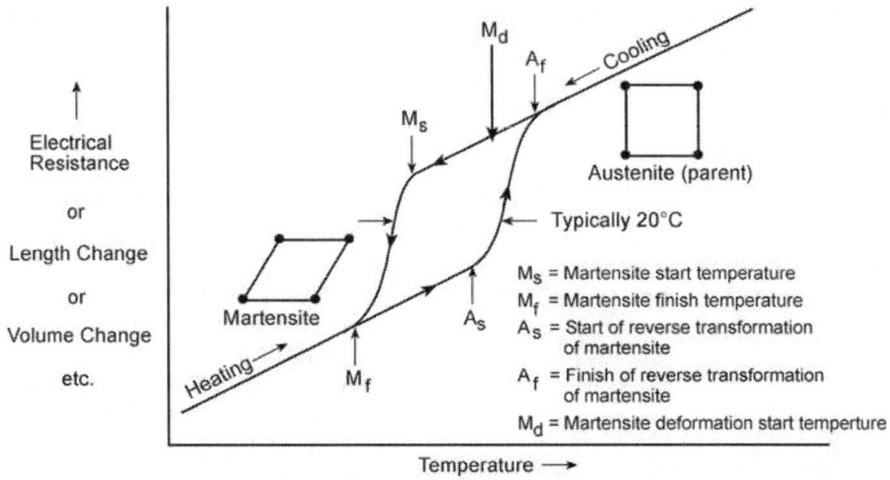

Fig. 10.8. Schematic diagram showing the transformation sequence of a TiNi alloy during cooling and reheating; the martensite deformation start temperature is also shown

The parent B2 phase in TiNi alloys is mechanically unstable at temperatures below a critical temperature known as the martensite deformation start temperature, M_d. If a TiNi shape memory alloy is deformed above A_s but below M_d, the material transforms to deformation induced martensite under stress but this martensite reverts to the parent phase (austenite) when the load is removed. This permits very large, fully recoverable strains to be realized. Because the strain is fully recoverable, the material behaves as if it were fully elastic but with a very low value of modulus of elasticity. For this reason, such behavior is called superelastic. The phenomenon of superelasticity is described in greater detail in several other articles [288-290]. Here it is sufficient to mention that this effect gives shape memory alloys a unique capability that can have engineering importance as discussed later.

There also is a martensite precursor effect in TiNi known as the R-phase transition. This is distinct from the martensite transformation and can account for maximum reversible strains under shape memory conditions of about 1%. The R-phase transition also leads to superelasticity. The R-phase transition has been discussed in detail for TiNi based alloys by Otsuka [291]. In general the larger recoverable strains associated with deformation induced martensite are of greater interest, so the R-phase transition will not be discussed further here.

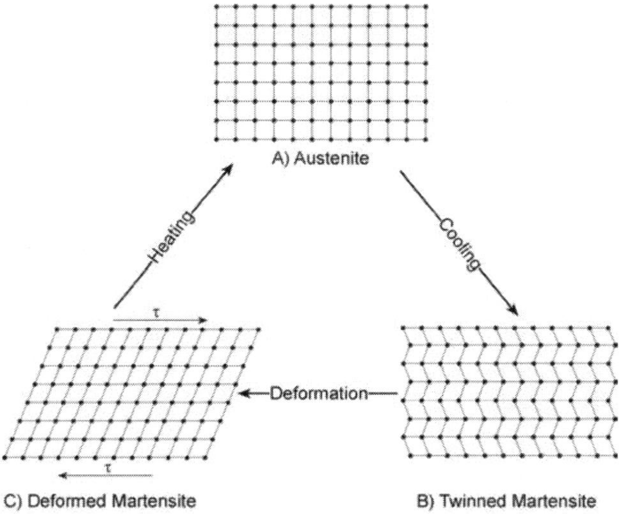

Fig. 10.9. Diagram showing the transformation from austenite to martensite and the shape change that occurs when the martensite untwins during deformation

The strain reversibility associated with the mechanical and thermal cycles depicted in Figs. 10.8 and 10.9 operates on a one time basis and is thus called "one-way" shape memory effect. If the material is processed in a special way, only some of the possible variants of the orientation relation between austenite and martensite form during deformation. In this case, formation of martensite also leads to a shape change. Clearly in these cases, the shape change becomes reversible during thermal cycling above and below M_s and A_s. This shape reversibility during thermal cycling is called a "two-way" shape memory effect. It has been shown that this capability can be incorporated into shape memory alloys using special thermomechanical treatments called "training treatments".

In summary, the shape memory effect in TiNi based alloys is the result of a reversible martensitic transformation coupled with a detwinning of the martensite during deformation. The shape memory effect in this system is possible because of several important metallurgical characteristics associated with this transformation. Included are a martensitic product that is internally twinned, a transformation that has a small accompanying volume change so as to exhibit a minimal temperature hysteresis, and a transformation temperature range that both is of practical interest and is low enough that competing decomposition reactions of the martensite or austenite are nonexistent or very sluggish.

The previous discussion of the basic behavior of shape memory alloys suggests that this effect can be beneficially used in a variety of devices, provided the temperature range over which the shape memory effect occurs is of practical interest. To broaden this temperature range, considerable effort has been expended in understanding the effect of exact alloy composition (Ti to Ni ratio, as well as interstitial impurity content) on the M_s, M_d, M_f, A_s and A_f temperatures. Similar studies of the effect of ternary additions of Cu, Pd, Pt, and Fe to TiNi on these critical temperatures also have been conducted. To this point, it has been shown that alloys with >50 at% Ni have a very strong M_s composition dependence [288]. This is shown in Fig. 10.10. This makes the production of commercial quantities of TiNi alloys with reproducible M_s temperature values difficult and potentially uneconomical. Moreover, Ni rich alloys are unstable when aged at temperatures low enough that could be encountered in service (e.g. 400°C). Long term exposure causes the formation of precipitates with attendant large increase in A_s temperature. This is shown in Fig. 10.11. Such variations in material characteristics can render devices based on the shape memory effect unserviceable.

The intent of this short summary is not to provide an exhaustive description of SME alloy compositions and their characteristics. Rather it is to call attention to this class of titanium based materials and to illustrate some of the aspects of this novel materials behavior.

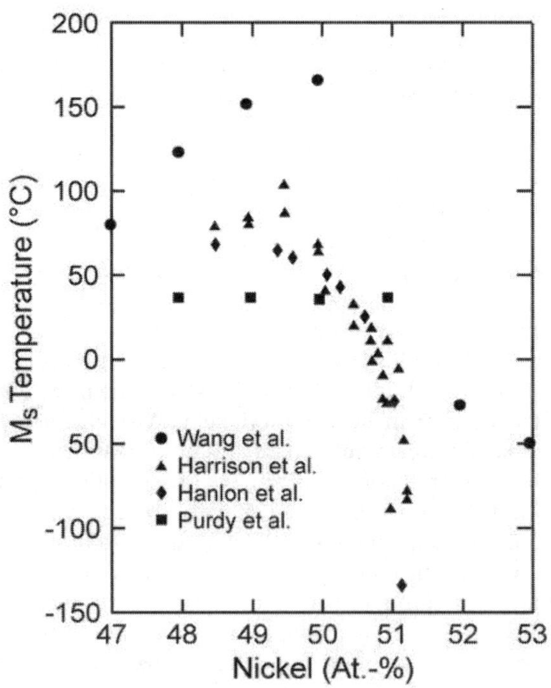

Fig. 10.10. Plot of the composition dependence of the M_s temperature in TiNi

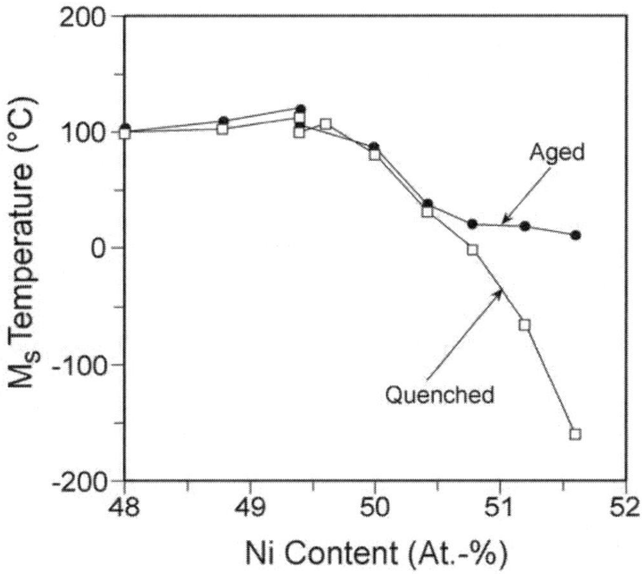

Fig. 10.11. Plot of the effect of aging on the M_s temperature of Ni rich TiNi alloys

10.4.2 Applications of Shape Memory Alloys

The foregoing discussion suggests that there are four distinct application areas or types for shape memory alloys. These are determined by the way the shape memory effect is utilized and include the following:

- Free recovery, where it is beneficial for an object to be capable of having its original shape restored by reheating to moderate temperatures. Basically this application derives its benefit from the motion associated with the shape recovery. Perhaps the most practical application of the SME is in TiNi alloys used for stents which are deployed in the human body to maintain flow in blood vessels. One such stent is shown in Fig. 10.12. This photo shows the TiNi mesh stent partially deployed. Here the stent is collapsed at low temperature but has an A_s temperature somewhat below body temperature so that it expands after it is placed in the body. Another example is the use of TiNi alloys in eyeglass frames allowing them to be "self-straightening" when inadvertently bent. Another is the use of TiNi SME alloy wires in women's undergarments to hold a particular shape which can be recovered after laundering, for example. These applications are examples where the "one-way" shape memory effect is useful.
- Constrained recovery, where a force is applied during recovery. This application benefits from the force generated when the recovery takes place. Typically this application uses the "one-way" shape memory effect. The use of TiNi alloys in tubing couplings is one example of constrained recovery. The relatively

high cost of TiNi alloys has limited the use of these couplings, but they represent a good conceptual example of an application where constrained recovery allows a desired functionality to be achieved which could not be accomplished in the absence of the SME.

- Another application type for shape memory alloys is in actuators that perform work. Temperature variations are required for the actuator to be effective. In this case the "two-way" shape memory effect is the more efficient means of obtaining the desired reversible motion, but "one-way" devices are in service which utilize a constant force bias spring to permit strain reversal during temperature cycling.
- Superelastic devices are used mainly for energy storage and to obtain large recoverable motions in devices. This application is limited to temperature regimes that lie between M_d and A_s but within this range, elastic displacements can be realized that are up to 15 times greater than from a comparable steel spring. Superelastic shape memory alloys also have the best fatigue capability in displacement limited applications. This is because the low effective stiffness allows large elastic strains to be accommodated at quite low stresses. A familiar use of supereleastic shape memory alloys is in the antennae of mobile telephones as shown in Fig. 10.13.

The use of shape memory alloys is growing as the special characteristics become more widely recognized. The cost of shape memory alloys will always be higher than ordinary metals because of the care required to maintain tight composition control and, in the case of "two-way" shape memory alloys, to perform the special processing needed. It is perhaps fortunate, in a sense, that shape memory alloys are used in limited quantities because producing them in large quantities, i.e. with large ingots, would be extremely challenging.

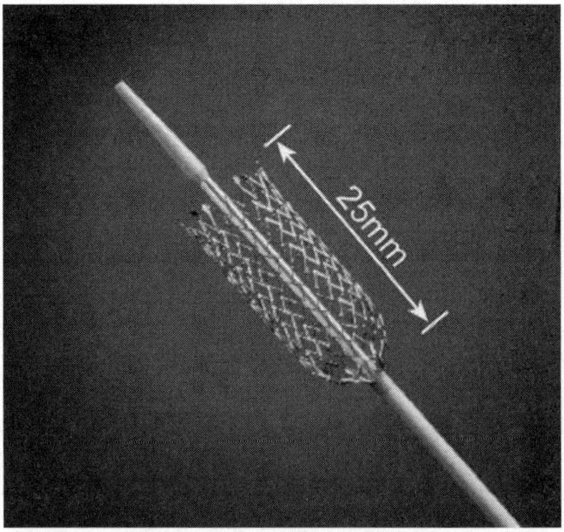

Fig. 10.12. Photo of a TiNi stent partially deployed from the cassette used to insert it

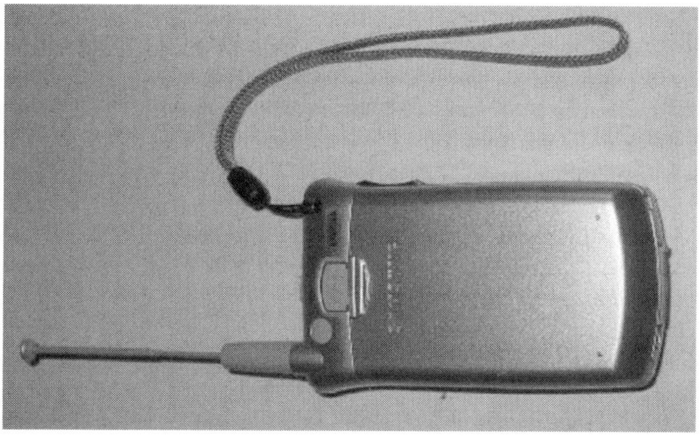

Fig. 10.13. Cellular telephone showing superelastic TiNi shape memory alloy autenna

10.5 Biomedical Applications

The use of titanium in the biomedical field has become a well established area because titanium fulfills the property requirements better than any competing material (stainless steels, CoCr-alloys, CP niobium, and CP tantalum) [292]. The properties which are of interest for biomedical applications are corrosion resistance, biocompatibility, bioadhesion (bone ingrowth), modulus of elasticity (should be as close as possible to the modulus of bone which is in the range of 10-30 GPa), fatigue strength, and good processibility including joining and casting. Especially the excellent corrosion resistance and biocompatibility make titanium the material of choice. Even the price of titanium, which is normally the drawback for its application, is only slightly higher as compared to the CoCr-alloys and to the types of stainless steel used. Compared to CP niobium and CP tantalum the price of titanium is much lower.

There are numerous different medical devices using titanium materials, e.g. bone plates, screws, hip joint implants, stents, heart valves, and various other kinds of fixtures also used in the dental area. The scope of this book permits only a few typical examples to be chosen to illustrate the materials selection criteria. For a broader description of the application of titanium in the biomedical area including all medical aspects, the reader is referred to a recent book on "Titanium in Medicine" [293].

Three groups of titanium (CP titanium, α+β titanium alloys, and β titanium alloys) are used in the biomedical area. In addition, there are devices that use shape memory alloys based on the compound TiNi. Traditionally, CP titanium and the Ti-6Al-4V alloy were the first titanium materials used for biomedical applications and even today these two materials are used in most applications. Because of the suspicion of long term toxicity problems with vanadium, the α+β alloys Ti-5Al-

2.5Fe and Ti-6Al-7Nb were developed in the 1980's. These two alloys have similar microstructures and properties as the Ti-6Al-4V alloy. Then in the 1990's, a variety of β titanium alloys were developed, mainly because of the higher fatigue strength and the lower modulus of elasticity as compared to α+β titanium alloys. These two properties are important for applications such as hip joint implants as will be discussed later.

The first example illustrates the use of titanium as a bone plate (Fig. 10.14). The two X-ray radiographs in the figure show the broken bone before and after attaching the bone plate. The schematic drawing in Fig. 10.14 illustrates more clearly the actual shape of the bone plate.

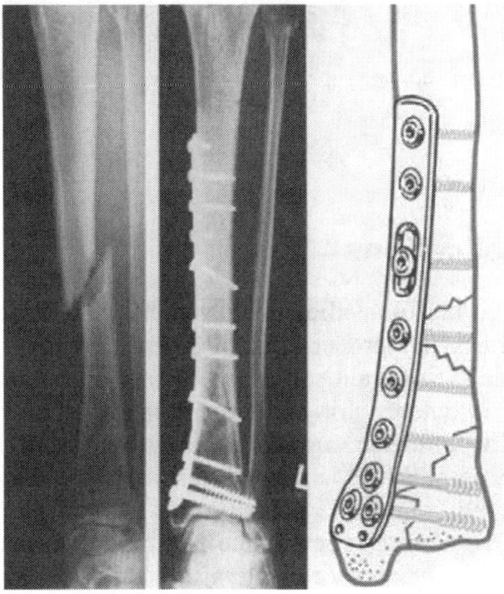

Fig. 10.14. Example of a bone plate implant, CP titanium grade 3 (courtesy Waldemar Link GmbH & Co)

For bone plates for which the strength requirement is not very high, the material of choice is CP titanium in the various grades 1 to 4 depending on the complexity in shape and the individual strength requirement. In most of these applications CP titanium grade 3 is the best compromise between very good formability and adequate fatigue strength. A typical HCF strength value at $R = -1$ is 280 MPa (see Table 4.3). CP titanium grade 3 allows easy manufacturing of more complex shaped bone plates. For example, for the close support of the backbone, a double curvature in the plates is needed and this contour can be readily obtained in CP titanium grade 3 by hot pressing in a contoured die below 600°C. Higher temperatures should be avoided because of the possibility of grain growth especially if a material with low Fe content is used (see Fig. 4.6, Sect. 4.1.1). If higher strength is required for bone plates, then either the Ti-6Al-4V alloy or the Ti-6Al-7Nb alloy is used [294]. Both alloys are used only for fairly straight plates because of their limited capability for contour creation. In most cases, the hot formed parts have to be machined after forming which increases the final product cost.

The most demanding application for biomedical implants requiring a high fatigue strength is the stem of a hip joint implant. A schematic of a complete artificial hip joint is shown in Fig. 10.15. The titanium alloy stem has a ceramic head (usually Al_2O_3 or ZrO_2) which can rotate in a cup made out of ultrahigh molecular weight polyethylene (UHMWPE). This combination of materials has a very low friction coefficient. The UHMWPE cup is usually held by a snap-in mechanism in a metal backed shell (not shown in Fig. 10.15) which in turn is fixed to the bone by screws. Both the metal backed shell and the screws also are made out of titanium (Ti-6Al-4V or CP titanium) [296]. The most common titanium alloy used for the hip joint stem is Ti-6Al-4V. The stem blanks are usually $\alpha+\beta$ forged and stress relieved, therefore the microstructure is either mill-annealed or fully equiaxed depending on the details of the processing route (see Sect. 5.1.3). Typical HCF strength values at $R = -1$ for these microstructures in the stress relieved condition are around 400 MPa (see Fig. 5.36).

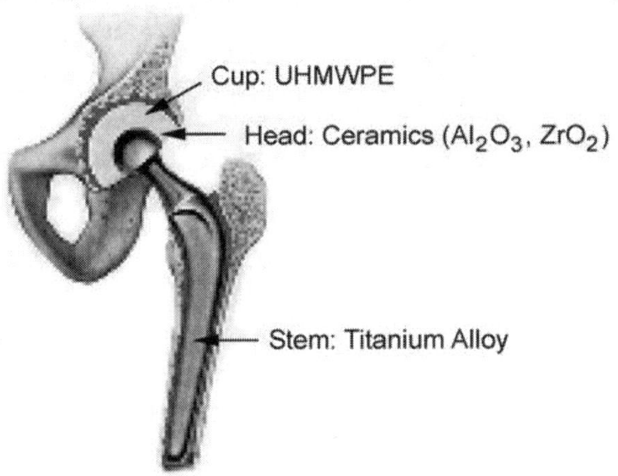

Fig. 10.15. Schematic of an artificial hip joint (courtesy M. Niinomi, Toyohashi University of Technology)

To facilitate better bioadhesion (bone ingrowth) the surface of the finished stem is very important. In general, the bone ingrowth behavior improves with increasing roughness of the stem surface. Various surface finishing methods are used (machining, etching, blasting with Al_2O_2, porous coatings, plasma spraying, etc.) [295-297]. One very popular surface treatment is plasma spraying of hydroxyapatite (main component of bone tissue) as bioactive coating on the titanium implant.

As an alternative manufacturing method to forging, investment casting of the hip stem can be used. In this case, the desired surface roughness of the stem can be directly obtained by designing it into the casting. Such a cast hip stem is shown in Fig. 10.16 [298]. The main disadvantage is that the fatigue strength of investment cast parts of $\alpha+\beta$ titanium alloys is usually lower than that of forged parts because of the fairly coarse fully lamellar microstructure typical of castings (see Chap. 5, Sect. 5.2.3). It is possible to create a so-called bi-lamellar microstructure (Fig. 5.44) in castings by a simple heat treatment. The increase in HCF strength (see Sect. 5.2.5) associated with this microstructure makes the usage of investment cast hip stems attractive. The S-N curves of specimens prepared from forged and cast hip stems are compared in Fig. 10.17. It can be seen that the forged hip stem has a higher HCF strength than the cast hip stem with the fully lamellar microstrucutre whereas the cast hip stem with the bi-lamellar microstructure even has a higher HCF strength than the forged material.

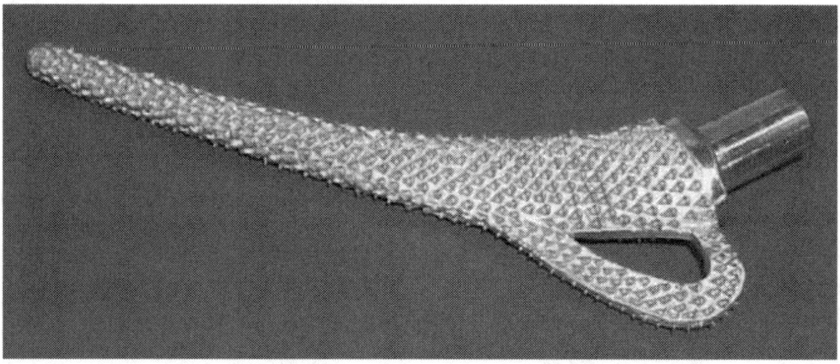

Fig. 10.16. Investment cast hip joint stem, Ti-6Al-4V

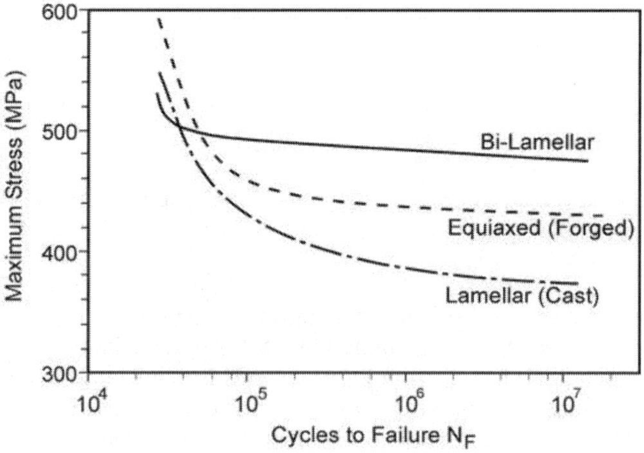

Fig. 10.17. S-N curves (R = –1) of specimens taken from forged and investment cast hip joint stems, bi-lamellar: special heat treatment of investment cast stems, Ti-6Al-4V

Besides Ti-6Al-4V, the newer α+β titanium alloy Ti-6Al-7Nb mentioned earlier is in commercial use for the hip stem [295]. As mentioned above, the microstructure as well as the mechanical properties of Ti-6Al-7Nb are similar to Ti-6Al-4V. A summary of processing, microstructure, and properties of the Ti-6Al-7Nb alloy and a summary of the experience with this alloy as joint replacements can be found elsewhere [299].

In the last 10 years, there has been a large effort to formulate new β titanium alloys as implant material using biocompatible alloying elements (Nb, Ta, Zr, Mo). The major performance advantage of β alloys are the higher strength (see Sect. 7.2) and the lower modulus of elasticity (see Sect. 2.3) as compared to α+β alloys. Another advantage for manufacturing is that investment castings of heavily stabilized β alloys have nearly identical properties compared to β annealed wrought products (see Sect. 7.3).

Howmedica developed the β alloy Ti-12Mo-6Zr-2Fe (TMZF) [300] and recommended the use of this alloy in the β annealed condition without any final aging treatment. This condition has the major advantage of a very low modulus of elasticity (74-85 GPa). If desired, the modulus value could be increased by aging (α phase precipitation) as discussed in Sect. 2.3 (see also Fig. 2.5). The disadvantage of the unaged condition is the lower yield stress with the attendant reduction in fatigue strength.

TIMET modified the Beta 21S alloy (Ti-15Mo-2.7Nb-3Al-0.2Si) by eliminating the 3% Al because aluminum is thought to be the only potentially harmful constituent. This modified Beta 21S alloy is called Beta 21SRx [301]. The oxygen content of Beta 21SRx was increased from 0.10-0.15% to 0.25-0.30% to obtain the same degree of α phase stabilization as in the original Beta 21S alloy. TIMET published mechanical property data for the unaged β annealed condition and after various final aging treatments. The modulus of elasticity varied between 83 and 94 GPa depending on heat treatment [301].

Other new β alloys for implants include Ti-15Mo-5Zr-3Al [302] and the Japanese alloys Ti-29Nb-13Ta-4.6Zr and Ti-29Nb-13Ta-4Mo [25, 303]. The microstructure and properties of these alloys are similar to the alloys described earlier.

The application of shape memory alloys based on the compound TiNi for stents has been already described in Sect. 10.4, see also Fig. 10.12.

10.6 Automotive Applications

Titanium is well suited for use in many components of a passenger car from a property standpoint (Fig. 10.18). But, as pointed out earlier in the introduction chapter of this book (Sect. 1.5), the high price of titanium has prevented realization of these applications in mass-produced passenger cars [304]. The different parts indicated in Fig. 10.18 can be assigned to different titanium alloy groups. CP titanium sheets would be the best candidate for the exhaust system because of the excellent formability and weldability, Ti-6Al-4V could be used for connecting rods because of high fatigue strength and adequate temperature capability, β titanium alloys are the best choice for suspension springs because of high strength and low modulus of elasticity, titanium based intermetallics could be used for valves as well as for turbocharger rotors because they combine high temperature capability and low density, even MMCs could find some applications (see for example Fig. 3.37 showing a particle reinforced titanium connecting rod fabricated by powder metallurgy).

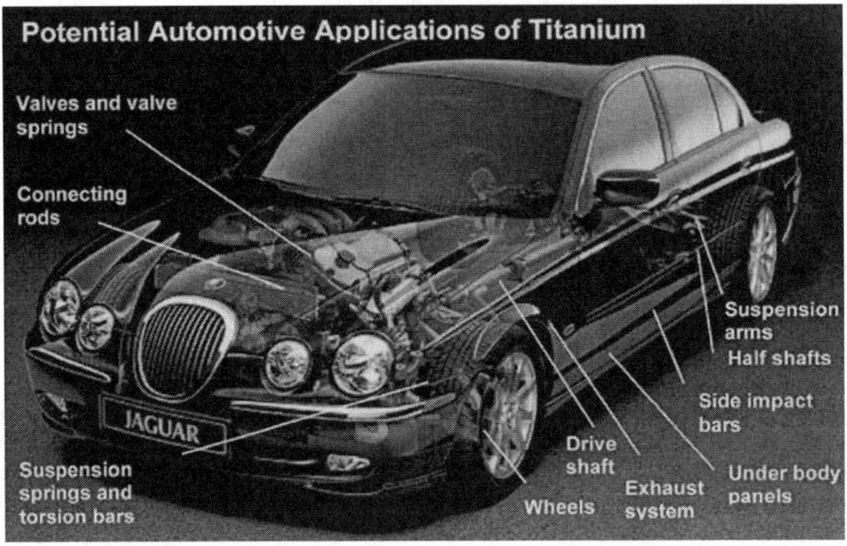

Fig. 10.18. Potential automotive applications of titanium demonstrated on a Jaguar S-type© car (courtesy Jaguar Cars Limited)

Recently, for the first time, a conventional titanium alloy was used in a mass-produced passenger car. Volkswagen equipped the latest Lupo FSI model with rear suspension springs made out of the low cost β alloy Ti-LCB [305]. The titanium springs replaced steel springs, both of these are shown together in Fig. 10.19. It can be seen that the titanium spring to the right has fewer windings than the steel spring on the left. This is possible because of the low modulus of elasticity of the titanium material combined with a high yield stress. The actual weight reduction using the Ti-LCB spring in the Lupo was about 40% as compared to the steel spring. The potential of Ti-LCB as spring material for mass-produced passenger cars has been previously pointed out [306]. In the same article, it was also pointed out that for the exhaust system CP titanium grade 2 would be the material of choice considering the whole manufacturing process. There are indications [307] that Volkswagen plans to use titanium exhaust system in the near future.

A few years ago, Mitsubishi introduced a titanium intermetallic component in a passenger car. The rotor head of the turbocharger rotor assembly in the Mitsubishi Lancer GSR model is a cast γ alloy. A photo of this rotor assembly is shown in Fig. 10.20. The rotor has an outside diameter of about 55 mm and is cast by Daido

Steel using the Levi cast method. Meanwhile, more than 10 000 cast γ turbocharger rotors are in service as was also pointed out in Sect. 8.2. The advantage of using γ material is the low weight of the rotor which minimizes the time required to spin up the turbocharger. This reduces the lag for increased power known in all turbocharged engines.

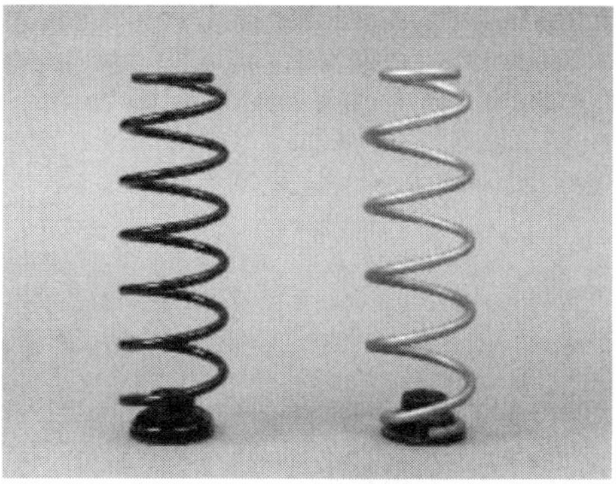

Fig. 10.19. Suspension springs: Ti-LCB spring on the right and steel spring on the left (courtesy O. Schauerte, Volkswagen)

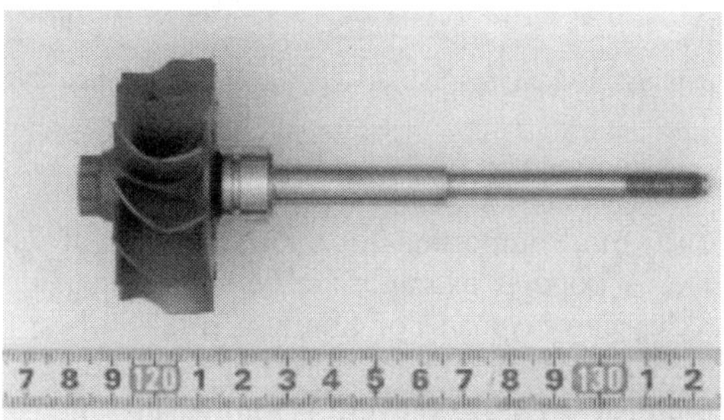

Fig. 10.20. Photo of a cast γ alloy turbocharger rotor installed in the Mitsubishi Lancer GSR model (courtesy Mitsubishi Automobile Co., Ltd.)

10.7 Sports Related Applications

As already pointed out in the introduction chapter (Sect. 1.5), there are numerous applications of titanium in the sporting equipment area [21-24]. The main reason for using titanium is the high strength to density ratio which allows lighter equipment to be made. In some cases, the low modulus of elasticity, the excellent corrosion resistance, and the elevated temperature capability of titanium are also important. Lighter sports equipment is an advantage for nearly all sports activities because the equipment is either moved by the athlete or the equipment is powered by an external source and is transporting the athlete. Popular examples for the last case are racing cars and racing motorcycles. For this group, the examples discussed in Sect. 10.6 for the application of titanium in mass-produced passenger cars also apply to racing cars and motorcycles. Since for racing cars the costs are not so dominant, it can be assumed that most of the parts shown in Fig. 10.18 and even more parts are made from titanium in racing cars. Detailed information is not publically available because it is perceived to create a competitive advantage. As example, an exhaust system for racing motorcycles using titanium tubes and sheets is shown in Fig. 10.21.

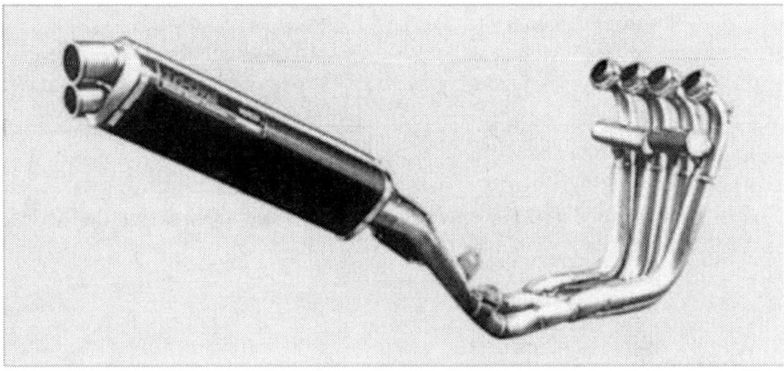

Fig. 10.21. Titanium exhaust system for racing motorcycles (courtesy Yoshimura Japan Co., Ltd.)

Low weight is even more important for sports equipment which is physically moved by the athlete. A popular example are bicycles. Nearly everything on expensive racing bicycles can be made of titanium (main tubes, handlebars, stem, axles, saddle rails, pedals, chainset, chain, and rim, hub, and spokes of the wheels, etc.) [22]. An example of a racing bicycle is shown in Fig. 10.22. In this case, even the whole gear shift mechanism is made from titanium. It should be emphasized that such titanium bicycles are far too expensive to be attractive for hobby bicycling or even for nonprofessional racing.

The same arguments apply for titanium racing wheelchairs used by some athletes at the Paralympic Games (Fig. 10.23). Such racing wheelchairs can cost up to 3000 US $ [308]. Conventional wheelchairs made from titanium also have become very popular because of low weight and good appearance. Further, scratches and chips are not a problem because they are not painted. These wheelchairs can be bought for as little as 1500 US $. The commonly used material is Ti-3Al-2.5V seamless tubing because it is strong and relatively easy to draw into tubing.

For sports that involve striking a ball, the hollow cast golf club head made from Ti-6Al-4V was already discussed in the introduction chapter (Sect. 1.5) and an example was shown in Fig. 1.9. Besides these Ti-6Al-4V cast heads, forged heads using various β alloys (Ti-15V-3Cr-3Al-3Sn, Ti-22V-4Al, Ti-20V-4Al-1Sn) have come on the market [309]. In both production routes (casting or forging) two pieces are produced and welded together. A powder metallurgical manufacturing route is used for softball bats. An example of such a bat is shown in Fig. 10.24. The material used is Ti-6Al-6V-2Sn offering higher annealed strength values than β annealed Ti-6Al-4V. The softball bat shown in Fig. 10.24 has a yield stress of 930 MPa and a tensile elongation of 12%.

Finally, a completely different type of sports equipment will be mentioned, namely a recently developed titanium air bottle for divers [310]. It has a weight advantage over similar bottles made from aluminum or steel. The air bottle can also serve for a complete different purpose, namely for firefighters [310].

Fig. 10.22. Titanium racing bicycle (courtesy The Japan Titanium Society)

Fig. 10.23. Titanium racing wheelchair (courtesy The Japan Titanium Society)

Fig. 10.24. Titanium softball bat made by PM method, Ti-6Al-6V-2Sn (courtesy S. Abkowitz, Dynamet Technologies, Inc.)

10.8 Appearance Related Applications

There are two different groups of appearance related applications. In the first group of applications, the natural gray-like appearance of the titanium metal is preserved. The preservation of the appealing titanium metal surface even after many years of outdoor exposure is due to the stable very thin titanium oxide surface layer which forms immediately upon exposure of the bare metal to air, for example after scratching. This stable oxide layer is the reason for the excellent corrosion resistance of titanium in many environments (see Sect. 2.9.2). In the second group of applications, the color of the titanium surface is deliberately changed by anodizing. This second group is a low volume application because it is only used for jewelry and decorative arts.

The application with the largest volume is the use of titanium for roofs, domes, and exterior walls of prestige buildings and monuments. A well-known example for using titanium in monuments is the famous Gagarin monument in Moscow built in 1961 in honor of the first cosmonaut Yuri Gagarin. A photo of this monument is contained in an overview article [311]. Using titanium for roofs and domes of prestige buildings has become very popular in Japan starting in 1973 [309]. One of the outstanding examples, the Fukuoka Dome, was shown in Fig. 1.8. Since the roof of this multi-role arena is retractable, this application also makes effective use of the high specific strength of titanium to create a lightweight structure, even for CP grades. The retractable roof used 48 500 m^2 of 0.3 mm thick CP titanium sheet [309]. There are many other examples for buildings in Japan [20, 309, 312, 313]. The annual quantity of titanium used for architectural purposes in Japan reached about 1300 metric tons in 1999 [313]. Examples outside Japan are the famous Guggenheim Museum in Bilbao, Spain [21] and the recently built (year 2000) Glasgow Science Centre using 6000 m^2 of 0.3 mm thick CP titanium sheet [310].

Other appearance related examples, using titanium in the natural gray-like color, are wrist watches and cameras. CP titanium grade 2 is mostly used for these applications. To prevent scratches and finger prints clear inorganic coatings are applied [309].

Very popular are also jewelry items using the natural gray-like appearance of titanium. Rings, ornaments, necklaces, bangles, earrings, etc. are on the market. A few examples are shown in Fig. 10.25.

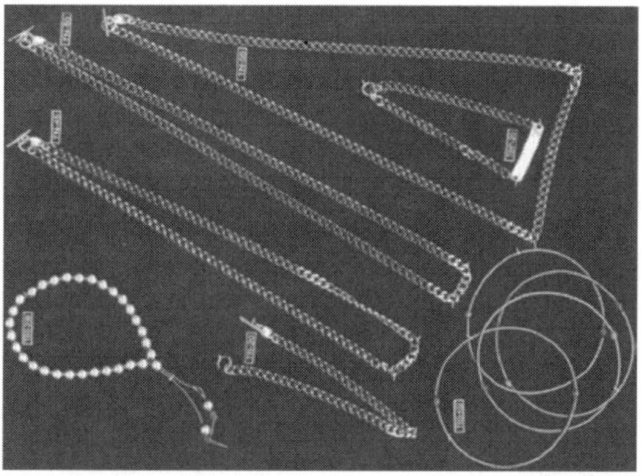

Fig. 10.25. Jewelry items made of titanium (courtesy Metal Surface Engineering-Horie Company, Japan)

In the area of jewelry and also in the area of arts the gray-like appearance of the titanium surface is selectively changed by growing oxide layers of varying thickness through anodizing. Anodizing under controlled conditions creates an oxide surface layer which imparts some very attractive colors. The relationship between applied voltage in the anodizing process, thickness of the oxide surface layer, and the resulting color of titanium is shown in Fig. 10.26 [314]. It can be seen that colors ranging from gold (thin oxide layer) over blue and green to pink (thick oxide layer) can be created. One example of decorative art is shown in Fig. 10.27. In the original piece of titanium art, the colors are quite striking and change from blue in the background to yellow, purple, and natural gray-like in the fish ornaments. There has been a growing public interest in this titanium based art and a growing acceptance.

It is fairly difficult to sell ordinary kitchen-ware made out of titanium even if advertised in combination with light weight arguments for camping and other carrying purpose. But in combination with decorative arts, "kichenware" items might be easier to sell. Some attractive examples (camping cutlery, ice cream spoons) created by an artist of the School of Art and Design at the University of Derby (UK) were illustrated in a recent publication [310].

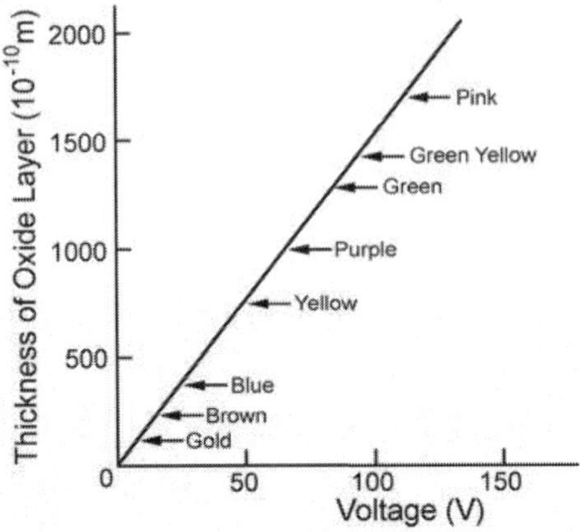

Fig. 10.26. Relationship between applied voltage in anodizing process, thickness of titanium oxide layer, and color of titanium [314]

Fig. 10.27. Example of decorative art using colored surface appearance of titanium (courtesy Metal Surface Engineering-Horie Company, Japan)

References

1. Jaffee R. I., Promisel N. E., eds.: *The Science, Technology and Application of Titanium,* Pergamon Press, Oxford, UK, (1970)
2. Jaffee R. I., Burte H. M., eds.: *Titanium Science and Technology*, Plenum Press, New York, USA, (1973)
3. Williams J. C., Belov A. F., eds.: *Titanium and Titanium Alloys*, Plenum Press, New York, USA, (1982)
4. Kimura H., Izumi O., eds.: *Titanium '80, Science and Technology*, AIME, Warrendale, USA, (1980)
5. Lütjering G., Zwicker U., Bunk W., eds.: *Titanium, Science and Technology*, DGM, Oberursel, Germany, (1985)
6. Lacombe P., Tricot R., Beranger G., eds.: *Sixth World Conference on Titanium*, Les Editions de Physique, Les Ulis, France, (1988)
7. Froes F. H., Caplan I. L., eds.: *Titanium '92, Science and Technology*, TMS, Warrendale, USA, (1993)
8. Blenkinsop P. A., Evans W. J., Flower H. M., eds.: *Titanium '95, Science and Technology*, The University Press, Cambridge, UK, (1996)
9. Gorynin I. V., Ushkov S. S., eds.: *Titanium '99, Science and Technology,* CRISM "Prometey", St. Petersburg, Russia, (2000)
10. Bomberger H. B., Froes F. H., Morton P. H.: *Titanium Technology: Present Status and Future Trends,* TDA, Dayton, USA, (1985) p. 3
11. Eylon D., Seagle S.R.: *Titanium '99, Science and Technology,* CRISM "Prometey", St. Petersburg, Russia, (2000) p. 37
12. Bania P. J.: *Titanium '92, Science and Technology*, TMS, Warrendale, USA, (1993) p. 2227
13. Seagle S. R.: Mater. Sci. Eng. A213, (1996) p.1
14. Gorynin I. V.: *Titanium '92, Science and Technology*, TMS, Warrendale, USA, (1993) p. 65
15. Yamada M: Mater. Sci. Eng. A213, (1996) p. 8
16. Boyer R. R.: J. of Metals 44, no. 5, (1992) p. 23
17. Combres Y., Champin B.: *Titanium '95, Science and Technology*, The University Press, Cambridge, UK, (1996) p. 11
18. Wilhelm H., Furlan R., Moloncy K. C.: *Titanium '95, Science and Technology*, The University Press, Cambridge, UK, (1996) p. 620
19. Schutz R. W., Watkins H. B.: Mater. Sci. Eng. A243, (1998) p. 305
20. Moriyasu T.: *Titanium '95, Science and Technology*, The University Press, Cambridge, UK, (1996) p. 21
21. Froes F. H., Allen P. G., Niinomi M.: *Non-Aerospace Applications of Titanium,* TMS, Warrendale, USA, (1998) p. 3
22. Blenkinsop P. A.: *Titanium '95, Science and Technology*, The University Press, Cambridge, UK, (1996) p. 1
23. Boyer R. R.: *Titanium '95, Science and Technology*, The University Press, Cambridge, UK, (1996) p. 41
24. Shira C., Froes F. H.: *Non-Aerospace Application of Titanium*, TMS, Warrendale, USA, (1998) p. 331

25. Niinomi M., Kuroda D., Morinaga M., Kato Y., Yashiro T.: *Non-Aerospace Application of Titanium*, TMS, Warrendale, USA, (1998) p. 217
26. Zarkades A., Larson F. R.: *The Science, Technology and Application of Titanium*, Pergamon Press, Oxford, UK, (1970) p. 933
27. Conrad H., Doner M., de Meester B.: *Titanium Science and Technology*, Plenum Press, New York, USA, (1973) p. 969
28. Fedotov S. G.: *Titanium Science and Technology*, Plenum Press, New York, USA, (1973) p. 871
29. James D. W., Moon D. M.: *The Science, Technology and Application of Titanium*, Pergamon Press, Oxford, UK, (1970) p. 767
30. Ivasishin O. M., Flower H. M., Lütjering G.: *Titanium '99, Science and Technology,* CRISM "Prometey", St. Petersburg, Russia, (2000) p. 77
31. Collings E. W.: *Materials Properties Handbook: Titanium Alloys*, ASM, Materials Park, USA, (1994) p. 1
32. Boyer R., Welsch G., Collings E. W., eds.: *Materials Properties Handbook: Titanium Alloys*, ASM, Materials Park, USA, (1994)
33. Partridge P. G.: Met. Rev. 12, (1967) p. 169
34. Yoo H. M.: Met. Trans. 12A, (1981) p. 409
35. Paton N. E., Williams J. C., Rauscher G. P.: *Titanium Science and Technology*, Plenum Press, New York, USA, (1973) p. 1049
36. Jones I. P., Hutchinson W. B.: Acta Met. 29, (1981) p. 951
37. Paton N. E., Baggerly R. G., Williams J. C.: Rockwell Report SC 526.7FR (1976)
38. Paton N. E., Williams J. C.: *Second International Conference on the Strength of Metals and Alloys*, ASM, Metals Park, USA, (1970) p. 108
39. Paton N. E., Backofen W. A.: Met. Trans. 1, (1970) p. 2839
40. Rosenberg H. W.: *The Science, Technology and Application of Titanium*, Pergamon Press, Oxford, UK, (1970) p. 851
41. Baker H., ed.: *Alloy Phase Diagrams,* ASM Handbook, Vol. 3, ASM, Materials Park, USA, (1992)
42. Hansen M.: *Constitution of Binary Alloys,* McGraw-Hill, New York, USA, (1958)
43. Burgers W. G.: Physica 1, (1934) p. 561
44. Newkirk J. B., Geisler A. H.: Acta Met. 1, (1953) p. 370
45. Otte H. M.: *The Science, Technology and Application of Titanium,* Pergamon Press, Oxford, UK, (1970) p. 645
46. Williams J. C.: *Titanium Science and Technology,* Plenum Press, New York, USA, (1973) p. 1433
47. Flower H. M., Davis R., West D. R. F.: *Titanium and Titanium Alloys,* Plenum Press, New York, USA, (1982) p. 1703
48. Peters M., Lütjering G., Ziegler G.: Z. Metallke. 74, (1983) p. 274
49. Benjamin D., ed.: *Properties and Selection: Stainless Steels, Tool Materials and Special-Purpose Materials*, Metals Handbook, 9th edn, Vol. 3, ASM, Metals Park, USA, (1980) p. 353
50. Béchel J., Hocheid B.: *Titanium, Science and Technology,* DGM, Oberursel, Germany, (1985) p. 1613
51. Blackburn M. J.: Trans. AIME 239, (1967) p. 1200
52. Williams J. C., Sommer A. W., Tung P. P.: Met. Trans. 3, (1972) p. 2979
53. Pearson W. B.: *Handbook of Lattice Spacings and Structures of Metals and Alloys*, Vol. 2, Pergamon Press, London, UK, (1967)
54. Williams J. C., Hickman B. S., Leslie D. H.: Met. Trans. 2, (1971) p. 477
55. Williams J. C.: *Titanium Technology: Present Status and Future Trends*, TDA, Dayton, USA, (1985) p. 75
56. Gysler A., Lütjering G., Gerold V.: Acta Met. 22, (1974) p. 901

57. Wagner L., Gregory J. K.: *Beta Titanium in the 1990's*, TMS, Warrendale, USA, (1993) p. 199
58. Zwicker U.: *Titan and Titanlegierungen*, Springer-Verlag, Berlin, Germany, (1974) p. 102
59. Liu Z., Welsch G.: Met. Trans. 19A, (1988) p. 121
60. Mishin Y., Hertig C.: Acta Mater. 48, (2000) p. 589
61. Schutz R. W., Thomas D. E.: *Corrosion,* Metals Handbook, 9th edn, Vol. 13, ASM, Metals Park, USA, (1987) p. 669
62. Davis J. R., ed.: *Stainless Steels*, ASM, Materials Park, USA, (1994) p. 139
63. Myers J. R., Bomberger H. B., Froes F. H.: *Titanium Technology: Present Status and Future Trends*, TDA, Dayton, USA, (1985) p. 165
64. Schutz R. W.: *Titanium '95, Science and Technology*, The University Press,, Cambridge, UK, (1996) p. 1860
65. Schutz R. W.: *Metallurgy and Technology of Practical Titanium Alloys,* TMS, Warrendale, USA, (1994) p. 295
66. Bania P. J., Parris W. M.: *Titanium 1990, Products and Applications*, TDA, Dayton, USA, (1990) p. 784
67. Smialek J. L., Nesbitt J. A., Brindley W. J., Brady M. P., Doychak J., Dickerson R. M., Hull D. R.: Mat. Res. Soc. Symp. Proc. 364, (1995) p. 1273
68. Leyens C., Peters M., Kaysser W. A.: *Titanium '95, Science and Technology,* The University Press, Cambridge, UK, (1996) p. 1935
69. Leyens C.: *Titan und Titanlegierungen*, DGM, Oberursel, Germany, (1996) p. 139
70. Johnson T. J., Loretto M. H., Kearns M. W.: *Titanium '92, Science and Technology*, TMS, Warrendale, USA, (1993) p. 2035
71. Kroll W. J.: Trans. El. Soc. 78, (1940) p. 35
72. Hunter M. A.: J. Amer. Chem. Soc. 32, (1910) p. 330
73. Cobel G., Fisher J., Snyder L. E.: *Titanium '80, Science and Technology,* AIME, Warrendale, USA, (1980) p. 1969
74. Rosenberg H. W., Green J. E.: *Titanium '92, Science and Technology,* TMS, Warrendale, USA, (1993) p. 2371
75. Chen G. Z., Fray D. J., Farthing T. W.: Nature 407, (2000) p. 361
76. Sears J. W., Young J. M., Kearns M.: *Titanium '92, Science and Technolgy,* TMS, Warrendale, USA, (1993) p. 2293
77. Mitchell A.: *Titanium '98,* International Academic Publisher, Beijing, China, (1990) p. 91
78. Buttrell W. H., Shamblen C. E.: *Titanium '95, Science and Technology,* The University Press, Cambridge, UK (1999) p. 1446
79. Adams R. T., Rosenberg H. W.: *Titanium and Titanium Alloys,* Plenum Press, New York, USA, (1982) p. 127
80. Chen C. C., Boyer R. R.: J. of Metals 31, no. 7, (1979) p. 33
81. Boyer R., Welsch G., Collings E. W., eds.: *Materials Properties Handbook: Titanium Alloys, Technical Note 4: Forging,* ASM, Materials Park, USA, (1994) p. 1083
82. Kuhlmann G. W.: *Forging Titanium Alloys,* Metals Handbook, 9th edn, Vol. 14, ASM, Metals Park, USA, (1988) p. 267
83. *Machining Data Handbook,* 2nd edn, Machinability Data Center, Metcut Research Associates, Inc., Cincinnati, USA, (1972)
84. Boyer R., Welsch G., Collings E. W., eds.: *Materials Properties Handbook: Titanium Alloys, Technical Note 7: Machining,* ASM, Materials Park, USA, (1994) p. 1119
85. Boyer R., Welsch G., Collings E. W., eds.: *Materials Properties Handbook: Titanium Alloys, Technical Note 3: Casting,* ASM, Materials Park, USA, (1994) p. 1079
86. Eylon D., Froes F. H., Gardiner R. W.: *Titanium Technology: Present Status and Future Trends,* TDA, Dayton, USA, (1985) p. 35
87. Savage S. J., Froes F. H.: *Titanium Technology: Present Status and Future Trends,* TDA, Dayton, USA, (1985) p. 60

88. Froes F. H., Eylon D.: *Titanium Technology: Present Status and Future Trends,* TDA, Dayton, USA, (1985) p. 49
89. Paton N. E., Hamilton C. H., eds.: *Superplastic Forming of Structural Alloys,* AIME, Warrendale, USA, (1982)
90. Mahoney M. W.: *Materials Properties Handbook: Titanium Alloys, Technical Note 5A: Superplastic Forming of Titanium Alloys,* ASM, Materials Park, USA, (1994) p. 1101
91. Winkler P.-J.: *Sixth World Conference on Titanium,* Les Editions de Physique, Les Ulis, France, (1988) p. 1135
92. Lee D., Backofen W. A.: Trans. AIME 239, (1967) p. 1034
93. Tisler R. J., Lederich R. J.: *Titanium '95, Science and Technology,* The University Press, Cambridge, UK, (1996) p. 596
94. Ashby M. F., Verrall R. A.: Acta Met. 21, (1973) p. 149
95. Kearns W. H., ed.: *Welding Handbook,* 7th edn, Vol. 4, American Welding Society, Miami, USA, (1982) p. 43
96. Boyer R., Welsch G., Collings E. W., eds.: *Materials Properties Handbook: Titanium Alloys, Technical Note 10: Welding and Brazing,* ASM, Materials Park, USA, (1994) p. 1159
97. Helm D., Lütjering G.: *Titanium '99, Science and Technology,* CRISM "Prometey", St. Petersburg, Russia, (2000) p. 1726
98. Juhas M. C., Viswanathan G. B., Fraser H. L.: *Friction Stir Welding,* CD-ROM, TWI, Cambridge, UK, (2000)
99. Wagner L., Lütjering G.: *Second International Conference on Shot Peening*, American Shot Peening Society, USA, (1984) p. 194
100. Niku-Lari A., ed.: *First International Conference on Shot Peening*, Pergamon Press, Oxford, UK, (1981)
101. Clauer A. H., Holbrook J. H., Fairand B. P.: *Shock Waves and High-Strain-Rate Phenomena in Metals,* Plenum Press, New York, USA, (1981) p. 675
102. Clauer A. H.: *Surface Performance of Titanium,* TMS, Warrendale, USA, (1996) p. 217
103. Krautkramer J., Krautkramer H.: *Ultrasonic Testing of Materials,* 4th edn, Springer-Verlag, Berlin, Germany, (1990)
104. Mester M. L., McIntire P., eds.: *Ultrasonic Testing,* Nondestructive Testing Handbook, 2nd edn, Vol. 7, ASNT, Columbus, USA, (1991)
105. Buck O., Thompson D. O., Paton N. E., Williams J. C.: *Internal Friction and Ultrasonic Attenuation in Crystalline Solids*, Springer-Verlag, Berlin, Germany, (1975) p. 451
106. Moyers J. C., Seagle S. R., Copley D. C., Gilmore R. S.: *Titanium '95, Science and Technology,* The University Press, Cambridge, UK, (1996) p. 1521
107. Libby H. L.: *Introduction to Electromagnetic Nondestructive Test Methods,* Krieger, Malabar, USA, (1979)
108. Birks A. S., Green R. E., eds.: *Electromagnetic Testing,* Nondestructive Testing Handbook, 2nd edn, Vol. 4, ASNT, Columbus, USA, (1986)
109. Thomas G.: *Transmission Electron Microscopy of Metals,* John Wiley and Sons, New York, USA, (1962)
110. Rhodes C. G., Williams J. C.: Met. Trans. 6A, (1975) p. 2103
111. Banerjee D., Williams J. C.: Scripta Met. 17, (1983) p. 1125
112. Blackburn M. J., Williams J. C.: ASM Quart. Trans. 60, (1967) p. 373
113. Blackburn M. J., Williams J. C.: Trans. AIME 239, (1967) p. 287
114. Spurling R. A., Rhodes C. G., Williams J. C.: Met. Trans. 5, (1974) p. 2597
115. Adams B. L., Wright S. I., Kunze K.: Met. Trans. 24A, (1993) p. 819
116. Cullity B. D.: *Elements of X-Ray Diffraction,* 2nd edn, Addison-Wesley, Reading, USA, (1978)
117. Kocks U. F., Tome C. N., Wenk H.-R.: *Texture and Anisotropy,* The University Press, Cambridge, UK, (1998)
118. Dieter G. E.: *Mechanical Metallurgy,* 2nd edn, McGraw-Hill, New York, USA, (1976)

119. Hertzberg R. W.: *Deformation and Fracture Mechanics of Engineering Materials,* 4th edn, John Wiley and Sons, New York, USA, (1996)
120. Lütjering G., Gysler A.: *Titanium Science and Technology,* DGM, Oberursel, Germany, (1985) p. 2065
121. Sakurai K., Itabashi Y., Komatsu A.: *Titanium '80, Science and Technology*, AIME, Warrendale, USA, (1980) p. 299
122. Boyer R., Welsch G., Collings E. W., eds.: *Materials Properties Handbook: Titanium Alloys*, ASM, Materials Park, USA, (1994) p. 228
123. Curtis R. E., Boyer R. R., Williams J. C.: Trans. ASM 62, (1969) p. 457
124. Margolin H., Williams J. C., Chesnutt J. C., Lütjering G.: *Titanium '80, Science and Technology*, AIME, Warrendale, USA, (1980) p. 169
125. Okazaki K., Conrad H.: Trans. JIM 13, (1972) p. 205
126. Okazaki K., Conrad H.: *Titanium and Titanium Alloys*, Plenum Press, New York, USA, (1982) p. 429
127. Finden P. T.: *Sixth World Conference on Titanium*, Les Editions de Physique, Les Ulis, France, (1988) p. 1251
128. Dieter G. E.: *Mechanical Metallurgy,* 2nd edn, McGraw-Hill, New York, USA, (1976) p. 685
129. Conrad H., Jones R.: *The Science, Technology and Application of Titanium,* Pergamon Press, Oxford, UK, (1970) p. 489
130. Fleischer R. L.: *The Strengthening of Metals*, Chapman and Hall, New York, USA, (1964) p. 93
131. Williams J. C., Baggerly R. G., Paton N. E.: Met. and Mater. Trans. 33A, (2002) p. 837
132. Boyer R., Welsch G., Collings E. W., eds.: *Materials Properties Handbook: Titanium Alloys*, ASM, Materials Park, USA, (1994) p. 247
133. Boyer R., Welsch G., Collings E. W., eds.: *Materials Properties Handbook: Titanium Alloys*, ASM, Materials Park, USA, (1994) p. 227
134. Jones R. L., Conrad H.: Trans. AIME 245, (1969) p. 779
135. Blackburn M. J., Williams J. C.: *Proc. Conf. on the Fundamental Aspects of Stress Corrosion Cracking*, NACE, Houston, USA, (1969) p. 620
136. Williams J. C., Thompson A. W., Rhodes C. G., Chesnutt J. C.: *Titanium and Titanium Alloys*, Plenum Press, New York, USA, (1982) p. 467
137. Boyer R., Welsch G., Collings E. W., eds.: *Materials Properties Handbook: Titanium Alloys*, ASM, Materials Park, USA, (1994) p. 238
138. Paton N. E., Williams J. C., Chesnutt J. C., Thompson A. W.: AGARD Conf. Proc., no. 185, (1976) p. 4-1
139. Boyd J. D.: *The Science, Technology and Application of Titanium,* Pergamon Press, Oxford, UK, (1970) p. 545
140. Paton N. E., Hickman B. S., Leslie D. H.: Met. Trans. 2, (1971) p. 2791
141. Williams J. C.: *Effect of Hydrogen on Behavior of Materials*, AIME, New York, USA, (1976) p. 367
142. Hall J. A., Banerjee D., Wardlaw T.: *Titanium, Science and Technology*, DGM, Oberursel, Germany, (1985) p. 2603
143. Lütjering G., Gysler A., Wagner L.: *Sixth World Conference on Titanium*, Les Editions de Physique, Les Ulis, France, (1988) p. 71
144. Lütjering G., Helm D., Däubler M.: *Fatigue 93,* EMAS, Warley, UK, (1993) p. 165
145. Frederick S. F.: AFML-TR-73-265, (1973)
146. Peters M., Lütjering G.: *Titanium '80, Science and Technology*, AIME, Warrendale, USA, (1980) p. 925
147. Lütjering G., Albrecht J., Invasishin O. M.: *Microstructure/Property Relationships of Titanium Alloys*, TMS, Warrendale, USA, (1994) p. 65
148. Lütjering G., Albrecht J., Ivasishin O. M.: *Titanium '95, Science and Technology*, The University Press, Cambridge, UK, (1996) p. 1163

149. Ivasishin O. M., Lütjering G.: *Titanium '99 , Science and Technology,* CRISM "Prometey", St. Petersburg, Russia, (2000) p. 441
150. Hines J. A., Peters J. O., Lütjering G.: *Fatigue Behavior of Titanium Alloys*, TMS, Warrendale, USA, (1999) p. 15
151. Lütjering G., Schmidt H.-J.: *Fatigue '87,* EMAS, Warley, UK, (1987) p. 1663
152. Tacke-Messing U., Wagner L., Lütjering G.: *Gefüge und Bruch,* Gebrüder Borntraeger, Berlin, Germany, (1990) p. 199
153. Torster F.: PhD Thesis, TU Hamburg-Harburg, Germany, (1995); Berichte aus der Werkstofftechnik, Shaker Verlag, Aachen, Germany, (1995)
154. Peters M., Gysler A., Lütjering G.: *Titanium '80, Science and Technology,* AIME, Warrendale, USA, (1980) p. 1777
155. Gray III G. T., Lütjering G.: *Titanium, Science and Technology,* DGM, Oberursel, Germany, (1985) p. 2251
156. Boyer R., Welsch G., Collings E. W., eds.: *Materials Properties Handbook: Titanium Alloys,* ASM, Materials Park, USA, (1994) p. 575
157. Boyer R., Welsch G., Collings E. W., eds.: *Materials Properties Handbook: Titanium Alloys,* ASM, Materials Park, USA, (1994) p. 584
158. Boyer R., Welsch G., Collings E. W., eds.: Materials Properties Handbook: Titanium Alloys, ASM, Materials Park, USA, (1994) p. 587
159. Wegmann G., Lütjering G., Albrecht J., Folkers K.-D., Liesner C.: *Titanium '95, Science and Technology,* The University Press, Cambridge, UK, (1996) p. 895
160. Wegmann G., Albrecht J., Lütjering G., Folkers K.-D., Liesner C.: Z. Metallkde. 88, (1997) p. 764
161. Schroeder G., Albrecht J., Lütjering G.: *Titanium '99, Science and Technology,* CRISM "Prometey", St. Petersburg, Russia, (2000) p. 545
162. Blackburn M. J., Smyrl W. H.: *Titanium Science and Technology,* Plenum Press, New York, USA, (1973) p. 2577
163. Adachi S., Wagner L., Lütjering G.: *Titanium, Science and Technology,* DGM, Oberursel, Germany, (1985) p. 2139
164. Lindemann J., Wagner L.: *Fatigue Behavior of Titanium Alloys,* TMS, Warrendale, USA, (1999) p. 47
165. Notkina E., Gysler A., Lütjering G.: *Fatigue in the Very High Cycle Regime,* Institute of Meteorology and Physics, Vienna, Austria, (2001) p. 149
166. Lütjering G.: *Compass 1999: Component Optimization from Materials Properties and Simulation Software,* EMAS, Warley, UK, (1999) p. 263
167. Helm D.: *Fatigue Behavior of Titanium Alloys,* TMS, Warrendale, USA, (1999) p. 291
168. Boyer R. R.: Mat. Sci. Eng. A213, (1996) p. 103
169. Boyer R. R., Williams J. C., Paton N. E.: *Titanium '99, Science and Technology,* CRISM "Prometey", St. Petersburg, Russia, (2000) p. 1007
170. Jaffee R. I., ed.: *Titanium Steam Turbine Blading,* Pergamon Press, New York, USA, (1990)
171. Puschnik H., Fladischer H., Lütjering G., Jaffee R. I.: *Titanium '92, Science and Technology,* TMS, Warrendale, USA, (1993) p. 131
172. Wells M. G. H., Roopchaud B., Montgomery J. S., Gooch W. S.: *Non-Aerospace Applications of Titanium,* TMS, Warrendale, USA, (1998) p. 289
173. Burkins M., Hansen J., Paige J., Turner P.: *Non-Aerospace Applications of Titanium,* TMS, Warrendale, USA, (1998) p. 273
174. Saal S., Wagner L., Lütjering G., Pillhöfer H., Daeubler M. A.: Z. Metallkde. 81, (1990) p. 535
175. Schauerte O., Gysler A., Lütjering G., Mailly S., Chabanne Y., Sarrazin-Baudoux C., Mendez J., Petit J.: *Fatigue Behavior of Titanium Alloys,* TMS, Warrendale, USA, (1999) p. 191

176. Petit J., Sarrazin-Baudoux C., Chabanne Y., Lütjering G., Gysler A., Schauerte O.: *Fatigue Behavior of Titanium Alloys,* TMS, Warrendale, USA, (1999) p. 203
177. Lütjering S., Smith P. R., Eylon D.: *Intermetallics and Superalloys,* Euromat Vol. 10, Wiley-VCH, Weinheim, Germany, (2000) p. 283
178. Daeubler M. A., Helm D., Neal D. F.: *Titanium 1990, Products and Applications,* TDA, Dayton, USA, (1990) p. 78
179. Gysler A., Lütjering G.: DFVLR-FB 79-24, (1979)
180. Daeubler M. A., Helm D.: *Titanium '92, Science and Technology,* TMS, Warrendale, USA, (1993) p. 41
181. Neal D. F.: *Titanium, Science and Technology,* DGM, Oberursel, Germany, (1985) p. 2419
182. Neal D. F.: *Titanium '95, Science and Technology,* The University Press, Cambridge, UK, (1996) p. 2195
183. Boyer R. R., Lütjering G.: *Titanium Alloy Processing,* TMS, Warrendale, USA, (1996) p. 349
184. Busongo F., Lütjering G.: *Proceedings of the 10th World Conference on Titanium,* Wiley-VCH, Weinheim, Germany, (2003)
185. Prandi B., Wadier J.-F., Schwartz F., Mosser P.-E., Vassel A.: *Titanium 1990, Products and Applications,* TDA, Dayton, USA, (1990) p. 150
186. Peters J. O., Lütjering G., Koren H., Puschnik H., Boyer R. R.: Mater. Sci. Eng. A213, (1996) p. 71
187. Combres Y., Champin B.: *Beta Ttianium Alloys in the 1990's,* TMS, Warrendale, USA, (1993) p. 477
188. Peters M., Lütjering G.: Z. Metallkde. 67, (1976) p. 811
189. Lütjering G., Gysler A.: *Aluminum, Transformation Technology and Application,* ASM, Metals Park, USA, (1980) p. 171
190. Sauer C., Busongo F., Lütjering G.: *Fatigue 2002,* EMAS, Warley, UK, (2002) p. 2043
191. Sauer C., Busongo F., Lütjering G.: Materials Science Forum, Vols. 396-402, Trans Tech Publications, Zuerich, Switzerland, (2002) p. 1115
192. Peters J. O., Lütjering G.: Z. Metallkde. 89, (1998) p. 464
193. Lütjering G., Gysler A.: *Aluminum Alloys, Physical and Mechanical Properties,* EMAS, Warley, UK, (1986) p. 1547
194. Lütjering G.: Mater. Sci. Eng. A263, (1999) p. 117
195. Albrecht J., Lütjering G.: *Titanium '99, Science and Technology,* CRISM "Prometey", St. Petersburg, Russia, (2000) p. 363
196. Peters J. O., Lütjering G., Koren M., Puschnik H., Boyer R. R.: *Titanium '95, Science and Technology,* The University Press, Cambridge, UK, (1996) p. 1403
197. Benedetti M., Peters J. O., Lütjering G.: *Proceedings of the 10th World Conference on Titanium,* Wiley-VCH, Weinheim, Germany, (2003)
198. Lindemann J.: PhD Thesis, TU Brandenburg, Cottbus, Germany, (1998); Fortschr.-Ber. VDI, series 5, no. 547, VDI Verlag, Düsseldorf, Germany, (1999)
199. Peters J. O., Lütjering G., Ivasishin O. M., Markovsky P. E.: *Third ASM Int. Conf. on Synthesis, Processing and Modelling of Advanced Materials,* ASM, Materials Park, USA, (1997) p. 269
200. Boyer R. R.: *Beta Titanium Alloys in the 1990's,* TMS, Warrendale, USA, (1993) p. 335
201. Froes F. H., Allen P. G., Niinomi M., eds.: *Non-Aerospace Applications of Titanium,* TMS, Warrendale, USA, (1998)
202. Bania P. J.: *Beta Titanium Alloys,* Société Francaise de Métallurgie et de Matériaux, Paris, France, (1994) p. 7
203. Boyer R. R.: *Beta Titanium Alloys,* Société Francaise de Métallurgie et de Matériaux, Paris, France, (1994) p. 253
204. Wheeler D. A., Vogt R. G., Cianci M. S.: *Beta Titanium Alloys in the 1990's,* TMS, Warrendale, USA, (1993) p. 375

205. Davies D. P., Gittos B. C., Terlinde G., Fischer G.: *Titanium '95, Science and Technology,* The University Press, Cambridge, UK, (1996) p. 1371
206. Terlinde G., Fischer G.: *Titanium '95, Science and Technology,* The University Press, Cambridge, UK, (1996) p. 2177
207. Dunlop D. C., Schutz R. W.: *Beta Titanium in the 1990's,* TMS, Warrendale, USA, (1993) p. 347
208. McAndrew J. B., Kessler H. D.: Trans. AIME 206, (1956) p. 1348
209. Yamaguchi M., Umakoshi Y.: Progress in Materials Science 34, (1990) p. 1
210. Kim Y.-W., Dimiduk D. M.: J. of Metals 43, no. 8, (1991) p. 40
211. Yamaguchi M., Inui H.: *Structural Intermetallics,* TMS, Warrendale, USA, (1993) p. 127
212. Huang S. C. Chesnutt J. C.: *Intermetallic Compounds: Principles and Practice,* Vol. 2, John Wiley and Sons, New York, USA, (1995) p. 73
213. Dimiduk D. M.: *Gamma Titanium Aluminides,* TMS, Warrendale, USA, (1995) p. 3
214. Dimiduk D. M., Martin P. L., Kim Y.-W.: Mat. Sci. Eng. A243, (1998) p. 66
215. Appel F., Wagner R.: Mat. Sci. Eng. R22, (1998) p. 187
216. Dimiduk D. M.: Intermetallics 6, (1998) p. 613
217. Dimiduk D. M.: Mat. Sci. Eng. A263, (1999) p. 281
218. Dimiduk D. M., McQuay P. A., Kim Y.-W.: *Titanium '99: Science and Technology,* CRISM "Prometey", St. Petersburg, Russia, (2000) p. 259
219. Kim Y.-W., Dimiduk D. M., Loretto M., eds.: *Gamma Titanium Aluminides,* TMS, Warrendale, USA, (1999)
220. Kim Y.-W., Wagner R., Yamaguchi M., eds.: *Gamma Titanium Aluminides,* TMS, Warrendale, USA, (1995)
221. Hemker K. J., Dimiduk D. M., Clemens H., Darolia R., Inui H., Larsen J. M., Sikka V. K., Thomas M., Whittenberger J. D., eds.: *Structural Intermetallics 2001,* TMS, Warrendale, USA, (2001)
222. Lipsitt H. A.: *High Temperature Ordered Intermetallic Alloys V,* Mat. Res. Soc. Symp. Proc. 288, (1993) p. 119
223. Banerjee D., Gogia A. K., Nandy T. K., Muraleedharan K., Mishra R. S.: *Structural Intermetallics,* TMS, Warrendale, USA, (1993) p. 19
224. Banerjee D.: *Intermetallic Compounds: Principles and Practice,* Vol. 2, John Wiley and Sons, New York, USA, (1995) p. 91
225. Banerjee D.: Progress in Materials Science 42, (1997) p. 135
226. Nandy T. K., Banerjee D.: *Structural Intermetallics,* TMS, Warrendale, USA, (1997) p. 777
227. Gogia A. K., Nandy T. K., Banerjee D., Carisey T., Strudel J. L., Franchet J. M.: Intermetallics 6, (1998) p. 741
228. Marcinkowski M. J.: *Electron Microscopy and Structure of Materials,* University of California Press, Berkeley, USA, (1972) p. 333
229. Blackburn M. J.: Trans. AIME 239, (1967) p. 660
230. Shechtman D., Blackburn M. J., Lipsitt H.: Met. Trans. 5, (1974) p. 1373
231. Banerjee D., Rowe R. G., Hall E. L.: *High-Temperature Ordered Intermetallic Alloys IV,* MRS, Pittsburgh, USA, (1991) p. 285
232. Ward C. H.: Intern. Mat. Rev. 38, (1993) p. 79
233. Ward C. H., Williams J. C., Thompson A. W., Rosenthal D. G., Froes F. H.: *Sixth World Conference on Titanium,* Les Editions de Physique, Les Ulis, France, (1988) p. 1103
234. Chesnutt J. C., Hall J. A., Lipsitt H. A.: *Titanium '95, Science and Technology,* The University Press, Cambridge, UK, (1996) p. 70
235. Strychor R., Williams J. C.: *Solid to Solid Phase Transformations,* AIME, Warrendale, USA, (1982) p. 249
236. Strychor R., Williams J. C., Soffa W. A.: Met. Trans. 19A, (1988) p. 225
237. Lütjering G., Proske G., Albrecht J., Helm D., Däubler M.: *Intermetallic Compounds (JIMIS-6),* The Japan Institute of Metals, Sendai, Japan, (1991) p. 537

238. Kumpfert J., Ward C. H., Peters M., Kaysser W. A.: *Synthesis/Processing of Lightweight Metallic Materials,* TMS, Warrendale, USA, (1995) p. 85
239. Banerjee D., Gogia A. K., Nandy T. K., Joshi V. A.: Acta Met. 36, (1988) p. 871
240. Rowe R. G., Banerjee D., Muraleedharan K., Larsen M., Hall E. L., Konitzer D. G., Woodfield A. P.: *Titanium '92, Science and Technology,* TMS, Warrendale, USA, (1993) p. 1259
241. Kelly T. J., Austin C. M.: *Titanium '95, Science and Technology,* The University Press, Cambridge, UK, (1996) p. 192
242. Blackburn M. J.: *The Science, Technology and Application of Titanium,* Pergamon Press, Oxford, UK, (1970) p. 633
243. Seeger J., Hartig C., Bartels A., Mecking H.: *High-Temperature Ordered Intermetallic Alloys IV,* MRS, Pittsburgh, USA, (1991) p. 157
244. Bhowal P. R., Merrick H. F., Larsen D. E.: Mater. Sci. Eng. A192/193, (1995) p. 685
245. Huang S. C., McKee D. W., Shih D. S., Chesnutt J. C.: *Intermetallic Compounds (JIMIS-6),*The Japan Institute of Metals, Sendai, Japan, (1991) p. 363
246. Proske G., Lütjering G., Albrecht J., Däubler M. A., Helm D.: *Titanium '92, Science and Technology*, TMS, Warrendale, USA, (1993) p. 1187
247. Rowe R. G.: *Titanium '92, Science and Technology,* TMS, Warrendale, USA, (1993) p. 343
248. Rowe R. G., Konitzer D. G., Woodfield A. P., Chesnutt J. C.: *High-Temperature Ordered Intermetallic Alloys IV,* MRS, Pittsburgh, USA, (1991) p. 703
249. Proske G., Lütjering G., Albrecht J., Helm D., Däubler M.: Mater. Sci. Eng. A152, (1992) p. 310
250. Lütjering G., Proske G., Terlinde G., Fischer G., Helm D.: *Titanium '95, Science and Technology,* The University Press, Cambridge, UK, (1996) p. 332
251. Takashima K., Cope M. T., Bowen P.: *Titanium '95, Science and Technology,* The University Press, Cambridge, UK, (1996) p. 340
252. Kim Y.-W., Boyer R., eds.: *Microstructure/Property Relationships in Titanium Aluminides and Alloys,* TMS, Warrendale, USA, (1990)
253. Pope D. P., Liu C. T., Whang S. H., Yamaguchi M., eds.: *High Temperature Intermetallics,* Elsevier, Amsterdam, The Netherlands, (1997)
254. Clemens H., Kestler H.: Advanced Engineering Materials 2, (2000) p. 551
255. Liu C. T., Maziasz P. J., Clemens D. R., Schneibel J. H., Sikka V. J., Nieh T. G., Wright J., Walku L. R.: *Gamma Titanium Aluminides,* TMS, Warrendale, USA, (1995) p. 679
256. Larsen J. M., Worth B. D., Balsone S. J., Jones J. W.: *Gamma Titanium Aluminides,* TMS, Warrendale, USA, (1995) p. 821
257. Prewo K. M.: *The Science, Technology and Application of Titanium,* Pergamon Press, Oxford, UK, (1970) p. 2333
258. Smith P. R., Gambone M. L., Williams D. S., Garner D. I.: Journal of Materials Science 33, (1998) p. 5855
259. Hanusiak W. M.: Atlantic Research Corporation – Advanced Materials Division, Wilmington, USA, (1999) private communication
260. Guo Z. X., Derby B.: *Titanium '92, Science and Technology*, TMS, Warrendale, USA, (1993) p. 2633
261. MacKay R. A., Brindley P. K., Froes F. H.: J. of Metals 43, no. 5, (1991) p. 23
262. Ward-Close M., Partridge P. G.: *Titanium '92, Science and Technology*, TMS, Warrendale, USA, (1993) p. 2479
263. Gigerenzer H., Kumnick A. J.: Textron Specialty Materials Report, Lowell, USA, (1990)
264. Brindley P. K.: *High Temperature Intermetallic Alloys II,* MRS, Pittsburgh, USA, (1987) p. 419
265. Vassel A., Indrigo C., Pautonnier F.: *Titanium '95, Science and Technology*, The University Press, Cambridge, UK, (1996) p. 2739

266. Metcalfe A. G.: J. Comp. Mat. 1, (1967) p. 356
267. Warren R., Anderson C. H.: Composites 15, (1984) p. 101
268. Clym T. W., Flower H. M.: *Titanium '92, Science and Technology*, TMS, Warrendale, USA, (1993) p. 2467
269. Das G.: *Titanium '92, Science and Technology*, TMS, Warrendale, USA, (1993) p. 2617
270. Larsen J. M., Revelos W. C., Gambone M. L.: *Intermetallic Matrix Composites II,* MRS, Pittsburgh, USA, (1992) p. 3
271. Jha S. C., Forster J. A., Pandey A. K., Delagi R. G.: ISI Japan 31, (1991) p. 1267
272. Bassi C., Peters J. A., Wittenauer J.: J. of Metals 41, no. 9, (1989) p. 18
273. Larsen J. M., Russ S. M., Jones J. W.: Met. and Mater. Trans. 26A, (1995) p. 3211
274. Larsen J. M., Russ S. M., Jones J. W.: *Characterization of Fibre Reinforced Titanium Metal Matrix Composites,* Specialized Printing Services, Loughton, UK, (1994) p. 1.1
275. Veeck S. J., Colvin G. N.: *Titanium '92, Science and Technology*, TMS, Warrendale, USA, (1993) p. 2495
276. Warner G. K.: B. F. Goodrich Landing Systems, Cleveland, USA, (2001) private communication
277. Onnes H. K.: Comm. Leiden 120b, (1911)
278. Meisner W., Ochsenfeld R.: Naturwissenschaften 21, (1933) p. 787
279. Buckel W.: *Super-conductivity*, VCH, Weinheim, Germany, (1991) p. 143
280. Kosing O. E., Scharl R., Schmuhl H. J.: *Proceedings of ASME TURBO EXPO 2001: Land, Sea, and Air,* CD-ROM, Paper no. 2001 GT 0283, ASME, Warrendale, USA, (2001)
281. Zhao Y., Zhou L., Deng J.: *Titanium '98,* International Academic Publishers, Beijing, China, (1999) p. 457
282. Advanced Materials and Processes, no. 9, (1993) p. 7
283. Outlook, Teledyne Wah Chang Albany, USA, Vol. 14, no. 1, (1993) p. 1
284. Shitazaki K., Tamura T., Kuriiwa T., Goto T., Kamegawa A., Takamura H., Okada M.: Materials Transactions 43, (2002) p. 1115
285. Tamura T., Hatakeyama M., Ebinuma T., Kamegawa A., Takamura H., Okada M.: Materials Transactions 43, (2002) p. 1120
286. Shirasaki K., Kuriiwa T., Tamura T., Kamegawa A., Okada M.: Materials Transactions 43, (2002) p. 1173
287. Chang L. C., Read T. A.: Trans. AIME 191, (1951) p. 49
288. Melton K. N.: *Engineering Aspects of Shape Memory Alloys,* Butterworth-Heinemann, London, UK, (1990) p. 21
289. Wayman C. M., Duerig T. W.: *Engineering Aspects of Shape Memory Alloys,* Butterworth-Heinemann, London, UK, (1990) p. 3
290. Duerig T. W., Zadno G. R.: *Engineering Aspects of Shape Memory Alloys,* Butterworth-Heinmann, London, UK, (1990) p. 369
291. Otsuka K.: *Engineering Aspects of Shape Memory Alloys,* Butterworth-Heinmann, London, UK, (1990) p. 36
292. Breme J.: *Sixth World Conference on Titanium,* Les Editions de Physique, Les Ulis, France, (1988) p. 57
293. Brunette D. M., Tengvall P., Textor M., Thomsen M., eds.: *Titanium in Medicine,* Springer-Verlag, Berlin, Germany, (2001)
294. Perren S. M., Pohler O. E. M.,, Schneider E.: *Titanium in Medicine,* Springer-Verlag, Berlin, Germany, (2001) p. 771
295. Windler M., Klabunde R.: *Titanium in Medicine,* Springer-Verlag, Berlin, Germany, (2001) p. 703
296. Hosonuma M., Shimamune T.: *Sixth World Conference on Titanium,* Les Editions de Physique, Les Ulis, France, (1988) p. 495
297. Breme J., Biehl V., Wack T., Eisenbarth E.: *Titanium '99, Science and Technoloy,* CRISM "Prometey", St. Petersburg, Russia, (2000) p. 1187

298. Albrecht J., Lütjering G., Nicolai H.-P., Liesner C.: *Fatigue 2002,* EMAS, Warley, UK, (2002) p. 2085
299. Semlitsch M., Weber H., Steger R.: *Titanium '95, Science and Technology,* The University Press, Cambridge, UK, (1996) p. 1742
300. Wang K., Gustavson L., Dumbleton J.: *Titanium '92, Science and Technology,* TMS, Warrendale, USA, (1993) p. 2697
301. Fanning J. C.: *Titanium '95, Science and Technology,* The University Press, Cambridge, UK, (1996) p. 1800
302. Steinemann S. G., Mäusli P.-A., Szmukler S., Semlitsch M., Pohler O., Hintermann H.-E., Perren S. M.: *Titanium '92, Science and Technology,* TMS, Warrendale, USA, (1993) p. 2689
303. Niinomi M., Kuroda D., Fukunaga K., Fukui H., Kato Y., Yashiro T., Suzuki A., Hasegawa J.: *Titanium '99, Science and Technology,* CRISM "Prometey", St. Petersburg, Russia, (2000) p. 1195
304. Froes F. H.: *Non-Aerospace Applications of Titanium,* TMS, Warrendale, USA, (1998) p. 317
305. Friedrich H., Kiese J., Haldenwanger H.-G., Stich A.: *Proceedings of the 10th World Conference on Titanium,* Wiley-VCH, Weinheim, Germany, (2003)
306. Sommer C., Peacock D.: *Titanium '95, Science and Technology,* The University Press, Cambridge, UK, (1996) p. 1836
307. Schauerte O., Kramer M., Kiese J., Walz W.: *Proceedings of the 10th World Conference on Titanium,* Wiley-VCH, Weinheim, Germany, (2003)
308. Froes F. H.: *Non-Aerospace Application of Titanium,* TMS, Warrendale, USA, (1998) p. 327
309. Moroishi T., Yamada M., Fujii M., Kusano A., Kawabe A.: *Titanium '95, Science and Technology,* The University Press, Cambridge, UK, (1996) p. 1662
310. Ward-Close C. M.: *Titanium '99, Science and Technology,* CRISM "Prometey", St. Petersburg, Russia, (2000) p. 27
311. Koryagin N. I.: *Sixth World Conference on Titanium,* Les Editions de Physique, Les Ulis, France, (1988) p. 49
312. Fukuhara Y.: *Sixth World Conference of Titanium,* Les Editions de Physique, Les Ulis, France, (1988) p. 381
313. Shimizu Y.: *Titanium '99, Science and Technology,* CRISM "Prometey", St. Petersburg, Russia, (2000) p. 19
314. Yamaguchi H., Miki K.: Alutopia 11, (1987) p. 41

Index

Druck: Strauss Offsetdruck, Mörlenbach
Verarbeitung: Schäffer, Grünstadt